On-Board Processing for Satellite Remote Sensing Images

On-board image processing systems are used to maximize image data transmission efficiency for large volumes of data gathered by Earth observation satellites. This book explains the methods, mathematical models, and key technologies used for these systems. It introduces the background, basic concepts, and the architecture of on-board image processing, along with on-board detection of the image feature and matching, ground control point identification, on-board geometric correction, calibration, geographic registration, etc.

- Describes algorithms and methodologies for on-board image processing with FPGA chips.
- Migrates the traditional on-ground computing to on-board operation and the image processing is implemented on-board, not on-ground.
- Introduces for the first time many key technologies and methods for on-board image processing.
- Emphasizes the recent progress in image processing by using on-board FPGA chips.
- Includes case studies from the author's extensive research and experience on the topic.

This book gives insights into emerging technologies for on-board processing and will benefit senior undergraduate and graduate students of remote sensing, information technology, computer science and engineering, electronic engineering, and geography, as well as researchers and professionals interested in satellite remote sensing image processing in academia, and governmental and commercial sectors.

On-Board Processing for Satellite Remote Sensing Images

Guoqing Zhou

CRC Press
Taylor & Francis Group
Boca Raton London New York

CRC Press is an imprint of the
Taylor & Francis Group, an **informa** business

Designed cover image: © Zhou et al. 2004

MATLAB® is a trademark of The MathWorks, Inc. and is used with permission. The MathWorks does not warrant the accuracy of the text or exercises in this book. This book's use or discussion of MATLAB® software or related products does not constitute endorsement or sponsorship by The MathWorks of a particular pedagogical approach or particular use of the MATLAB® software.

First edition published 2023
by CRC Press
6000 Broken Sound Parkway NW, Suite 300, Boca Raton, FL 33487-2742

and by CRC Press
4 Park Square, Milton Park, Abingdon, Oxon, OX14 4RN

CRC Press is an imprint of Taylor & Francis Group, LLC

© 2023 Taylor & Francis Group, LLC

ISBN: 978-1-032-32964-2 (hbk)
ISBN: 978-1-032-33429-5 (pbk)
ISBN: 978-1-003-31963-4 (ebk)

DOI: 10.1201/9781003319634

Typeset in Times
by KnowledgeWorks Global Ltd.

Contents

Preface

Due to the temporal-spatial limitations of Earth observation with a single satellite, the author in early 2000 proposed "Intelligent Earth Observation Satellite Network System (IEOSNS)". The envisioned IEOSNS is a space-based architecture for the dynamic and comprehensive on-board integration of Earth observing sensors, data processors and communication systems. It is intended to enable simultaneous, global measurements and timely analysis of the Earth's environment for a variety of users. It should be said that the visionary idea proposed in early 2000, which is radically different from traditional one, was really innovative.

On-board image processing is one of the most important components of the envisioned IEOSNS. The objective of this book is therefore to describe the methods, mathematical models, and key technologies on on-board processing for remotely sensed images. This book consists of ten chapters. The order of chapters has been arranged. The early chapters introduce the background, the basic concepts, and the architecture of on-board image processing. The later chapters discuss more specialized aspects on methods regarding on-board detection of the image feature point and matching, and on-board ground control point (GCP) identification and matching. Afterward, this book in detail describes the key technologies on on-board space resection, on-board geometric correction, on-board geometric calibration, on-board georeferencing without ground control point, and on-board orthorectification using FPGA. Finally, a case of application for flooding change detection with on-board image processing is exampled. The coverage of the later chapters would be particularly useful for those advanced readers or in the advanced courses. In addition, only a limited number of key references is given at the end of each chapter, and readers can expand their knowledge on subjects of particular interest.

Several of the apparent features of the book that I believe can enhance its value as a textbook and as a reference book are: First, the algorithms and methodologies described in this book are for on-board image processing using FPGA. Second, the key technologies for on-board image processing using FPGA are different from the traditional methods, such as image feature extraction, ground control points identification, image georeferencing, orthorectification, etc. Third, either the linear readers or the sporadic readers are appropriate, i.e., the sporadic readers can completely understand the essence of any chapter but must not read all the previous chapters. These chapters with a logical order are designed to especially deliberate for various interested readers. Readers therefore read the chapters relevant to his/her own subjects.

This book is based on my early research work and the articles published in journals. A gathering of these unconnected subjects is annealed to finally become lecture notes for a one-semester, 3-credit graduate course. This book can therefore be used as a textbook in universities for upper level undergraduate and graduate students, and serve as a reference book for researchers in academia, governmental, and industrial sectors. Prerequisites for upper level undergraduate and/or graduate course are "Introduction to Remote Sensing", or "Remote Sensing and Digital Image Processing", or "Digital Image Processing with Remote Sensing Perspective", or "Photogrammetry".

I wish to acknowledge my sincere appreciation to my Ph.D. students and M.S. students, Dr. Huang Jingjin, Dr. Zhang Rongting, Dr. Liu Dequan, Mr. Jiang Linjun, Mr. Liu Yilong, Mr. Wang fan, Mr. ShuLei, for their contributions to these chapters.

I especially wish to express my appreciation to the spacecraft manufacturers, government agencies, and private firms who supplied spacecraft figures and relevant information for this book. They are NAVY NEMO satellite for the ORASIS, DLR for the BIRD mission, ESA for PROBA and for the Satellite On-Board Processing (OBP) system, NASA GSFC for the SpaceCube 2.0 system,

Europe for Dubbed ø-sat, NASA for EO-1 satellite, NASA JPL for the IPEX spacecraft. Although I have endeavored to cite their descriptions from the published articles or web, a few individuals may be missed for citation.

Finally, I wish to thank my wife, Michelle M. Xie and two sons, Deqi Zhou and Mitchell Zhou, for their patience, understanding, endurance, support and love.

Guoqing Zhou

Author Biography

Guoqing Zhou received his first Ph.D. degree from Wuhan University, Wuhan, China, in 1994, and his second Ph.D. degree from Virginia Tech, Blacksburg, VA, USA, in 2011.

He was a Researcher with the Department of Computer Science and Technology, Tsinghua University, Beijing, China, from 1994 to 1995, and a Post-Doctoral Researcher with the Institute of Information Science, Beijing Jiaotong University, Beijing, China, from 1995 to 1996. He continued his research as an Alexander von Humboldt Fellow with the Technical University of Berlin, Berlin, Germany, from 1996 to 1998, and was a Post-Doctoral Researcher with The Ohio State University, Columbus, OH, USA, from 1998 to 2000. He was an Assistant Professor, an Associate Professor and a Full Professor with Old Dominion University, Norfolk, VA, USA, in 2000, 2005 and 2010, respectively. He is currently with Guilin University of Technology, Guilin, China. He has authored 11 books, 7 book chapters, and more than 450 peer-refereed publications with more than 265 journal articles.

1 Introduction

1.1 BACKGROUND

In 2000, the US National Aeronautics and Space Administration (NASA) presented NASA's vision "to improve life here on Earth" (Ramapriyan et al. 2002). To achieve this goal, Earth Science Enterprise (ESE) at NASA combines the NASA-unique capabilities for spaceborne observations with research in various Earth science disciplines to address a number of scientific questions (https://www.nasa.gov/topics/earth/index.html). Some of the key science and application goals are to improve predictive capabilities for the following (Ramapriyan et al. 2002):

- Weather forecasts, such as
  - Five-day forecasts with more than 90% accuracy,
  - Seven-to-ten-day forecasts with 75% accuracy.
- Environmental warnings, such as
  - Two-day warning of air quality,
  - Two-day warning of wet.
- Climate predictions, such as
  - Routine 6 to 12-month seasonal predictions,
  - Experimental 12–24-month predictions.
- Natural hazards forecast, such as
  - Continuous monitoring of surface deformation in vulnerable regions with millimeter accuracy,
  - Improvements in earthquake and volcanic eruption forecasts.

Over the past decades, previous efforts in the development of spaceborne-based remote sensing had focused on the efforts for offering a wide range of spatial, spectral and temporal resolutions (Zhou et al. 2004). Significant advances in collecting, archiving and disseminating data have been achieved. It has been widely thought that a considerable progress has been made in ingesting, archiving and distributing large volumes of data using distributed databases with Earth Observing System Distributed Information System (EOSDIS) since the 1980s (Ramapriyan et al. 2002; Moore and Lowe 2001; Moore and Lowe 2002), discovery of the existence of datasets and services through the Global Change Master Directory (Olsen 2000; Smith and Northcutt 2000), the area of interoperability through the EOS Data Gateway (EDG) (Nestler and Pfister 2001), Data and Information Access Link (DIAL) (McDonald et al. 2001), Distributed Oceanographic Data System (DODS) (Cornillon 2000), Alexandria Digital Library (Frew et al. 1999), EOSDIS Clearing House (ECHO) (Pfister and Nestler 2003), and other efforts, such as image fusion with different resolutions, differently sensed images (Zhou et al. 2004). However, with increasing widths of applications from different users, the different requirements presented are various. For example, some users may require highly frequent and highly repetitive coverage with relatively low spatial resolution (e.g., meteorology); some users may desire the high spatial resolution as possible with the lowly frequent repeat coverage (e.g., terrestrial mapping); while some users need both spatial resolution and frequent coverage as high as possible, plus rapid image delivery (e.g., military surveillance) (Figure 1.1). In order to meet the various requirements from different uses, many spaceborne-based imaging systems for remote sensing have been proposed, such as satellite constellation for Earth observations, format flying for Earth observations, intelligent Earth observing satellite (EOS) systems, etc. One of them is typically the "Intelligent Earth Observation Satellite Network System (IEOSNS)" proposed by Zhou et al. (2004).

DOI: 10.1201/9781003319634-1

Various Users		Illustration
Mobile user	A real-time user, e.g., a mobile GIS user, requires a real-time downlink for geo-referenced satellite imagery with a portable receiver, small antenna and laptop computer.	
Real-time user	A mobile user, e.g., a search-and-rescue pilot, requires a real-time downlink for geo-referenced panchromatic or multispectral imagery in a helicopter.	
Lay user	A lay user, e.g., a farmer, requires geo-referenced, multispectral imagery at a frequency of 1-3 days for investigation of his harvest.	
Professional user	A professional user, e.g., a mineralogist, requires hyperspectral imagery for distinguishing different minerals.	
Professional user	A topographic cartographer, e.g., a photogrammetrist, requires panchromatic images for stereo mapping.	

FIGURE 1.1   Examples for different end-users. (Courtesy from Zhou et al. 2004; Astronautix 2021)

1.2   ARCHITECTURES OF IEOSNS

With the development of five generations of spacecraft evolutions, many scholars have con-jectured about the future generation of EOSs (Zhou et al. 2004). It is widely accepted that the EOS has passed the threshold of maturity as a commercial space activity, and the future generation of satellites should be intelligent (Zhou et al. 2004). Zhou et al. (2004) envisioned an Intelligent Earth Observation Satellite Network System (IEOSNS), which is a space-based configuration for the dynamic and comprehensive on-board integration of earth observing sen-sors, data processors and communication systems. Zhou et al. (2004) also thought that "[it] enables simultaneous, global measurement and timely analysis of the Earth's environment for real-time, mobile, professional and common users in the remote sensing communities. This is because user's demands in the mapping, natural resources, environmental science, Earth moni-toring, etc., communities have migrated from basic imagery to temporal, site-specific, updated mapping products/image-based information. Data and information revisions would be requested more frequently, which is, in many ways, analogous to today's weather updates". In addition, layer users are less concerned with the technical complexities of image processing, requiring imagery providers directly provide users with value-added images (e.g., orthorectified images, enhanced edges of features, etc.) and value-added products (e.g., flooding extent, disaster area, etc.) (Campbell et al. 1998). Therefore, timely, reliable and accurate information, with the capa-bility of direct downlink of various bands of satellite data/information and operation as simple as selecting a TV channel, is highly preferred (Zhou et al. 2004).

1.2.1   ARCHITECTURE OF IEOSNS

It is apparent that no single satellite can reach the goal of meeting the requirements of the various users above. The design of the intelligent satellite system would use a multi-layer satellite web with high-speed data communication (cross-link, uplink and downlink) and multiple satellites with on-board data processing capability (Figure 1.2) (Zhou et al. 2004; Schetter et al. 2003; Campbell and Böhringer. 1999; and Zetocha, 2000).

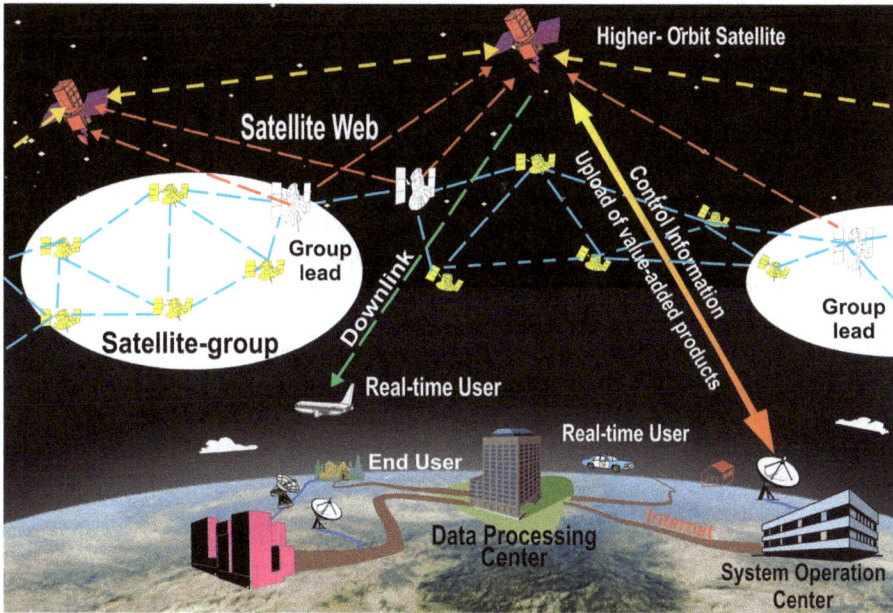

FIGURE 1.2   The architecture of a future intelligent Earth Observing Satellite system. (Courtesy from Zhou et al. 2004.)

1.2.2   MULTI-LAYER SATELLITE NETWORKS

The envisioned IEOSNS consists of two layers of networked satellites (see Figure 1.2). Zhou et al. (2004) described: "The first layer, which consists of hundreds of Earth Observing satellites viewing the entire earth, is distributed in low orbits ranging from 300 km to beyond. Each EOS is small, lightweight and inexpensive relative to current satellites. These satellites are divided into groups called *satellite groups*. Each EOS is equipped with a different sensor for collection of different types of data and an on-board data processor that enables it to act autonomously, reacting to significant measurement events on and above the Earth. They collaboratively work together to conduct the range of functions currently performed by a few large satellites today (Zhou et al. 2004). There is a lead satellite in each group, called *group-lead*; the other satellites are called *member-satellites*. The group-lead is responsible for management of the member-satellites and communication with other group-leaders in the network (constellation) in addition to communication with the geostationary satellites. Such a design can reduce the communication load and ensure effectiveness of management and coverage of data collection".

The second layer is composed of geostationary satellites because not all EOSs are in view of or in communication with global users. The second layer satellite network is responsible for communication with end-users (e.g., data downlink) and ground stations, and ground data processing centers, in addition to the further processing of data from group-lead satellites (see Figure 1.2) (Zhou et al. 2004).

All the satellites are networked together into an organic measurement system with high-speed optical and radio frequency communications. User requests are routed to specific instruments maximizing the transfer of data to archive facilities on the ground and on the satellite (Prescott et al. 1999). Thus, all group-leads must establish and maintain a high-speed data cross-link with one another in addition to uplinking with one or more geostationary satellites, which in turn maintain high-speed data cross-links and downlinks with end-users and ground control stations and processing centers.

1.2.3 Performance of Satellite Constellation

The normal operating procedure is for each EOS to independently collect, analyze and interpret data using its own sensors and on-board processors. These collected data will not be transmitted to ground users, the ground station, or geostationary satellites unless they detect changed data. When an EOS detects an event, e.g., a forest fire, the sensing-satellite rotates its sensing system into position and alters its coverage area via adjusting its system parameters in order to bring the event into focus (Schoeberl et al. 2001). Meanwhile, the sensing-satellite informs member-satellites in its group, and the member-satellites adjust their sensors to acquire the event, resulting in multi-angle, multi-sensor, multi-resolution and multispectral observation and analysis of the event. These data sets are merged onto a geostationary satellite that assigns priority levels according to the changes detected (Panagiotis et al. 2020). Following a progressive data compression, the data is then available for transmission to other geostationaries. The links between the geostationary satellites provide the worldwide real-time capability of the system. Meanwhile, the geostationary satellite further processes the data to develop other products, e.g., predictions of fire extent after 5 days, weather influence on a fire, pollution caused by a fire, etc. These value-added products are then also transmitted to users (Zhou et al. 2004).

If the geostationary satellite cannot analyze and interpret the data, the "raw" data will be transmitted to the ground data processing center (GDPC). The GDPC will interpret these data in terms of the user's needs and then upload the processed data back to the geostationary satellites. In the constellation, all satellites can be independently controlled by either direct command from a user on the ground or autonomously by the integrated satellite-network system itself.

The satellite transmits the image in an order of priority, the more important parts of the data first. For example, the multispectral imagery of a forest fire may have a higher priority than the panchromatic imagery. Panchromatic imagery for 3D mapping of a landslide may have priority over the multispectral imagery. Of course, the autonomous operation of the sensors, processors and prioritization algorithms can be subject to override by system controllers or authorized users. This concept of performance is similar to the *sensor-web* concept as envisioned by the *Earth Science Vision Initiative*, and *Earth Science Vision Enterprise Strategic Plan of NASA* (Zhou, 2001; Earth Science Vision Initiative, http://staac.gsfc.nasa.gov/esv.htm; The Earth Science Enterprise Strategic Plan, http://www.earth.nasa.gov/visions/stratplan/index. html; NASA Space Technology, Instrument and Sensing Technology, http://ranier.hq.nasa.gov/ Sensors_page/InstHP.html).

1.2.4 Event-Driven Observation

The normal operating procedure is for each programmed sensor to independently collect a different data type, such as temperature, humidity or air pressure, land use, etc., like http://sensorwebs. jpl.nasa.gov/overview/pictures/instrument.html an electronic nose (Enose) for sniffing out different types of gases, analyzing and interpreting data using its own sensors and on-board processors. These collected data will *not* be transmitted to ground users, the ground station, or geostationary satellites unless they detect changed data. When an EO satellite detects an event, e.g., a forest fire, the sensing-satellite rotates its sensing system into position and alters its coverage area by adjusting its system parameters in order to bring the event into focus (Schoeberl et al. 2001). Meanwhile, the sensing-satellite informs member-satellites in its group, and the member-satellites adjust their sensors to acquire the event, resulting in a multi-angle, multi-sensor, multi-resolution and multispectral observation and analysis of the event (Figure 1.3). These data sets are merged to a geostationary satellite that assigns priority levels according to the changes detected (Zhou et al. 2004). Following a progressive data compression, the data is then available for transmission to other geostationary satellites. The links between the geostationary satellites provide the worldwide real-time capability

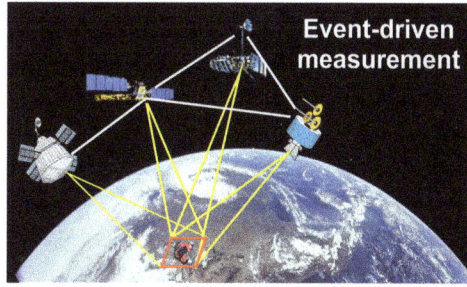

FIGURE 1.3   Event-driven Earth observation.

of the system. Meanwhile, the geostationary satellites further processes the data to develop other products, e.g., predictions of a fires extent after five days, weather influence on a fire, pollution caused by a fire, etc. These value-added products are then also transmitted to users (Panagiotis et al. 2020).

1.2.5 END-USER OPERATION

End-users expect directly downlinked satellite data, where the concept of data in fact means image-based knowledge products and information products, rather than traditional remotely sensed data, using their own receiving equipment. The operation appears to the end-users as simple and easy as selecting a TV channel by using a remote control (Figure 1.4) (Zhou et al. 2004). Moreover, all receivers are capable of uploading the user's command, and mobile and hand-held receivers have GPS receivers installed, i.e., mobile user's position in the geodetic coordinate system can be real-time determined and uploaded to geostationary satellite. The onboard data distributor will retrieve an image (block) from its database according to the user's position (ELS 2018).

In this fashion, an ordinary and layer user on the street is able to use a handheld wireless device to downlink/access the image map of his surroundings from a geostationary satellite or from the Internet. The intelligent satellite system will enable people not only to see their environment, but also to "shape" their physical surroundings (Courtesy of https://earthobservatory.nasa.gov/features/ColorImage). The downlinked data that users receive is not an actual image; instead, it receives a signal, much like a TV antenna receiving a TV signal, rather than direct picture and sound. This signal must be transformed into picture and sound by TV set. Similarly, the IEOSNS signal (which

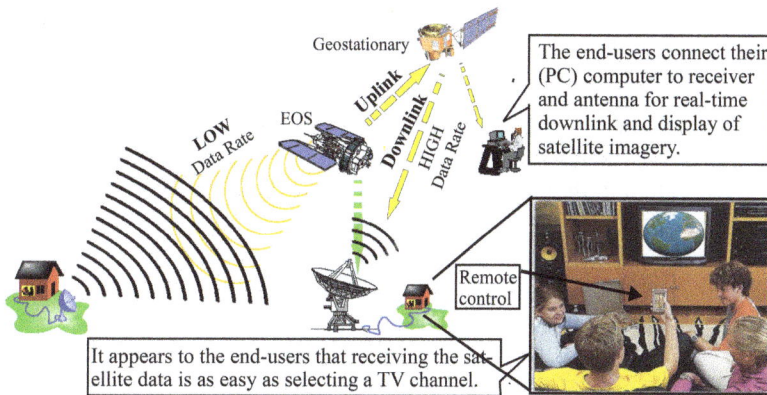

FIGURE 1.4   End-user operation like selecting a TV channel. (Courtesy from Zhou et al. 2004.)

we call a *special signal*) is absolutely different from the signal of current EOS. Thus, IEOSNS satellite signal must be transformed into an image by the users receiving equipment (Zhou 2020). Therefore, users need:

1. **User Software for Data Downlink:** The *special signal* is transformed by software, which is provided by the ground control center so that real-time and common users can easily use it. For a lay user (e.g., a farmer) complicated application software is unnecessary because the user analyzes and interprets the images using their perceptual faculties. For more advanced users (e.g., a professor), advanced software will still be necessary because they use "imagery" in different ways.

2. **Accessible Frequency:** Different users need different imagery, e.g., a photogrammetrist needs forward and afterward stereo panchromatic imagery for stereo mapping; a biologist needs hyperspectral imagery for flower research. Thus, different types of satellite images are assigned with different broadcast frequencies, which the ground control station provides access to for authorized users.

1.3   CHARACTERISTICS OF IEOSNS

The architecture for the envisioned IEOSNS in Section 1.2 is flexible because any additional satellites can easily be inserted without risk to the infrastructure, and the instruments and platforms are organically tied together with network information technology. The constellation (multi-layer satellite network) ensures that global data is collected on a frequency of decade minutes' base or shorter; event-driven data are collected with multi-angle, multi-resolution, multi-bands, and users can acquire images of any part of the globe in real-time. This design concept provides a plug-and-play approach to the development of new sensors, measurement platforms and information systems, permitting smaller, lighter, standardized satellites with independent functions to be designed for shorter operational lifetimes than today's large systems so that the instrument technology in space can be kept closer to the state-of-the-art.

The envisioned IEOSNS will perform much of the event detection and response processing that is presently performed by ground-based systems through the use of high-performance processing architectures and reconfigurable computing environments (Alkalai, 2001; Armbruster and Wijmans 2000; Bergmann and Dawood 1999). The envisioned IEOSNS will act autonomously in controlling instruments and spacecraft, while also responding to the commands of the user interested to measure specific events or features. So, users can select instrument parameters on demand and control on-board algorithms to preprocess the data for information extraction (Dong et al. 2011).

1.4   KEY TECHNOLOGIES FOR IEOSNS

The envisioned IEOSS consisting of a multi-layer satellite-network will produce large amounts of scientific data, which creates significant challenges in the processing, transmission, storage and distribution of data and data products. Thus, the envisioned IEOSS will require the fastest processors, the highest communication channel transfer rates and the largest data storage capacity, as well as real-time software systems to ensure that on-board data processing and the post-processed data flows smoothly from the satellite network to the global users (Prescott et al. 1999; Schetter et al. 2003). The key technologies to realize this capability are:

- Various types of intelligent and smart sensors and detectors,
- High data rate transmission and high-speed network communication and
- Most powerful on-board data processing capabilities.

1.4.1 INTELLIGENT AND SMART SENSORS AND DETECTORS FOR DATA COLLECTION

Many current event detections through cost-effective image analysis on the ground, e.g., air pollution detection using hyperspectral image analysis, will be replaced by on-board detector or sensor. Thus, the IEOSNS requires various smart and efficient sensors/detectors so that sudden events on the ground can easily be detected and observed in a timely manner (Digital Libraries 1999).

- **Biological Sensors:** This type of sensor is mainly used for environmental science investigation, such as in terrestrial and freshwater sciences, e.g., the monitoring of toxic chemicals and pollutants both in waters and in soils.
- **Chemical Sensors:** The chemical sensors can analyze atmospheric particles, their size and chemistry, the transport, dispersion and deposition of heavy metals, etc.
- **Microwave Sensors:** Microwave radiometers sense far infrared radiation emitted by the Earth with wavelengths in the vicinity of 1.5 cm. A microwave detector can penetrate clouds and distinguish between ground and ice or snow surfaces (Hopkins 1996 https://aos.wisc.edu/~hopkins/aos100/satpltfm.htm).
- **Neural Network Sensor:** Neural network technology is advancing on several fronts. The type of sensor can simulate the human visual system, revealing the composition of information contained in a single pixel. The technique, which uses the human visual system as a model, reveals the composition of information contained in a single pixel (Heaven 2019). Thus, neural network sensors essentially increase the resolution of satellite images (Zhou 2001, http://www.globaltechnoscan.com/3may_9may/onr_visual_system.htm).
- **Smart Dust Sensor:** These minute, inexpensive devices are self-powered and contain tiny on-board sensors and a computer on a scale of just five square millimeters (Zhou 2001, http://www.rand.org/scitech/stpi/ourfuture/Internet/sec4_sensing.html).
- **Computerized Sensor:** This sensor uses digital computers to control and analyze reactors and other processes (Zhou 2001, http://www.ornl.gov/ORNLReview/meas_tech/sensor.htm). The computers will be provided with direct sensing capabilities in order to interact directly with their environment.
- **Smart Sensor:** These sensors can automatically detect a specific, sudden event, such as a forest fire, volcanic activity, an oil spill or a burning coal seam (Zhou 2001, http://www.unifreiburg.de/fireglobe/iffn/tech/tech_9.htm).

1.4.2 HIGH DATA RATE TRANSMISSION AND HIGH-SPEED NETWORK COMMUNICATION

In the envisioned IEOSNS constellation, the satellites are in different orbits, and their relative velocities vary significantly. Hence, the establishment and maintenance of real-time network communication, including high-speed data crosslink of EOSs, uplink of user/ground control station and geostationary, downlink of user and geostationary (see Figure 1.2), is **NOT** a simple problem (Welch et al. 1999; Surka et al. 2001). Obviously, the technology for high-speed wireless (optical or RF) data linking to connect satellite to satellite, and satellite to ground for high data rate transmission and the network management are vital elements for this concept.

Therefore, developments in information sciences and technology are critical to the achievement of the envisioned IEOSNS goal for its deployment for operational applications.

1.4.3 ON-BOARD DATA PROCESSING CAPABILITIES

The success of on-board data processing is crucial to realize the envisioned IEOSNS. On-board data processing includes, such as an image data processor, data distributor, data management processor, housekeeping, resource management, on-board command planning, platform/sensor control, etc. One of the essential capabilities provided by On-Board Processing (OBP) is satellite autonomy

(Prescott et al. 1999; Ramachandran et al. 1999). This autonomy requires the mission operations and data processing/interpretation activities to evolve from ground-based control/analysis toward on-board control/analysis. The following is only a part of on-board data processing (Kazutami et al. 2005).

1. **On-board Image Processing:** Some image processing, such as image filtering, enhancement, compression, radiometric balance, edge detection and feature extraction, could be automatically processed on-board with techniques currently available and to be developed within the next ten years. However, higher-level intelligent image processing, like classification, spatial information extraction, change detection, image interpretation, and pattern recognition, may need several generations of development. It has been demonstrated that full-automation of image analysis and image interpretation is quite difficult, particularly in complex areas such as wetlands and urban environments. In particular, for the envisioned IEOSS, the important function is its change detection capability, i.e., the envisioned IEOSS only transmits those data that have been changed when compared with images stored on a database system (Tomar and Verma 2007).

2. **Data Storage and Distribution:** The envisioned IEOSNS requires huge data storage capabilities on-board and autonomous operation of data distribution; thus, some advanced and novel data handling technologies, such as data compression, data mining, advanced database design, data and/or metadata structures, etc., will be required to support autonomous data handling (Caraveo et al. 1999).

3. **On-Board Software:** Real-time software systems for integrating all of the components of the satellite network and completing the flow of data from collection and transmission, to information extraction and distribution will be one of the key elements in the envisioned IEOSS. Additionally, in order to produce the value-added data products useful to common users, the current application software, algorithm, dynamic searching, etc., will need to be improved. In order to directly downlink to common users, some advanced concepts, such as dynamic and wireless interaction technology will need to be designed for handling the huge data computational requirements of dynamic interaction (Defencetalk 2020; Krebs 2021).

1.5   TYPICAL SYSTEMS OF ON-BOARD IMAGE DATA PROCESSING

1.5.1   NAVAL EARTH MAP OBSERVER (NEMO)

The NEMO satellite was developed by US Navy (see Figure 1.5), aiming at demonstration of real-time hyperspectral imagery (HIS) processing in space and providing unclassified, space-based HIS at moderate resolution for direct use by Naval forces and the civil sector (Zhou, 2001, http://nemo.nrl.navy.mil/concept.html). The key technologies relevant to on-board data processing in NEMO are (Davis et al. 2000):

1. ORASIS stands for Optical Real-Time Adaptive Signature Identification System. It was developed by the Naval Research Laboratory's (NRL's) for OBP, analysis and feature extraction. The ORASIS is housed in NEMO's Imagery On-Board Processor (IOBP), an advanced high-performance computer consisting of a highly parallel array of commercial Analog Device ADSP21060 Super Harvard Architecture RISC Computer (SHARC) digital signal processor (DSP) and a unique radiation-tolerant architecture resulting in a relatively low power, low cost, and high-performance computing platform. Major technology efforts include developing radiation-tolerant design, extensive redundancy and reconfigurability, thermal control, and packaging applicable to spaceflight avionics applications.

FIGURE 1.5 NEMO spacecraft. (Courtesy of https://space.skyrocket.de/doc_sdat/nemo.htm.)

The ORASIS algorithm and the IOBP are used to demonstrate the real-time HSI processing as follows:

- Real-time feature extraction and classification with greater than 10× data reduction,
- High-performance IOBP with greater than 2.5 giga FLOPS of sustained computational power and
- On-board data storage (56 gigabit)

1.5.2 BIRD Mission (Fire Monitoring)

The BIRD (Bi-spectral Infrared Detection) mission is a small satellite mission funded by DLR (Deutsches Zentrum für Luft- und Raumfahrt (DLR)) (see Figure 1.6), aiming at demonstration

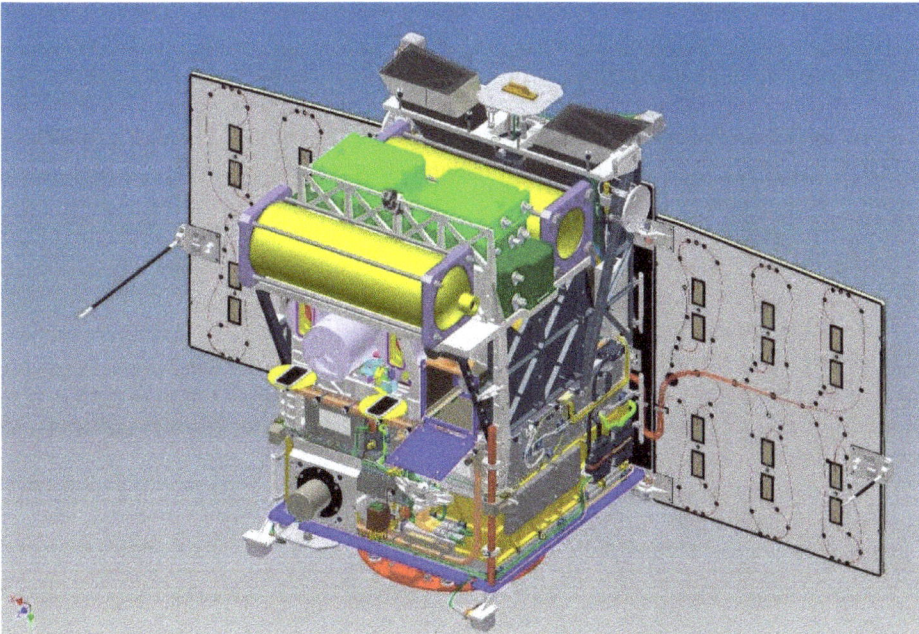

FIGURE 1.6 BIRD spacecraft. (Courtesy of https://directory.eoportal.org/web/eoportal/satellite-missions/b/biros.)

of the different innovative technologies related to detection and assessment of hot spot events like vegetation fires, volcanic activities, burning oil wells and coal seams using push-broom sensors (Zhou, 2001, http://www.uni-freiburg.de/fireglobe/iffn/tech/tech_9.htm). The interesting capabilities relevant to on-board data processing are (Oertel et al. 1998; Halle et al. 2000).

1. **On-board data processing capabilities**

   The purpose of the on-board data processing system is to reduce the downlink data stream by generation of thematic maps. This reduction was implemented through on-board classificator, which was developed by a neural network implemented in the neuro-chip "Recognition Accelerator NI1000" from Nestor Inc.

   The on-board processing is needed in cases when users must obtain information with strong time restrictions, for example, fire warning or managing, gale warning or detection of pollution of the environment. The classificator of BIRD is with the features below (see Figure 1.6).
   - Trained to detect fire directly on-board,
   - A thematic on-board classificator for disaster warning and monitoring,
   - Radiometric and geometric on-board correction of sensor signals,
   - Geometric on-board correction of systematic alignment errors, and
   - Geometric on-board correction of spacecraft attitude.
2. **On-board geocoding of thematically processed data with real-time downlink**
   - Immediate downlink of regional data,
   - Downlink of an alert message if required and
   - Store-and-forward, data downlink to low-cost payload ground stations.

   At the end of the classification process on-board BIRD, images with so-called "class maps" are the results. They will be labeled with the time-code of the data-take of the corresponding row-data image. These class maps are directly available for the user at the ground station (see Figure 1.7).

1.5.3   PROBA ON-BOARD DATA PROCESSING

PROBA (**PR**oject for **O**n-**B**oard **A**utonomy) is an ESA mission (see Figure 1.8) aiming at demonstrating capability of on-board data processing and the benefits of on-board autonomy through the advanced technologies in orbit. Especially, the following autonomy functionalities are expected to be demonstrated (European Space Agency Signature 2021; Teston et al. 1997; Zhou 2001, http://telecom.esa.int/artes/artes2/fileincludes/multimedia/multi.cfm)

1. **On-board housekeeping**: decision-making including failure detection, failure identification and first-level recovery actions.
2. **On-board data management:** data handling, storage and downlinks.
3. **On-board resources usage:** power and energy usages.
4. **On-board instrument commanding:** planning, scheduling, resource management, software exchanges, navigation and instrument pointing, downlinks of the processed data.
5. **On-board science data distribution:** automatic direct data distribution to different users without human involvement. Minimum possible delay.
6. **On-board platform control:** platform control of the PROBA is performed autonomously on-board by a high-accuracy autonomous double-head star tracker, a GPS receiver and a set of reaction wheels (Eoportal Directory 2021).

1.5.4   SPACECUBE 2.0

The SpaceCube system was developed by from NASA Goddard Space Flight Center. The SpaceCube system is considered as "next generation" on-board science-data processors for space missions

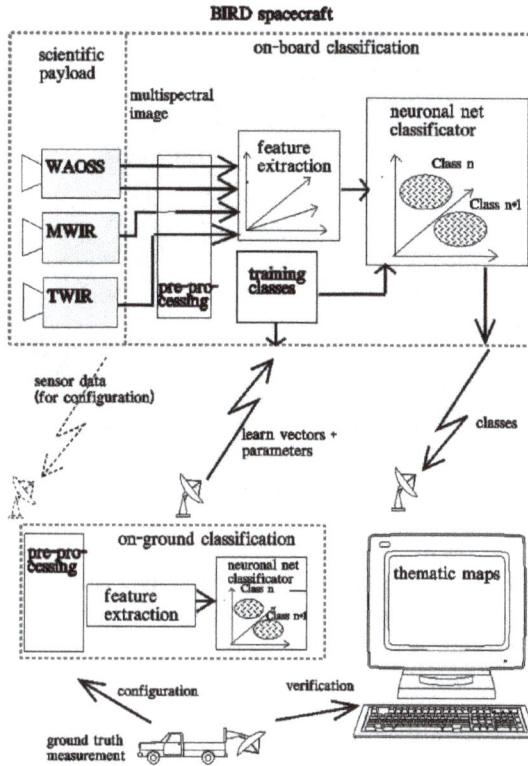

FIGURE 1.7 Architecture of BIRD on-board classification process. (Courtesy of https://www.explica.co/firebird-the-german-mission-that-detects-fires-from-space/.)

FIGURE 1.8 PROBA spacecraft. (Courtesy of ESA 2021, https://earth.esa.int/web/guest/missions/esa-operational-eo-missions/proba.)

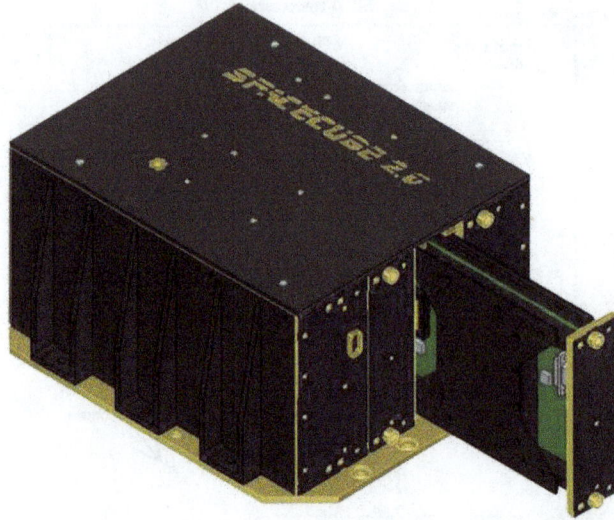

FIGURE 1.9 NASA's SpaceCube 2.0 Hybrid Science Data Processor (Flight Model). (Courtesy of https://www.businesswire.com/news/home/20130930005547/en/NASA-SpaceCube-2.0-Sports-Peregrine-UltraCMOS%C2%AE-Power.)

(Figure 1.9). The SpaceCube provides 10–100 times the standard on-board computing power while lowering relative power consumption and cost. The extra performance and smaller size enable SpaceCube 2.0 to support four HD cameras and the "FireStation" heliophysics instrumentation package, which detects and measures terrestrial gamma-ray flashes from lightning and thunderstorms (Petrick et al. 2015, and from https://www.psemi.com/newsroom/press-releases/613939-nasa-space-cube-2-0-sports-peregrine-ultracmos-power-management-solution).

Petrick et al. (2014) wrote: "The NASA GSFC's SpaceCube 2.0 uses four of Peregrine's PE99155 point of load (POL) power management chips to provide regulated power at multiple supply voltages for the processing elements, memory and associated circuitry like oscillators and gigabit transceivers. Peregrine's flexible and small-form-factor design allowed all of the regulation functions to be placed on a single PCB, realizing area and overall power-consumption savings over the SpaceCube 1.0 power design" (also from https://www.businesswire.com/news/home/20130930005547/en/NASA-SpaceCube-2.0-Sports-Peregrine-UltraCMOS%C2%AE-Power). Therefore, the SpaceCube™ v2.0 system is a high performance, reconfigurable, hybrid data processing system that can be used in a multitude of applications, including those that require a radiation hardened and reliable solution (Inoue 2003).

1.5.5   Ø-Sat System for Space Mesh

Esposito et al. (2019) described: "Dubbed ø-sat, or PhiSat is the first European satellite to demonstrate how on-board artificial intelligence can improve the efficiency of sending Earth observation data back to Earth (see Figure 1.10). This revolutionary artificial intelligence technology has flown on one of the two CubeSats. The two CubeSats, each about the size of a shoebox, collects data, which can be made available through the Copernicus Land and Marine Environment services, using state-of-the-art dual microwave and hyperspectral optical instruments. They also carry a set of intersatellite communication technology experiments. To avoid downlinking these less than perfect images back to Earth, the ø-sat artificial intelligence chip is capable of filtering cloud cover out so that only usable data are returned". (from https://www.esa.int/Applications/Observing_the_Earth/Phsat/First_Earth_observation_satellite_with_AI_ready_for_launch)

FIGURE 1.10   Ø-SAT. (Courtesy of https://directory.eoportal.org/web/eoportal/satellite-missions/p/phisat-1.)

1.5.6   SATELLITE ON-BOARD PROCESSING (OBP) SYSTEM

The Satellite OBP system, designed and partially developed by the European Space Agency (ESA) provides for each user terminals (UTs) a high data rate reception capability, up to 32.784 Mbit/s transmit capability (Figure 1.11). The OBP System is designed in such a way that a large number of UTS's (up to 150,000 for a 262 Mbit/s on-board switch capacity) could be served by the OBP Network.

In broadcast mode, satellite systems provide a tremendous amount of capacity and in principle supply every human on Earth with several gigabits per second of information. In addition, satellites

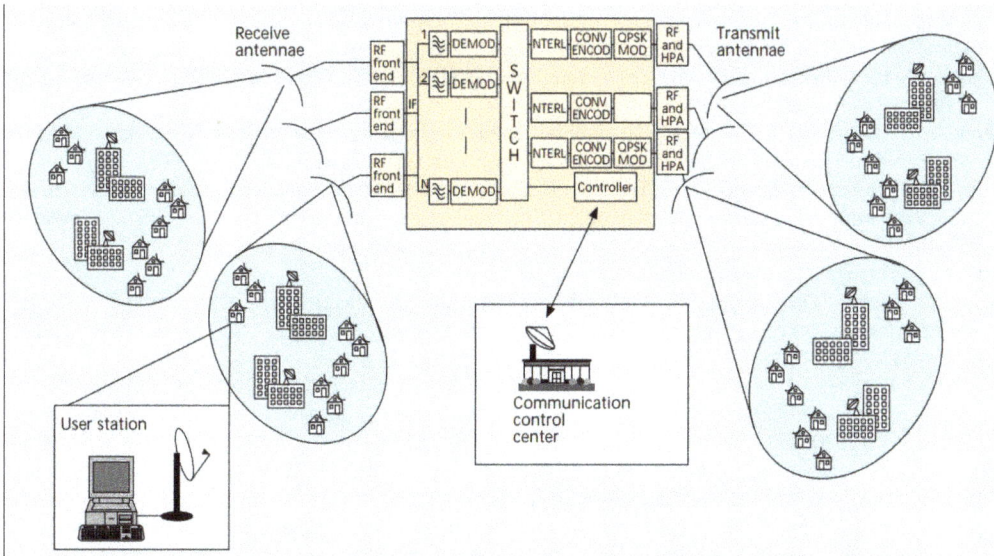

FIGURE 1.11   The On-board Processing System. (Courtesy of Manfred 2000.)

are also capable, in principle, of collecting a large amount of information from very many sources and delivering this information to a few sinks. To increase the capacity of satellites in orbit, new technologies below are needed (Wittig 2000):

- Multiple-beam antennas to cover a large area and to allow low-cost user stations to transmit toward the satellite.
- On-board regeneration and decoding to improve the link budget.
- On-demand on-board switching to provide unlimited connectivity and make the most efficient use of the capacity.

1.6 CONCLUSION

This chapter introduced the architectures of an envisioned IEOSNS. The envisioned IEOSNS is a space-based architecture for the dynamic and comprehensive on-board integration of Earth observing sensors, data processors and communication systems.

This chapter also overviewed the several existing typical "intelligent" satellites with on-board data processing capabilities. The future is promising. A thorough feasibility study addressing the key technologies of each of the components, benefits and issues are necessary in the near future.

REFERENCES

Alkalai, L., An overview of flight computer technologies for future NASA space exploration missions, *3rd IAA Symposium on Small Satellites for Earth Observation*, Elsevier BV, Berlin, Germany, 2001, April 2–6.

Armbruster, P., Wijmans, W., Reconfigurable on-board payload data processing system developments at the European Space Agency, *Proceedings of ESA-presentation at SPIE 2000*, 2000, SPIE, Vol. 4132–12, 91–98

Astronautix, Photo Gallery, http://www.astronautix.com/a/astroe.html, last accessed on September 21, 2021.

Bergmann, N.W., Dawood, A.S., Reconfigurable computers in space: Problems, solutions and future directions, *The 2nd Annual Military and Aerospace Applications of Programmable Logic Devices (MAPLD'99) Conference*, Laurel, Maryland, USA, 1999.

Businesswire, NASA SpaceCube 2.0 Sports Peregrine UltraCMOS® Power Management Solution, https://www.businesswire.com/news/home/20130930005547/en/NASA-SpaceCube-2.0-Sports-Peregrine-UltraCMOS%C2%AE-Power, September 30, 2013.

Campbell, M. E., Böhringer, K.F., Vagners, J. Intelligent satellite teams, *Proposal to the NASA Institute for Advanced Concepts*, July 1998.

Campbell, M., Böhringer, K.F., Intelligent satellite teams for space systems, *2nd Int. Conf. on Integrated Micro/Nanotechnology for Space Applications*. Pasadena, California, April 11–15, 1999.

Caraveo, P. A., Bonati, A., Scandelli, L., Denskat, U., NGST on-board data management study, *NGST Science and Technology Exposition*, Astronomical Soc./Pacific, Hyannis, Massachusetts, USA, September 13–16, 1999.

Cornillon, P., Personal Communication, December 2000.

Davis, C.O., Horan, D., Corson, M., On-orbit calibration of the naval Earth Map observer (NEMO) coastal ocean imaging spectrometer (COIS). *Proceedings of SPIE*, November 11–15, 2000, Vol. 4132.

Defencetalk, Lockheed Launches First Smart Satellite Enabling Space Mesh Networking, https://www.defencetalk.com/lockheed-launches-first-smart-satellite-enabling-space-mesh-networking-73155/, January 16, 2020.

Digital Libraries, *Proceedings of the Third IEEE METADATA Conference*, 1999, http://www.computer.org/proceedings/meta/1999/papers/55/jfrew.htm

Dong, Y., Chen, J., Yang, X., Improved energy optimal OpenMP static scheduling algorithm. *Journal of Software*, 2011, 22(09): 2235–2247.

ELS, Tyvak 0129 Narrative v3, Technical Report, Washington DC, 2018. 21 pp.

Eoportal Directory, BIROS (Bi-spectral InfraRed Optical System). https://directory.eoportal.org/web/eoportal/satellite-missions/b/biros, May 17, 2014.

Eoportal Directory, Spacecraft, https://directory.eoportal.org/web/eoportal/satellite-missions/p/phisat-1, June 19, 2020.

ESA, First Earth Observation Satellite with AI ready for launch, https://www.esa.int/Applications/Observing_the_Earth/Ph-sat/First_Earth_observation_satellite_with_AI_ready_for_launch, September 12, 2019.

ESA, Missions/Ph-sat, https://www.esa.int/ESA_Multimedia/Missions/Ph-sat/(sortBy)/view_count/(result_type)/images, April 13, 2020.

Esposito, M., et al, Highly integration of hyperspectral, thermal and artificial intelligence for the ESA PhiSat-1 mission, *Proceedings of the 70th IAC (International Astronautical Congress)*, Washington DC, USA, October 21–25, 2019.

European Space Agency Signature, PROBA-1. https://earth.esa.int/web/guest/missions/esa-operational-eo-missions/proba, September 23, 2021.

Francesc, M.M.J., et al. In-orbit validation of the FMPL-2 instrument—The GNSS-r and l-band microwave radiometer payload of the FSSC at mission. *Remote Sensing*, 2020, 13(1): 121. doi:10.3390/rs13010121

Frew, J., Freeston, M., Hill, L., Janée, G., Larsgaard, M., Zheng, Q., Generic query metadata for geospatial digital libraries, *Computer Science and Engineering*, January 1, 1999, https://www.researchgate.net/publication/243491744_Generic_query_metadata_for_geospatial_digital_libraries.

Gunter's Space Page, NEMO, https://space.skyrocket.de/doc_sdat/nemo.htm, September 21, 2021.

Halle, W., Venus, H., Skrbek, W., Thematic data processing on-board the satellite BIRD, 2000. http://www.spie.org/Conferences/Programs/01/rs/confs/4540A.html.

Headquarters, N.. NASA's Earth Science Enterprise Research Strategy for 2000–2010 [R/OL]. http.www.earth.nasa.gov, 2000.

Heaven, D., Why deep-learning AIs are so easy to fool. *Nature*, 2019, 574: 163–166.

Hopkins, E.J., Weather satellite platforms, 1996. https://aos.wisc.edu/~hopkins/aos100/satpltfm.htm, last accessed September 21, 2021.

Inoue, H., The Astro-E mission. *Advances in Space Research*, 2003, 32(10): 2089–2090. doi:10.1016/S0273-1177(03)90649-1

Kazutami, M., et al. X-ray telescope on-board Astro-e. III. Guidelines to performance improvements and optimization of the ray-tracing simulator. *Applied Optics*, 2005, 44(6): 916–940. doi:10.1364/AO.44.000916.

Krebs, Gunter D. Gunter's Space Page. Retrieved January 31, 2023, from https://space.skyrocket.de/index.html, September 21, 2021.

Krebs, Gunter D. Pathfinder risk reduction (Tyvak 0129, Pony Express 1). Gunter's Space Page, Retrieved October 7, 2021, from https://space.skyrocket.de/doc_sdat/pathfinder-risk-reduction.htm

McDonald, K. R., Suresh, R., Di, L., The data and information access link (DIAL). *In 17th International Conference on Interactive Information and Processing Systems for Meteorology, Oceanography, and Hydrology*, Albuquerque, NM, May 2001.

Moore, M., Lowe, D., Leveraging open-source development in large-scale science data management systems. *Proceedings of SPIE – The International Society for Optical Engineering*, January 22, 2001, 4483: 291–300.

Moore, M., Lowe, D., Providing rapid access to EOS data via data pools, *Proceedings of SPIE, Earth Observing Systems VII*, Seattle, WA, July 7–10, 2002.

NASA, Earth Science Vision Initiative, http://staac.gsfc.nasa.gov/esv.htm.

NASA, How to Interpret a Satellite Image: Five Tips and Strategies. https://earthobservatory.nasa.gov/features/ColorImage

NASA, NASA Space Technology, Instrument and Sensing Technology. http://ranier.hq.nasa.gov/Sensors_page/InstHP.html.

NASA, The Earth Science Enterprise Strategic Plan. http://www.earth.nasa.gov/visions/stratplan/index.html.

Nestler, M.S., Pfister, R., The Earth Observing System Data Gateway. In AGU Spring Meeting Abstracts, 2001, 2001: U21A–02.

Oertel, D., Zhukov, B., Jahn, H., Briess, K., Lorenz, E., Space-borne autonomous on-board recognition of high temperature events, *The Int. IDNDR Conference on Early Warning Systems for the Reduction of Natural Disasters*, in Potsdam, Germany, September 7–11, 1998.

Olsen, L.M., Discovering and using global databases, In *Global Environmental Databases: Present Situation; Future Directions*, R. Tateishi, D. Hastings. (Editors), International Society for Photogrammetry and Remote Sensing (ISPRS), 2000.

Panagiotis, B., et al. A review on early forest fire detection systems using optical remote sensing. *Sensors (Basel, Switzerland)*, 2020, 20(22), 6442.

Petrick, D., et al. Adapting the SpaceCube v2.0 data processing system for mission-unique application requirements, *2015 NASA/ESA Conference on Adaptive Hardware and Systems (AHS). IEEE*, 2015: 1–8. doi:10.1109/AHS.2015.7231153

Petrick., D., Geist, A., Albaijes, D., et al. SpaceCube v2.0 space flight hybrid reconfigurable data processing system//*2014 IEEE Aerospace Conference*. *IEEE*, 2014: 1–20. doi:10.1109/aero.2014.6836226

Pfister, R.G., Nestler, M. S., Sharing community data, services and tools using the EOS Clearinghouse (ECHO). *In AGU Fall Meeting Abstracts*, 2003. 2003: U41B–0006.

Prescott, E.G., Smith, A.S., Moe, K., Real-time information system technology challenges for NASA's Earth Science Enterprise, *Proceedings of the 20th IEEE Real-Time Systems Symposium, Phoenix, Arizona*. December 1999.

Ramachandran, R., Conover, H.T., Graves, S.J., Keiser, K., Pearson, C., Rushing, J., A next generation information system for Earth science data, *The International Symposium on Optical Science, Engineering and Instrumentation*, SPIE, September 24, 1999, SPIE, Denver, Colorado.

Ramapriyan, H., McConaughy, G., Lynnes, C., Harberts, R., Roelofsb, L., Kempler, S. and McDonnald K., Intelligent archive concepts for the future, *The First International Symposium on Future Intelligent Earth Observing Satellite (FIEOS)*, International Society of Photogrammetry and Remote Sensing (ISPRS), Nov. 10–11, 2002, Denver, Colorado, USA

Schetter, T., Campbell, M., Surka, D., Multiple agent-based autonomy for satellite constellations, *Artificial Intelligence*, 2003, 145 : 147–180.

Schoeberl, M., Bristow, J., Raymond, C., Intelligent distributed spacecraft infrastructure, *Earth Science Enterprise Technology Planning Workshop*, NASA, January 23–24, 2001.

Smith, S.G., Northcutt, R.T., Improving multiple site services: The GCMD proxy server system implementation, EOGEO 2000. *Earth Observation (EO) & Geo-Spatial (GEO) Web and Internet Workshop*, Im Selbstverlag des Instituts für Geographie der Universität Salzburg, April 17–19, 2000.

Surka, D.M., Brito, M.C., Harvey, C.G., Development of the real-time object agent flight software architecture for distributed satellite systems, *IEEE Aerospace Conference Proceedings*, Big Sky, Montana, USA, March 10–17, 2001.

Teston, F., Creasey, R., Bermyn, J., Mellab, K., PROBA: ESA's autonomy and technology demonstration mission, *48th International Astronautic Congress*, October 6, 1997, Turin, Italy.

Tomar, G.S., Verma, S., Gain & size considerations of satellite subscriber terminals on Earth in future intelligent satellites, *IEEE International Conference on Portable Information Devices*, Orlando, Florida, USA, June 4, 2007.

Welch, L., Brett, T., Pfarr, B.B., Adaptive management of computing and network resources for real-time sensor webs, *AIST NRA99*, 1999. http://www.esto.nasa.gov:8080/programs/aist/NRA99-selections/AIST63.html.

Wittig, Manfred, Satellite on-board processing for multimedia applications. *IEEE Communications Magazine,* 2000, 38(6): 134–140.

Zetocha, P., Intelligent agent architecture for on-board executive satellite control, *Intelligent Automation and Control, TSI Press Series on Intelligent Automation and Soft Computing*, TSI Press, Albuquerque, NM, 2000, 9: 27–32.

Zhou, G., *Urban High-Resolution Remote Sensing: Algorithms and Modelling*, Taylor & Francis/CRC Press, 2020, ISBN: 978-03-67-857509, 465 pp.

Zhou, G., Architecture of future intelligent Earth observing satellites (FIEOS) in 2010 and beyond, technical report, *National Aeronautics and Space Administration Institute of Advanced Concepts (NASA-NIAC)*, NASA Institute for Advanced Concepts, USA, 2001, 60 pp.

Zhou, G., Baysal, O., Kaye, J., Concept design of future intelligent Earth observing satellites, *International Journal of Remote Sensing*, July 2004, 25(14): 2667–2685.

2 On-Board Processing for Remotely Sensed Images

2.1 INTRODUCTION

It is generally expected that spaceborne-based mission operation activities for earth observing would evolve from ground-based toward on-board ones (Teston et al. 1997; Altinok et al. 2016). Many basic and complex operations are expected to gradually migrate onto the on-board computer (Zhou et al. 2004). This is because the traditional on-ground operations by humans are essentially limited for rapidly response to time-critical events, such as life rescue in ocean. To this end, the new generation of satellites, associated with the ground-based operations are expected to be performed together in an overall system-level architecture rather than in isolation modes (Bernaerts et al. 2002).

Additionally, the trend in remote sensing community is on one hand to increase the number of spectral bands, frequency of earth observation and the geometric resolution of the imaging sensors, which leads to higher data volumes. On the other hand, the user is often only interested in special information of the received sensor data and not in the whole data mass (Halle et al. 2000). Therefore, image data processing must be dedicated to on-board tasks for a satellite. Thus, the new generation of satellite must have strong on-board data processing capabilities, whose device has been given various names, such as on-board data processor, or on-board signal processor, or on-board computer, or on-board data processing software, or on-board detector, or on-board classifier, etc. (Halle et al. 2000). All these on-board processing systems should meet the following basic demands:

- Reducing the data volumes and data downlinking rate without loss of any information,
- User can get image knowledge and information, rather than traditional images, immediately without large processing and distribution efforts and
- User can directly downlink the useful information and knowledge extracted from images with a real-time mode without time delay.

Especially, for those rapidly response to time-critical events, such as life rescue, fire spot monitoring, gale warning or damage monitoring of the environment indeed are to migrate to the on-board performance from ground-based image processing. The conventional image processing includes, but not limited to, should be the first step to migrate to on-board processing:

- Pre-processing of satellite images (e.g., noise removal, enhancement and segmentation),
- Geo-coding of satellite images,
- Change detection (e.g., land cover, land use, shoreline erosion and glacier moving) and
- Object recognition (e.g., wildfire, nuclear facilities and moving objects).

On-board processing is therefore expected to be applied in several critical satellites and instrument functions. One of the most essential capabilities provided by on-board processing is autonomous. Autonomous data collection allows smart instruments to adjust their data collection scheme configuration in order to optimize a specific measurement.

2.2 ARCHITECTURE OF ON-BOARD IMAGE PROCESSING

The European Space Agency (ESA) has proposed an architecture for on-board data system for image processing and analysis. The basic architecture consists of telecommand and telemetry

DOI: 10.1201/9781003319634-2

modules, on-board computers, data storage and mass memories, remote terminal units, communication protocols and busses (Figure 2.1). The elements in Figure 2.1, which is described by the SAVOIR Advisory Group, are common to all on-board image processing projects, but subject to modification for different requirements from users in accordance with different missions from science, exploration, earth observation and so on (ESA 2021).

The ESA's mission considered that the major functions in the on-board architecture in Figure 2.1 have the following:

- *On-board computers*, which is a central core of the spacecraft-based on-board mission and is considered as a spacecraft management unit (SMU) or a command and a data handling management unit (CDMU) in the ESA mission.
- *Remote terminal units (RTU)*, which is considered carrying on a medium-large size spacecraft and as a remote interface unit (RIU) in the ESA mission.
- *Platform solid state mass memories*, which is considered for the mass memory for the payload data in ESA's Earth observation and science missions.
- *Telemetry (TM)/telecommand (TC) units*, which are considered as being multiplexed to the intended internal addresses. Two categories of commands are the high priority commands (HPCs) and the normal commands in the ESA mission (see Figure 2.1).
- *Busses and communication protocols*, which is for the use of communications protocols (see Figure 2.1).
- *Internal command and control busses,* which are commonly considered for command, control and data transfers of a spacecraft platform (Wang et al. 2017).

2.3   TOPICS OF ON-BOARD DATA PROCESSING

The topics of on-board image or data processing are very wide, different missions may contain different on-board processing methods and algorithms. Per the experiences over decades in on-board image processing, and from a perspective of on-board data processing levels, the three categories of on-board data processing are the following.

2.3.1   Low-Level On-Board Data Processing

The low-level of on-board data processing should have the following basic capabilities:

- ✓ On-board image filtering, enhancement and radiometric balance,
- ✓ On-board data compression,
- ✓ On-board geometric correction of sensor signals,
- ✓ On-board geometric correction of systematic alignment errors and
- ✓ On-board geometric correction of spacecraft attitude.

- **On-board image filtering**

    Actually, even current common cameras have the functions of filtering on-board. For example, the Image Processing & Storage subsystem (IPS) of the Spanish Earth Observation Minisatellite (MINISOB) can receive the (analog) signal from the detector electronics ("proximity electronics"), to process it for temporary on-board storage and to transmit it to ground using ESA standards.
- **On-board image compression**

    Earth observation usually collects in most cases very large volumes of image data; it is therefore essential to minimize the volume for increasing on-board storage capacity and the on-board transmission rate. One of the typical on-board image processing tasks is on-board image compression, which is usually implemented by on-board digital signal processor (DSP). Different missions and tasks usually require different compression rate.

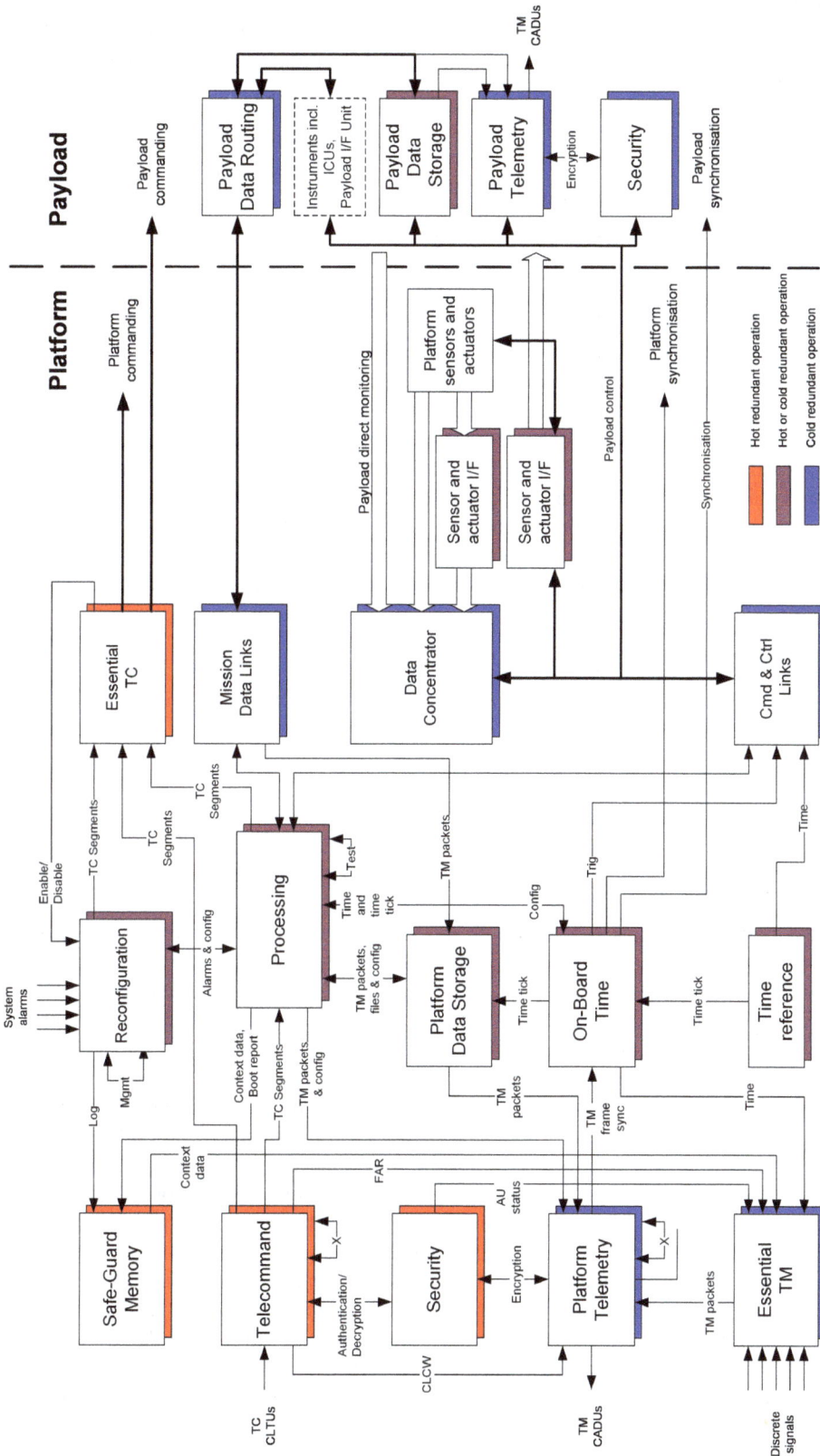

FIGURE 2.1 Functional architecture of on-board data system. (Courtesy of https://www.esa.int/Enabling_Support/Space_Engineering_Technology/Onboard_Computers_and_Data_Handling/Architectures_of_Onboard_Data_Systems.)

In recent years, many methods and algorithms have been presented for spaceborne applications using highly integrated custom circuits (ASIC) technology, or field-programmable gate arrays (FPGA) technology, which makes it possible to implement, in high pixel rate systems, sophisticated real-time compression schemes (Lambert-Nebout and Moury 1999; Lambert-Nebout et al. 2001).

- **On-board storage**

  In addition to the memory requirements arising from the compression and processing functions, there are massive memory requirements from mission requisites. This is performed based on a big solid-state mass memory.

- **Data encryption**

  In order to prevent those unauthorized use of the data, all of the data are usually required for an optional encryption.

- **On-board navigation and control**

  Navigation of the spacecraft is performed autonomously on-board by a combination of GPS measurements and orbit propagation. Three-axis attitude control and fine pointing are provided by a high-accuracy autonomous double-head star tracker, a GPS receiver and a set of reaction wheels (Teston et al. 1999). Autonomous star tracker (AST) is one of the most important attitude determination sensors, and provides the full-sky coverage and achieves the high-pointing accuracy required in Earth observation (Gantois et al. 2006). A set of gyroscopes may be accommodated for improving the short-term pointing stability. Multi-antennae GPS can determine the position with a decimeter accuracy level.

2.3.2   MIDDLE-LEVEL ON-BOARD DATA PROCESSING

The middle-level on-board data processing should have, e.g., the following capabilities:

- ✓ Completely autonomous housekeeping,
- ✓ Completely autonomous data management,
- ✓ Completely autonomous sensor and platform control and
- ✓ Autonomous resource management.

- **On-board image data management**

  On-board data management provides the allocation of on-board resources and in scheduling of operations with a considerable flexibility relative to traditional ground-based data management. For example, PROBA (PRoject for On-Board Autonomy) developed by ESA mission proposed the following operations functions for on-board image data management (Bernaerts et al. 2002; Gantois et al. 2006):

  1. **On-board housekeeping:** Satellite autonomously take care of all routine housekeeping and resource management tasks. This includes the decision-making process in the case of anomalies, failure detection, failure identification and first-level recovery actions, as well as software loading, unloading and management (Teston et al. 1999). A summary of the information available on-board is downlinked to the control center at regular intervals (https://earth.esa.int/web/eoportal/satellite-missions/p/proba-1). More detailed information may be downlinked on specific request by the control center.

  2. **On-board data management:** Satellite should have enough capabilities to autonomously perform all conceivable manipulations of data to meet the various user's tasks on-board, e.g., data handling, data storage, data downlink, data distribution, usually called *"distributor"*, etc. (Zhou et al. 2004).

  3. **On-board resources usage:** Satellite is expected to be capable of autonomous management and assignment of resources usage, such as power (Teston et al. 1997). The allocation should be performed on a dynamic basis, resolving task constraints and priorities.

- **On-board platform control**

    Spaceborne platforms are expected to be controlled intelligently and autonomously, including, but not limited to, the following several aspects (Bernaerts et al. 2002; Gantois et al. 2006):

    ✓ On-board platforms adjust their positions and attitude in space relative to the constellation of sensors in response to collaborative data gathering,
    ✓ On-board event-driven autonomous observation with single satellite and multiple satellites in a networked satellite system and
    ✓ On-board decision support, schedule and planning for Earth observations.

    The on-board data processing and intelligent sensor control relies on technologies that support the configuration of sensors, satellites and sensor webs of space-based resources (https://sbir.nasa.gov/content/board-science-decisions-and-actions-1).

2.3.3 High-Level On-Board Data Processing

A high-level data processing is required for generation of value-added products, for which many robust algorithms (less human interaction) are used. This level of processing should be cost effectively performed on the ground at present. Such a processor should have an advanced capability for high-level on-board image processing, such as change detection capability. Several examples of typical high-level on-board data processing include (Zhou et al. 2004):

✓ On-board completely autonomous mission planning and schedule,
✓ On-board thematic classifier for disaster warning and monitoring,
✓ On-board change detection so that only specified change data are transmitted and
✓ On-board predication via specific model.

- **On-board mission planning and schedule**

    The required image data processing rate is high and beyond the capability of the most current on-board digital processing capability. For example, Future Intelligent Earth Observation (IEOS) proposed by Zhou et al. (2004) will resolve the planning and scheduling of missions on-board using a combination of a constraints solver and optimizer to achieve the best possible mission data return possible. Ideally, a completely autonomous mission planning, i.e., the schedules are programmed in on-board software, is feasible in principle. When required, the on-ground and the OBMM (on-board mission manager) mission planning tools will be used for coordinating the schedule of activities, whose resulting schedule must be confirmed on-ground prior to its execution on-board (Teston et al. 1999).

- **On-board classification**

    Associating with increasing of on-board processing for the images acquired by satellite and spaceborne, on-board classification is considered as one of the most important tasks for on-board image processing. So-called on-board classification is indeed classification and segmentation for remotely sensed images onto orbital platforms without ground interference, *i.e.*, image classification for previously computed at the ground will now be computed on-board (Lopez et al. 2013).

    The on-board classification algorithms majorly consist of

    ✓ Expert-derived decision tree classifiers,
    ✓ Machine-learned classifiers such as SVM classifiers and regressions, classification and regression trees (CART),
    ✓ Bayesian maximum-likelihood classifiers,
    ✓ Spectral angle mappers and
    ✓ Direct implementations of spectral band indices and science products.

Several of the matureal examples can be seen in public domain after their demonstration. For instance,

a. **NASA EO-1 satellite (Earth Observing-1)**

The NASA EO-1 satellite (Earth Observing-1), launched on November 21, 2000, is as part of a one-year technology validation/demonstration mission of the Autonomous Sciencecraft Experiment (ASE) (Chien et al. 2009). Classifiers on EO-1 include a manually constructed classifier, a Support Vector Machine (SVM), a Decision Tree and a classifier derived by searching over combinations of thresholded band ratios. Classifiers can be used on-board a spacecraft to identify high-priority data for downlink to Earth, providing a method for maximizing the use of a potentially bandwidth-limited downlink channel. On-board analysis can also enable rapid reaction to dynamic events, such as flooding, volcanic eruptions or sea ice break-up. The manual and SVM classifiers have been uploaded to the EO-1 spacecraft (Castano et al. 2006).

- **Thermal classification and summarization:** The on-board thermal classification algorithm on EO-1 finds all pixels matching conditions H1, H2 and H3 listed and labels them as "HOT". All remaining pixels that satisfy conditions E1, E2 and E3 are labeled as "EXTREME". The thermal algorithm downlinks a map of the hot and extreme pixels as well as the extracted 12 spectral bands of the hot and extreme pixels (of requested). Figure 2.2 shows sample thermal products (Chien et al. 2009).

- **Flood classification:** The on-board flood classification algorithm is intended to enable on-board recognition of major flooding events (Ip et al. 2006). Through iterative analysis of data from test sites two algorithms were developed for a range of sediment loads. The first of these flood detection algorithms uses the ratio of 0.55 μm/0.86 μm Hyperion data. The second utilizes the ratio between the 0.99 μm/0.86 μm data. One complication in utilizing the flood classifier is accurate estimation of cloud cover for masking the flood scene as clouds are also H2O and thus likely to be classified as surface water. In order to address this issue, prior to flood detection the images are screened for clouds using a cloud classifier developed by MIT Lincoln Laboratory (Griggin et al. 2003). Figure 2.3 shows a sequence of Hyperion flood scenes with the corresponding derived classification maps, which demonstrates the EO-1's 16-day-repeat cycle is sufficient to capture the onset and retreat of a flood event such as this one:

- **Cryosphere classification:** EO-1 and ASE also demonstrated on-board classification of Snow, Water, Ice and Land (SWIL). A classifier was manually derived (Doggett et al. 2006) and later a SVM algorithms was automatically learned by training on expert-labeled data. Figure 2.4 shows a sequence of Hyperion images and derived classification. The cryosphere algorithm uses the Normalized Difference Snow Index (NSDI) defined as NSDI = (0.56 μm − 1.65 μm)/(0.56 μm+1.65 μm). Another cryosphere classifier was developed using Support Vector Machine (SVM) learning techniques and is described in (Castano et al. 2006).

b. **The BIRD mission (Bi-spectral Infrared Detection)**

The BIRD mission (Bi-spectral Infrared Detection) is a small satellite mission funded by Deutsches Zentrum für Luft- und Raumfahrt (DLR, i.e. German Aerospace Center) on October 22, 2001. The classification module on the satellite, called "On-Board-Classification", and the classification test and verification module at the ground station, called "On-Ground-Classification". Each module contains a unit for the data pre-processing like a feature extraction from the available sensor image data. For the classification experiment, the features are the radiation values of the sensors, and are input for the classification process. The center of the classification process is the Neuro Chip NI1000, which works internally with so-called "Radial Basis Function Networks" (Halle et al. 2002).

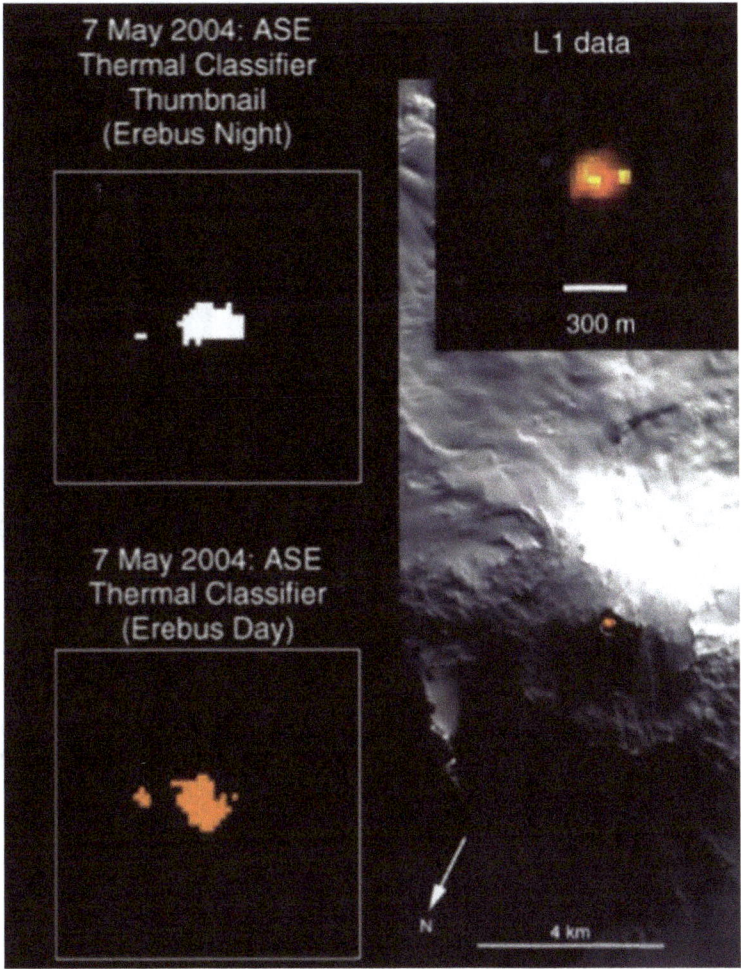

FIGURE 2.2   Hot and extreme pixel classification maps as well as Level One full data for EO-1/ASE trigger and autonomous response data acquired of the Mount Erebus volcano on May7th, 2004, on two overflights. (Courtesy of Davies et al. 2006.)

FIGURE 2.3   Dlamantina flooding sequence. False color and classified scenes of prime sites (256 × 1024 pixels/7.7km × 30km center subset of a Hyperion scene) along the Diamantina River, Australia (January 5, January 30, February 6, February 15, February 22 and February 29 in 2004). (Courtesy of Ip et al. 2006.)

FIGURE 2.4   Sequence of false color and classified images as developed on-board EO-1 tracking sea ice breakup at Prudhub Bay, Alaska February29, 2004 to June27, 2004. (Courtesy of Chien et al. 2009.)

At the end of the classification process on-board BIRD, images with so called "class maps" are the results. They will be labeled with the time-code of the data-take of the corresponding row-data image. These class maps are directly available for the user at the ground station. An example of the on-board classification capability is the result after the data-take in South-West Australia on July 1, 2002 (Figure 2.5(a)). The on-board classifier has detected a huge fire frond in a forest near the Vancouver Bay. (Figure 2.5(b)) (Halle et al. 2002).

c.   **The IPEX spacecraft (Intelligent Payload Experiment)**

The IPEX (Intelligent Payload Experiment) spacecraft is a small spacecraft (10 × 10 × 10 cm) that was launched into Earth orbit on December 6, 2013. It carries a random forest classifier that performs on-board analysis of images collected by its

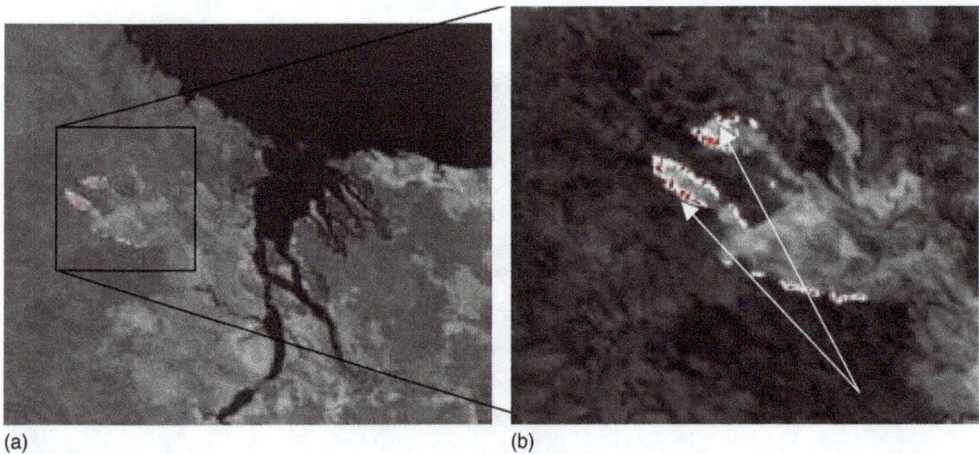

(a)   (b)

FIGURE 2.5   Area of bushfires in the area of Sydney in Australia obtained by BIRD on January 4, 2002; MIR with overlay bushfire. (Courtesy of Halle et al. 2002.)

five 3-megapixel cameras (Altinok et al. 2016). The classifier architecture was adapted from the TextureCam instrument, which integrates image acquisition, processing and classification into a single system (Bekker et al. 2014).

This is the first instance of a random forest running in space, and it is the first instance of computer vision on-board a small CubeSat spacecraft (Thompson et al. 2015). Full classification maps for five well-illuminated, noncorrupted scenes are in Figure 2.6:

- **On-board detection**

  For example, satellite imagers gathered by on-board video cameras can be processed on-board the satellite and compared with normal status images stored on a database system. Identification of changes in these images will result in transmission of early warning messages to ground about these disasters.

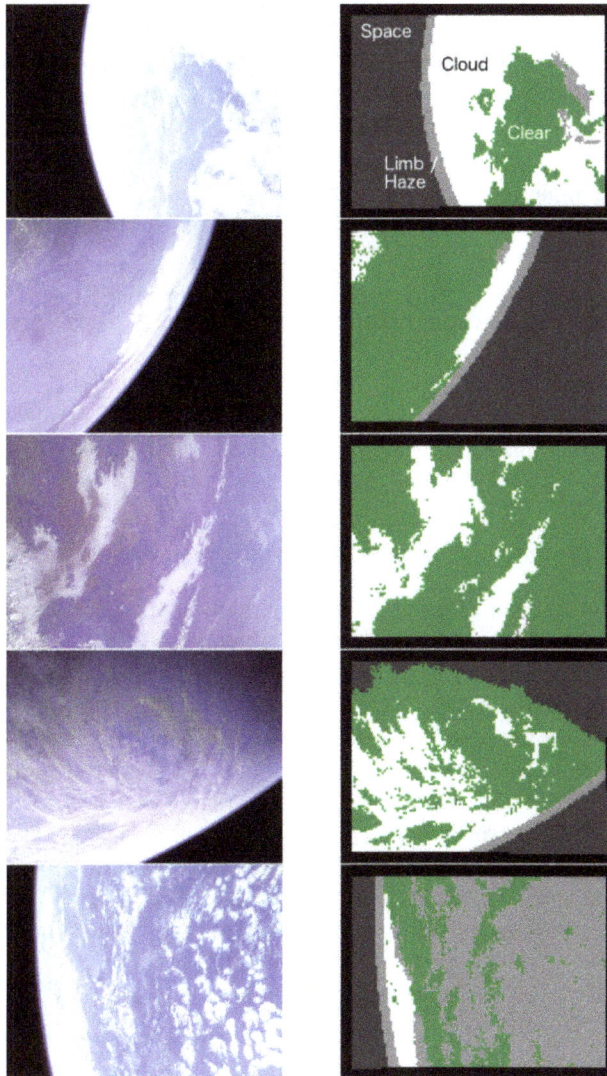

FIGURE 2.6 All returned IPEX classification maps. Top to bottom: 3746b, 866a3, ae0d6, 574ce, cebdf. Color coding: Black: unclassified. Dark gray: outer space. Light gray: horizon/haze. White: opaque cloud. Green: clear. (Courtesy of Altinok et al. (2016).)

• **On-board data distributor**

Intelligent Earth Observation Satellite Network System (IEOSNS) will automatically and directly distribute data to different users upon their request without other human involvement and with minimum delay. The optimal downlink times should be uploaded in the form of a file from a ground control center or calculated on-board the geostationary satellites. The more important parts of the data are sent first, followed by the less important parts of the data.

• **On-board instrument commanding**

The typical instrument commands contain planning, scheduling, resource management, navigation, and instrument pointing, downlinks of the processed data, etc. (Zhou et al. 2004). Specific pre-defined operations procedures are called up as required by specifying items (2) and (3). These procedures will be resident on-board, or may be uplinked from ground. Therefore, a spacecraft command language will need to be implemented on-board on top of the basic telecommand execution module (Teston et al. 1999).

One of two radically different options (which are still being evaluated) may be selected for implementing operations procedures, while taking account of the fact that complex operations procedures as well as on-board software updates need to be executed:

✓ Complex operations are considered as on-board software updates so that any complex ground interaction is executed by the uplink of an on-board software task which is scheduled by the on-board scheduler and which has access to a library of on-board services.

✓ Software updates are seen as complex operational procedures, and any ground interaction is executed through a dedicated spacecraft command language and an on-board interpreter.

For what concerns the lower-level commanding design, packet telecommanding is supported using the Coda Optimistic Protocol-1(COP-1) protocol. Commands are interpreted at spacecraft computer level or at instrument level. Direct priority commands are supported for emergency recovery situations.

2.4   CHALLENGES OF ON-BOARD DATA PROCESSING

The on-board processing is promising in the future for many applications, but many challenges will in practice be encountered. The two typical challenges are below.

1. **On-board computer**

   High-performance on-board computer must provide sufficient performances to support spacecraft autonomy, especially, must support the processing normally performed on-ground and the on-board scientific data processing (Teston et al. 1999). Thereby, a high-performance processor is required to carry out the spacecraft management including guidance, navigation, control, housekeeping and monitoring, on-board scheduling and resource management (Teston et al. 1999). The development of such a processor, which is a space version of standard commercial processors, has been initiated by ESA in later 1990s (Teston et al. 1997). In addition, the capability to meet various users' requirements for data distribution through the selective use of the downlink and the on-board mass memory is highly required.

2. **On-board software**

   On-board software must reach the functionality to meet the user's needs from traditional image-based data to advanced image-based information/knowledge (Zhou et al. 2004). This means that many performances done on-ground have been migrated to on-board processing, thus, an autonomy requires larger and more reactive software than is the case on conventional spacecraft to allow the support of custom, re-usable, and automatically generated software (Teston et al. 1999). Especially, on-board software is required to have a true capability of deployment of "knowledge building systems (KBSs)". This capability can be envisioned as an end-to-end system starting with sensors (spaceborne, airborne or Earth-bound) and ending with users who derive knowledge through scientific research and/or exploit the knowledge in real-time mode. The knowledge itself is preserved for posterity. Knowledge building involves a dynamic interplay between people and technology that transforms observations into data initially. The dynamic interplay then transforms data into information and information into knowledge (Ramapriyan et al. 2002; Zhou 2003).

2.5 INTRODUCTION TO FPGA

The FPGA has been on the market for more than a decade; its principal use is for discrete logic integration and board-level integrated circuit (IC) count reduction. It is itself a digital IC and represents the highest transistor count of all, up to 100 million transistors 9 (Caffrey et al. 1999). For comparison, the Intel Pentium III is in the neighborhood of 9.5 million transistors 10 (https://en.wikipedia.org/wiki/Pentium_III). Recent advances yielding this density and improving speed performance have made these parts appropriate for hardware DSP. An historic concern for DSPs has been technology obsolescence. The FPGA market is not driven by DSP applications, so this technology will not lose technical support or future product development as many custom DSP ASICs have in the past.

FPGA has been widely used in aerospace field. FPGA chips with aerospace-grade include one-time programming anti-fuse FPGA launched by Actel company and reconfigurable Virtex series FPGA based on SRAM launched by Xilinx company (Song, 2010). In this chapter, the FPGA chip of the Xilinx company is selected as the hardware platform of developing the algorithm. Figure 2.7 shows the internal structure of Xilinx Spartan-2 series FPGA chip. The FPGA chip of Xilinx company is mainly composed of programmable input and output block (IOB), configurable-logic block (CLB), digital clock manager (DCM), embedded block RAM, rich wiring resources, embedded bottom functional units and embedded special hardware modules (Tian et al. 2008).

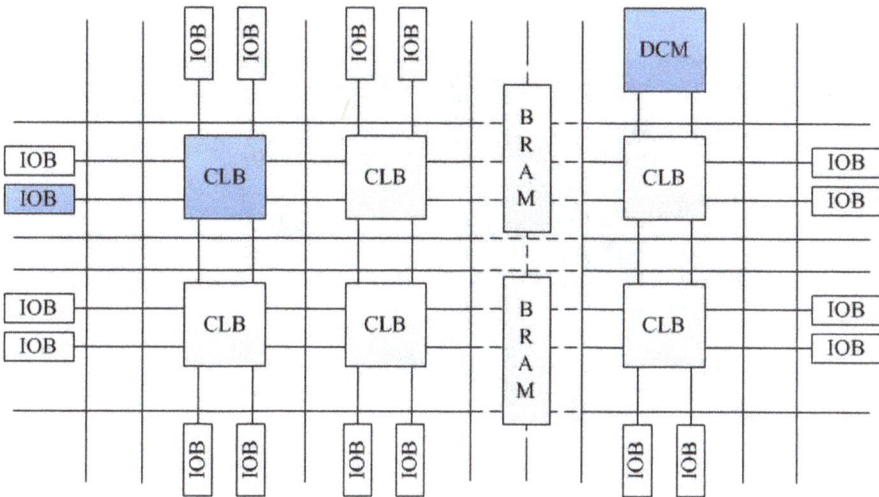

FIGURE 2.7 Internal structure of a FPGA chip. (Tian et al. 2008.)

Because different series of FPGA chips are aimed at different application fields, its internal structure should be different (Zhou et al. 2015a, 2015b, 2015c, 2015d).

The Xilinx has four series of FPGA chip devices, namely Spartan, Virtex, Kinex and Artix. The latest generation of products is 7 series FPGA, including Virtex-7, Kinetex-7 and Artix-7, of which the latter two are the latest product series. Although the new generation of 7 series FPGA is similar to Virtex-6 series in logic, it has many innovations, including adopting 28nm high-performance and low-power technology, creating a unified and scalable architecture, improving tools and intellectual property (IP), etc. Compared with previous products, 7 series FPGA not only improves performance, but also reduces power consumption. The main difference between the three 7 series FPGAs is the size of the device and the target market: Artix-7 series is optimized for low cost and low power consumption, uses small package design and is positioned for a large number of embedded applications. Virtex-7 series, as before, is positioned as the highest performance and maximum capacity; Kinex-7 series lies between the other two series, finding a balance between cost and performance (Bailey 2013).

Considering the complexity of a single image space resection algorithm and the cost of design and development, Artix-7 series FPGA chip (XC7A200T-2FBG676C) with low cost and low power consumption is selected as the hardware processor. The FPGA chip of Artix-7 series adopts 6-input LUT (look-up table). This type of FPGA chip has rich logic resources and more I/O pins, which can basically meet the experimental needs. The hardware experimental platform is the AC701 evaluation development board containing the FPGA chips of Artix-7 series. The layout of FPGA chips and peripheral circuits on the development board is shown in Figure 2.8. FPGA development board provides a good hardware development environment. AC701 board has similar characteristics with many embedded processing systems. It provides a DDR3 SODIMM memory (storage capacity of 1G and bandwidth of 533MHz/1066Mbps), a PCI Express 4-channel edge connector, an Ethernet PHY (physical interface transceiver) with three modes, general I/O and a UART to USB bridge. The hardware resources of the evaluation, or called "trial" FPGA chip are shown in Table 2.1. The CLB of Artix-7 is composed of four Slices and additional logic. One Slice is composed of multiple Logic Cells. The internal structure of a Slice is shown in Figure 2.9. Among the I/O pins of 500 users, there are 400 high-performance I/O pins supporting 1.2V to 3.3V voltage, which is the number of pins that users can freely allocate.

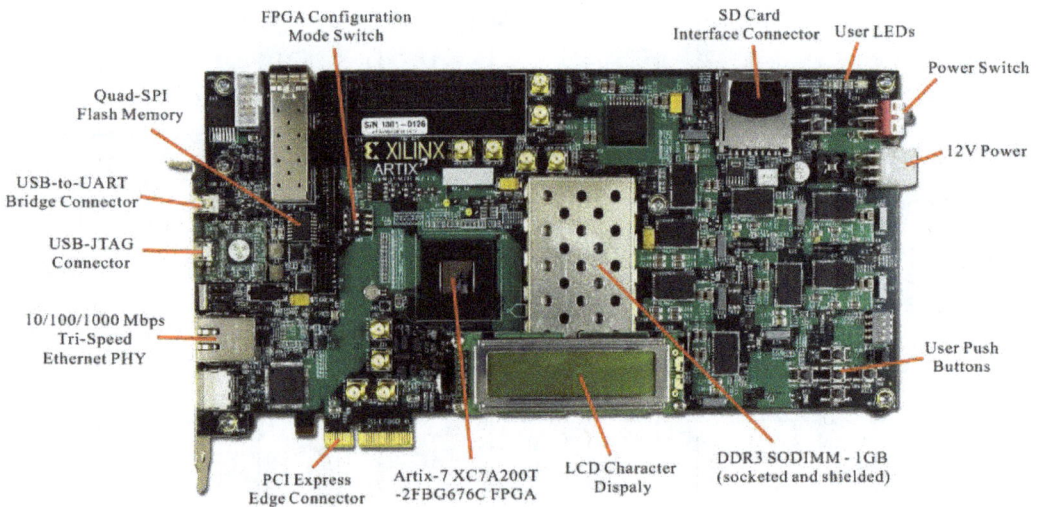

FIGURE 2.8   AC701 evaluation development board.

TABLE 2.1

Hardware Resources of Artix-7 XC7A200T-2FBG676C FPGA Chip (Xilinx 2014)

Components		Quantity (PCs)	Components		Quantity (PCs)
Logic Cells		215,360	PCIe		1
CLBs	Slices[1]	33,650	Block RAM Blocks[4]	18 kB	763
	Max Distributed RAM	2888		36 kB	365
				Max(kB)	13,140
DSP48E1 Slices[2]		740	GTPs		16
CMTs[3]		10	XADC Blocks		1
Total I/O Banks		10	Max User I/O[5]		500

Note: (1) Slice of 7 series FPGA includes 4 LUTS and 8 triggers; Only 25–50% of the LUTS of Slice can be used as distributed RAM or shift register (SRL). (2) Each DSP Slice contains a pre-adder, a 25×18 multiplier and an accumulator. (3) Each CMT (clock management tile) contains an MMCM (mixed mode clock manager) and a PLL (phase-locked loop). (4) Block RAMs are essentially 36 kB in size, and each block can also be used as two independent 18 kB blocks. (5) Only 400 of them can be assigned by users.

2.5.1 FPGA-BASED DESIGN AND IMPLEMENTATION FOR ISE SOFTWARE

It is necessary to design circuit functions and to select a FPGA device for on-board geometric calibration. The design of the circuit functions usually adopts the top-down hierarchical method. Starting from the system level, the system is divided into multiple secondary modules, and then the secondary modules are subdivided into tertiary modules, which can be subdivided until they can

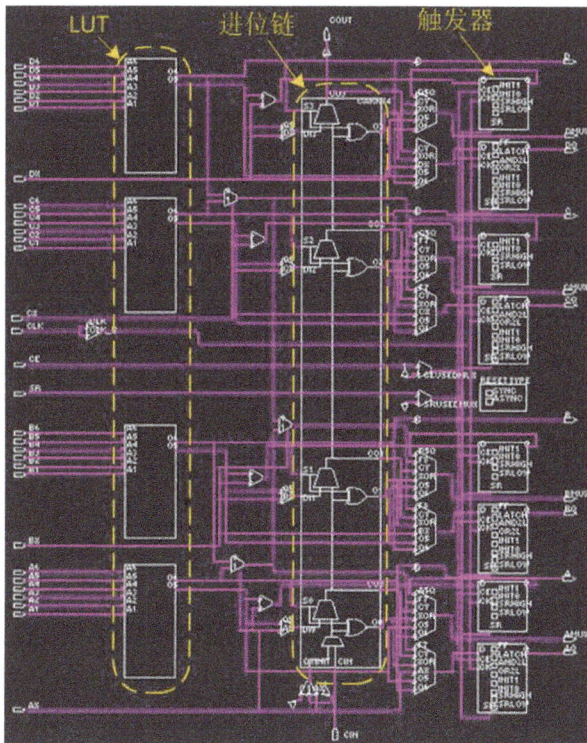

FIGURE 2.9   Internal structure of Artix-7 slice unit.

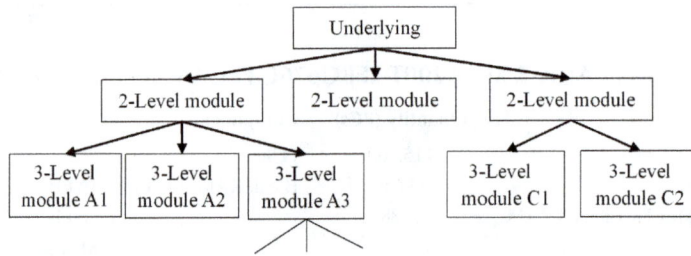

FIGURE 2.10   The top-down hierarchical design method when using a FPGA chip. (Tian et al. 2008.)

be directly realized by using the basic computing unit or IP core, as shown in Figure 2.10. Different types of FPGA have different hardware resources. For a complex design, many devices should be selected. The selection of the devices has been described in the section above.

After designing circuit functions and selecting devices, we can enter a specific design and development process. The software platform using FPGA-based design and development is a called "integrated software environment (ISE)" (version 14.7), which is from Xilinx company. The basic process of FPGA development using ISE software can refer to the five steps proposed by Tian and Xu (2008), *i.e.*,

1. Create engineering and design inputs,
2. Create Testbench and simulate RTL level (Register Transfer Level) code,
3. Add timing and pin constraints (as shown in Figure 2.11),
4. Integrate and implement them and
5. Generate bitstream file and program FPGA chip.

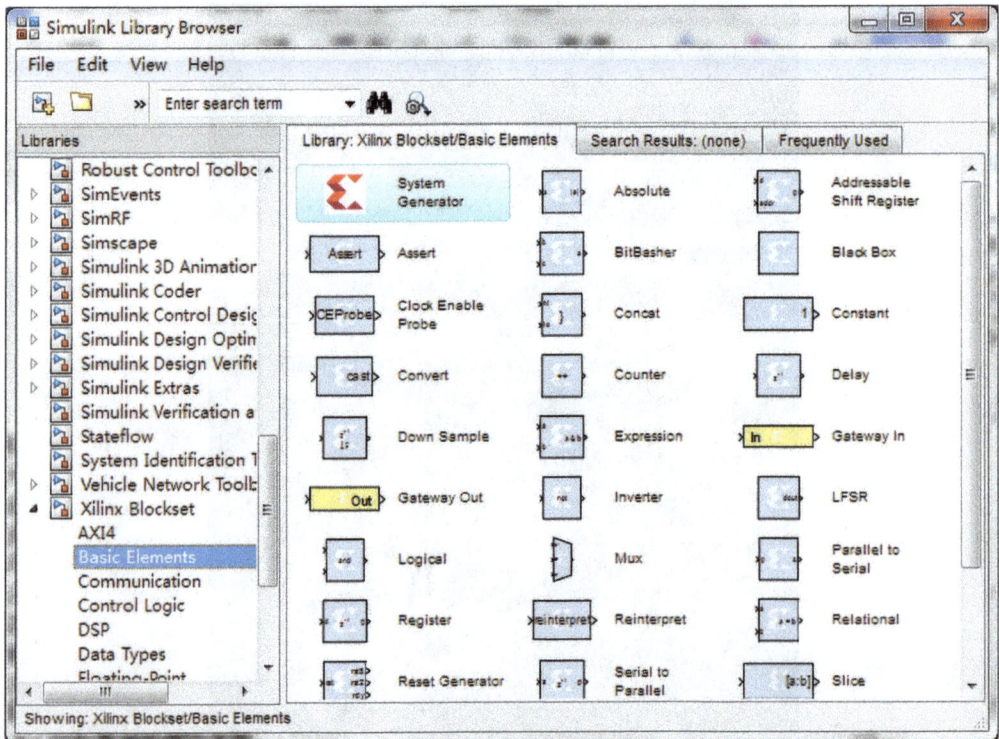

FIGURE 2.11   Composition of the Xilinx basic module library.

An optimal design circuit usually needs repeated modification and verification. To this end, Steps 1–4 above are usually repeated. In addition to the functional simulation in Step 2, after the synthesis and implementation steps in Step 4, the post-synthesis simulation and timing simulation are carried out respectively to further verify and evaluate whether the design circuit meets the requirements. After obtaining the bitstream file of the designed circuit, the simulation test based on the hardware platform can also be carried out (Zhou and Kaufman Hall 2001).

2.5.2 FPGA DEVELOPMENT METHOD BASED ON SYSTEM GENERATOR

Since the single-image space rear intersection involves many computational steps, to select the traditional FPGA digital system design and development method based on the Hardware Description Language (HDL) for the design and development of the entire translation computing system, the implementation is quite difficult even with a large number of IP cores (encapsulated by HDL code). A simple FPGA design and development is to use System Generator for DSP, an integrated development tool for FPGA digital processing system launched by Xilinx. System Generator enables seamless connectivity with Simulink tools in MATLAB software, allowing designers to drag Xilinx's IP kernel module into Simulink and then connect them to build the FPGA computing system. Because the whole design process does not need to write any code and is completed in a visual environment, it can easily and quickly modify any part of the system design, and the operation is very simple and flexible. The System Generator tool enables the rapid construction of the FPGA system and automatically generates HDL code, allowing designers not to spend too much time getting familiar with the HDL language, thus it could greatly reduce the difficulty of design development and shorten the cycle of design development (Tian and Xu 2008). The research in this chapter will use this tool to complete the construction of the translation computing system to accelerate the hardware design and verification of the algorithm FPGA implementation.

The System Generator tool provides all the basic modules needed to complete the design of an FPGA system, or called the IP core module. These basic modules constitute the Xilinx module library, which is divided into three major categories: basic module library, parametric module library and module library controlling peripheral circuits. The system constructed by these Xilinx modules can be implemented on FPGA in MATLAB/Simulink environment. The basic module library (Xilinx blockset library on the left in Figure 2.11) contains more than 100 underlying modules, which are specifically divided into 12 sub-categories, which is enough to meet the needs of design and development. An essential module for the System Generator model includes the System Generator module, the Gateway In and Gateway Out Module which are both in the Basic Elements. The System Generator module is used to set the model of the FPGA chip and convert the design system to ISE engineering-related documents. The Gateway In and Gateway Out modules also belong to data type modules, which respectively connect the entry and exit between Simulink and System Generator tools to realize the conversion between fixed-point numbers and floating-point numbers. In the basic module library, common mathematical operations such as addition, subtraction, multiplication, division, cosine, opening, opposite number, size comparison can find the corresponding IP kernel modules in the Basic Elements, Math and Float-Point libraries. The counter, data selector, delay unit, register, ram, data format conversion, logic operation and other functional modules often used in operation have corresponding IP cores. These basic IP kernel modules can be used for constructing complex computational systems.

The general process of FPGA design and development using System Generator is shown in Figure 2.12. Specific design and development steps are: (1) MATLAB/Simulink is used for algorithm modeling and behavior level simulation, and the Simulink file with suffix. SLX is generated; (2) Automatically generate RTL level code (i.e. HDL code file) and IP core code further processed in ISE through System Generator, and generate testbench test file used in function simulation at the same time; (3) Use Modelsim software or ISIM simulation tool for functional simulation to verify

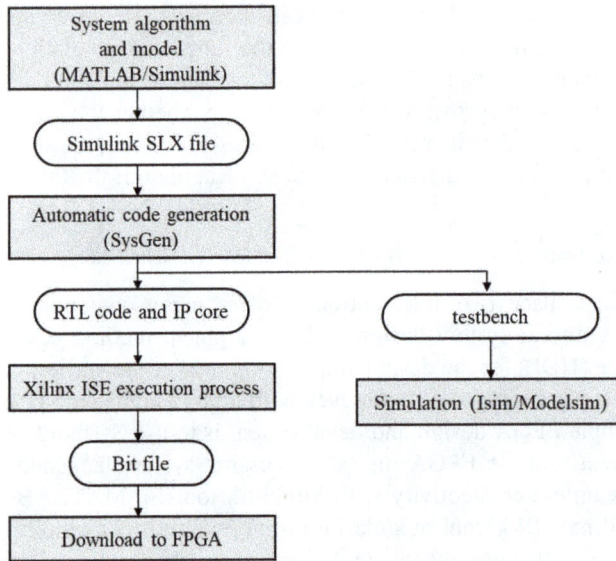

FIGURE 2.12 FPGA development process based on System Generator (Tian and Xu 2008)).

whether the system meets the design requirements; (4) Complete the execution process of setting constraints, synthesis and implementation;(5) The bitstream file is generated, and then the bitstream file is downloaded to FPGA to complete the whole design process.

2.6 INTRODUCTION TO RECONFIGURABLE COMPUTING TECHNOLOGY

Reconfigurable computing (RCC) uses high-gate-density FPGAs as general-purpose processing elements. RCC accelerates applications by exploiting fine-grained parallelism in a given algorithm. A typical FPGA clock speed is an order-of-magnitude slower than typical microprocessors; thus, any acceleration is accomplished with parallel execution. In its simplest form, the goal is to have tens, hundreds or more arithmetic computations operating simultaneously. This parallelism puts the FPGA in a unique computational position of being able to maintain throughput while increasing the number of operations. For instance, a microprocessor processes fewer samples in time as filter length increases, while an FPGA exhibits increased logic utilization, but maintains total data throughput by increasing the number of multiply-accumulate units in the design.

Disadvantages of RCC implementations include limited bit precision of operations, fixed-point arithmetic and the need to program RCC processors with either a hardware description language or with specialized extended high-level languages that contain calls to RCC implementations of core processing routines. The first two limitations require algorithm designers to truly understand the precision needs of the algorithm and how fixed-point arithmetic will impact the stability and precision of results. This challenge is frequently overlooked when designing for microprocessors that deliver 32-bit floating-point performance, but in an FPGA 32-bit floating-point is exceedingly inefficient. Computing techniques also have a large impact on efficiency and performance for FPGAs. For instance, when taking the magnitude of a complex number, the obvious approach uses the square root operation, but the square root operation is inefficient in the hardware implementation. It has been demonstrated that the CORDIC (Coordinate Rotation Digital Computer) algorithm for coordinate transformation is far more efficient in hardware for performing this computation (Volder 1959). Designing processing systems in an FPGA often requires creative solutions and understanding of algorithms originally developed, in some cases,

30 years ago. It is well worth the effort, though, because the computation available for given weight, space and power is so much greater that it is feasible to do substantial in situ processing of remotely sensed data.

DAPS has produced implementations of three general-purpose spectral processing algorithms. The first work was done on the RCA-2 board and was reported by Caffrey et al. (1999). The algorithm was a simple dot product operator that represents the application of a matched-filter to a spectral data cube. The data cube had 16 spectral channels in a 512×512 scene. Twelve spectral signatures were applied to the data cube, which required 2.0s (averaged over multiple trials) on a 400 MHz Pentium II. The RCA-2 implementation required 0.12s (Szymanski et al. 2000), for a factor of 17 speedups. This design took a portion of a single RCC processor. If the full RCA-2 were used, 32 spectral signatures could be run with no reduction in data rates, resulting in a factor of 45 speedups. Even for a simple algorithm, this speedup illustrates the scalable nature of RCC. Furthermore, the architecture can be scaled at the module level, with multiple modules running in parallel and processing a large number of spectral signatures.

2.7 CONCLUSION

This chapter introduced the architectures of on-board data processing and emphasized the typical topics of on-board data processing by categorizing the on-board image data processing into three levels, i.e., low-level, middle-level and high-level on-board image data processing. Especially, this chapter presents the challenges of on-board image data processing, including hardware and software challenges. Finally, this chapter introduces the FPGA, a chip for data processing, and the potential applications of FPGA in on-board image processing.

REFERENCES

Altinok, A., Thompson, D.R., Bornstein, B., et al., Real-time orbital image analysis using decision forests, with a deployment onboard the IPEX spacecraft. *Journal of Field Robotics*, 2016, 33(2): 187–204. doi: 10.1002/rob.21627.

Bailey, D.G., Design for embedded image processing on FPGAs. *Library of Congress Cataloging-in-Publication Data*. 2013, Wiley Press, ISBN 978-0-470-82849-6.

Bekker, D.L., Thompson, D.R., Abbey, W.J., et al., Field demonstration of an instrument performing automatic classification of geologic surfaces. *Astrobiology*, 2014, 14(6): 486–501. doi: 10.1089/ast.2014.1172

Bernaerts, D., Bermyn, J., Teston, F., PROBA (project for on-board autonomy). *In Smaller Satellites: Bigger Business*, Springer, Dordrecht, 2002, pp. 53–68.

Caffrey, M.P., Szymanski, J.J., Begtrup, A., et al., High-performance signal and image processing for remote sensing using reconfigurable computers//Advanced signal processing algorithms, architectures, and implementations IX. *International Society for Optics and Photonics*, 1999, 3807: 142–149. doi:10.1117/12.367629.

Castano, R., Mazzoni, D., Tang, N., et al., Onboard classifiers for science event detection on a remote sensing spacecraft. *Proceedings of the 12th ACM SIGKDD international conference on Knowledge Discovery and Data Mining*, 2006: 845–851. doi: 10.1145/1150402.1150519.

Chien, S., Tran, D., Schaffer, S., et al., Onboard classification of hyperspectral data on the Earth Observing One mission. *IEEE 2009 First Workshop on Hyperspectral Image and Signal Processing: Evolution in Remote Sensing*, 2009: 1–4, doi: 10.1109/WHISPERS.2009.5289012.

Davies, A.G., Chien, S., Baker, V., et al., Monitoring active volcanism with the Autonomous Sciencecraft Experiment on EO-1. *Remote Sensing of Environment*, 2006, 101(4): 427–446. doi: 10.1016/j.rse.2005.08.007.

Doggett, T., Greeley, R., Chien, S., Castano, R., Cichy, B., Davies, A.G., Rabideau, G., Sherwood, R., Tran, D., Baker, V., Autonomous detection of cryospheric change with hyperion on-board earth observing-1, *Remote Sensing of Environment*, 2006, doi: 10.1016/j.rse.2005.11.014.

ESA. Architectures of Onboard Data Systems. https://www.esa.int/Enabling_Support/Space_Engineering_Technology/Onboard_Computers_and_Data_Handling/Architectures_of_Onboard_Data_Systems, June 19, 2021.

7

Gantois, K., Teston, F., Montenbruck, O., et al., PROBA-2 mission and new technologies overview. *Small Satellite Systems and Services – The 4S Symposium*, December 2006, ESA Publications Division ESTEC, Postbus 299 2200 AG Noordwijk, The Netherlands.

Griggin, M., Burke, H., Mandl, D., et al., Cloud cover detection algorithm for EO-1 Hyperion imagery. *IGARSS 2003. 2003 IEEE International Geoscience and Remote Sensing Symposium*. Proceedings (IEEE Cat. No. 03CH37477). IEEE, 2003, 1: 86–89, doi: 10.1109/IGARSS.2003.1293687.

Halle, W., Brie, K., Schlicker, M., et al., Autonomous onboard classification experiment for the satellite BIRD. *International Archives of Photogrammetry Remote Sensing and Spatial Information Sciences*, 2002, 34(1): 63–68.

Halle, W., Venus, H., Skrbek, W., Thematic data processing on-board the satellite BIRD, 2000, http://www.spie.org/Conferences/Programs/01/rs/confs/4540A.html.

Ip, F., Dohm, J.M., Baker, V.R., et al., Development and testing of the autonomous spacecraft experiment (ASE) floodwater classifiers: Real-time smart reconnaissance of transient flooding. *Remote Sensing Environment*, 2006, 101: 463–481.

Lambert-Nebout, C., Latry, C., Moury, G., On-board image compression for space missions. *Annales Des Telecommunications*, 2001, 56(11/12): 632–645.

Lambert-Nebout, C., Moury, G., A survey of on-board image compression for CNES space missions. *IEEE International Geoscience & Remote Sensing Symposium*, 1999. doi:10.1109/IGARSS.1999.775023.

Lopez, S., Vladimirova, T., Gonzalez, C., et al., The promise of reconfigurable computing for hyperspectral imaging onboard systems: A review and trends. *Proceedings of the IEEE*, 2013, 101(3): 698–722, doi: 10.1109/JPROC.2012.2231391.

On-Board Science for Decisions and Actions. Oct. 31 2013, NASA, https://sbir.nasa.gov/content/board-science-decisions-and-actions-1

Pentium III. Launched on February 28, 1999, https://en.wikipedia.org/wiki/Pentium_III

PROBA-1 (Project for On-Board Autonomy – 1). Jun 12, 2012, https://earth.esa.int/web/eoportal/satellite-missions/p/proba-1

Ramapriyan, H., McConaughy, G., Lynnes, C., Harberts, R., Roelofsb, L., Kempler, S., McDonnald, K., Intelligent archive concepts for the future, *The First International Symposium on Future Intelligent Earth Observing Satellite (FIEOS)*, Denver, Colorado, Nov. 10–11, 2002.

Song, K.F., Application of FPGA in aerospace remote sensing systems. *Ome Information*, 2010, 27(12): 49–55.

Szymanski, J.J., Blain, P.C., Bloch, J.J., et al., Advanced processing for high-bandwidth sensor systems// Imaging spectrometry VI. *International Society for Optics and Photonics*, 2000, 4132: 83–90. doi:10.1117/12.406575.

Teston, F., Creasey, R., Bermyn, J., Mellab, K., PROBA: ESA's autonomy and technology demonstration mission, *48th International Astronautic Congress*, Turin, Italy, October 6, 1997.

Teston, F., Creasey, R., Bermyn, J., et al., PROBA: ESA's autonomy and technology demonstration mission. *Small Satellite Conference*, 1999, https://digitalcommons.usu.edu/smallsat/1999/all1999/41/.

Thompson, D.R., Altinok, A, Bornstein, B, et al., Onboard machine learning classification of images by a CubeSat in earth orbit. *AI Matters*, 2015, 1(4): 38–40, doi:10.1145/2757001.2757010.

Tian, Y., Xu, W.B., *Practical Course of Xilinx FPGA Development. Beijing: Tsinghua University Press*, 2008.

Tian, X., Zhou, F., Chen, Y.W., Liu, L., Chen, Y., Design of field programmable gate array based real-time double-precision floating-point matrix multiplier. *Journal of Zhejiang University (Engineering Science)*, 2008, 42(9): 1611–1615.

Volder, J.E., The CORDIC trigonometric computing technique. *IRE Transactions on Electronic Computers*, 1959 EC-8(3): 330–334. doi:10.1109/tec.1959.5222693.

Wang, W.P., Wang, X.H., Han, B., et al., Design of OBDH subsystem for remote sensing satellite based on onboard route architecture//*MATEC Web of Conferences*. EDP Sciences, 2017, 139: 00189. doi:10.1051/matecconf/201713900189.

Xilinx. 7 Series FPGAs Overview v1.15, February 2014. URL: http://china.xilinx.com/support/documentation/data_sheets/ds180_7Series_Overview.pdf.

Zhou, G., *Real-Time Information Technology for Future Intelligent Earth Observing Satellite System*, Hierophantes Press, ISBN: 0-9727940-0-X, February 2003.

Zhou, G., Baysal, O., Kaye, J., Concept design of future intelligent earth observing satellites, *International Journal of Remote Sensing*, 2004, 25(14): 2667–2685.

Zhou, G., Kaufman Hall, R. Architecture of future intelligent earth observing satellites (FIEOS) in 2010 and beyond. *Earth Observing Systems VIII*, San Diego, CA, 2001.

Zhou, G., Li, C., Yue, T., Jiang, L., Liu, N., Sun, Y., Li, M., An overview of in-orbit radiometric calibration of typical satellite sensors, *The 2015 Int. Workshop on Image and Data Fusion (IWIDF)*, Kona, Hawaii, USA, July 21–23, 2015a. doi: 10.5194/ISPRS, Archives-XL-7-W4-235-2015.

Zhou, G., Li, C., Yue, T., Jiang, L., Liu, N., Sun, Y., Li, M., FPGA-based data processing module design of onboard radiometric calibration in visible/near infrared band, *The 2015 International Workshop on Image and Data Fusion (IWIDF2015)*, Kona Hawaii, USA, July 21–23, 2015b.

Zhou, G., Li, C., Yue, T., Liu, N., Jiang, L., Sun, Y., Li, M., FPGA-based data processing module design of onboard radiometric calibration in visible and near infrared bands, *Proceedings of SPIE on 2015 International Conference on Intelligent Earth Observing and Applications, 0277-786X, Vol. 9808*, Guilin, China, October 23–24, 2015c.

Zhou, G., Liu, N., Li, C., Jiang, L., Sun, Y., Li, M., Zhang, R., FPGA-based remotely sensed imagery denoising, *2015 IEEE International Geoscience and Remote Sensing Symposium (IGARSS)*, Milan, Italy, July 26–31, 2015d.

3 On-Board Detection and On-Board Matching of Feature Points

3.1 INTRODUCTION

The detection and matching of feature points are used as the first step of satellite image processing such as attitude estimation, geometrical calibration, object detection and tracking and ortho-rectification (Mair et al. 2010). While the traditional method whose implementation of detection and matching on ground is not suitable for the increasing requirement for real-time or near real-time remotely sensed imagery applications (Zhou et al. 2004; Zhang 2011). To meet the requirement of real-time processing, various detection and matching algorithms are implemented on field programmable gate array (FPGA), which can offer highly flexible designs and scalable circuits (Pingree 2010). Meanwhile, the parallel processing characteristic and the pipeline structure of FPGA allow data to be processed more quickly with a lower power consumption than a similar microprocessor implementation and/or CPU (Dawood et al. 2002; Mueller et al. 2009). In addition, several FPGA-based implementations are proposed with soft cores microprocessors to carry out the complex algorithms (Ball 2005), such as, motion estimation algorithms (González et al. 2012; 2013; 2015) and epsilon quadratic sieve algorithm (Meyer-Bäse et al. 2010). Hence, the FPGA-based implementation of detection and matching algorithms are widely researched.

Among the detection and matching algorithms, the scale-invariant feature transform (SIFT), speed-up robust feature (SURF), oriented-fast and rotated brief (ORB) and theirs modified algorithms have an excellent performance (Rublee et al. 2011), and these algorithms also have been implemented using FPGAs to achieve various applications. For example, Yao et al. (2009) proposed a Xilinx Virtex-5 FPGA implementation of optimized SIFT feature detection. The proposed FPGA implementation can carry out the feature detection of a typical image of 640×480 pixels within 31 ms. Svab et al. (2009) presented a hardware implementation of the SURF on FPGA. The implementation achieves about 10 frames per second (fps) at image of 1024×768 pixel^2 and the total power consumption is less than 10 W. Schaeferling and Kiefer (2010) proposed a hardware architecture to accelerate the SURF algorithm on Virtex-5 FPGA. Lentaris et al. (2013) proposed a hardware-software co-design scheme using Xilinx Virtex-6 FPGA to speed-up the SURF algorithm for the ExoMars Programme. Schaeferling and Kiefer (2011) implemented a complete SURF-based system on Xilinx Virtex-5 FX70T FPGA for object recognition. The average time was 481 ms for one frame. Sledevič and Serackis (2012) proposed an FPGA-based implementation of a modified SURF algorithm. The proposed architecture achieves real-time orientation and the descriptor calculation can achieve on 60 fps 640×480 video stream only on a 25 MHz clock. Battezzati et al. (2012) proposed a FPGA architecture for implementation SURF algorithm. The entire system is about 340 ms for one frame. Zhao et al. (2013) proposed a real-time SURF-based traffic sign detection system by exploiting parallelism and rich resources in FPGAs. The proposed hardware design is able to accurately process video streams of 800×600 pixel^2 at 60 frame fps. Fan et al. (2013) proposed a high-performance hardware implemented using OpenSURF (Evans 2009) algorithm on XC6VSX475T FPGA. The proposed implementation achieved 356 fps with 156 MHz clock. Krajnik et al. (2014) presented a complete hardware and software solution of an FPGA-based computer vision embedded module capable of carrying out SURF image features extraction algorithm. Chen et al. (2015) proposed an FPGA architecture of OpenSURF. The result discovered that the

DOI: 10.1201/9781003319634-3

architecture can detect features and extract descriptors from video streams of 800×600 pixel^2 at 60 fps. Gonzalez et al. (2016) proposed an FPGA implementation of an algorithm for automatically detecting targets in remotely sensed hyperspectral images. Weberruss et al. (2015) proposed a hardware architecture of ORB on FPGA, which offer lower power consumption and higher frame rates than general hardware does. Among all of them, the SIFT algorithm, which has characteristics on such as scale invariance, rotational invariance and affine invariance, is the most popular and achieves the best performance. However, a few characteristics of SIFT descriptor, such as large computational burden, floating point arithmetic, poor real-time performance, limit its applications with FPGA-based implementation. While SURF algorithm has less computation burden and quicker calculating speed, especially, its detector is easy to implement using hardware with fixed-point arithmetic and parallel characteristic. The feature points also can be detected at different scales. In contrast, its descriptor is poor for a real-time performance and with a large consumption of hardware resources. Therefore, the SURF algorithm is not suitable for on-board processing directly. On the other hand, the ORB algorithm consists of an oriented FAST detector and a rotated binary robust independent elementary features (BRIEF) descriptor (Calonder et al. 2010), which are easy to be implemented in FPGA. While the FAST detector has no scale invariant and need to be further considered, especially when a series images are not at the same scale.

With the analysis above, to implement on-board detection and matching with a high fps, a modified SURF detector and a BRIEF descriptor and the modified SURF detector and the BRIEF descriptor are implemented using FPGA, are proposed in this chapter (Huang et al. 2018; Zhou 2020).

3.2　FEATURE DETECTOR AND DESCRIPTOR ALGORITHM

3.2.1　PC-BASED SURF FEATURE DETECTOR AND ALGORITHM

3.2.1.1　Surf Detector

The basic idea and steps of the SURF detector, which was first proposed by Bay et al. (2006), are summarized as follows. More details can be referenced to Bay et al. (2006).

3.2.1.1.1　*Integral Image Generation*

Integral image is a novel method to improve the performance of the subsequent steps for SURF detector. It is defined by

$$I(x,y) = \sum_{r=1}^{y} \sum_{c=1}^{x} i(r,c) \tag{3.1}$$

where $I(x, y)$ represents the integral value at location (x,y) of the image, $i(r,c)$ represents the gray value at the location (r,c) of the image. When using the integral image, it is efficient to calculate the summation of pixels in an upright rectangle area of image.

3.2.1.1.2　*Hessian Matrix Responses Generation*

The expression of Hessian matrix and the determinant of Hessian matrix can be expressed by, respectively

$$H(x,\sigma) = \begin{bmatrix} L_{xx}(x,\sigma) & L_{xy}(x,\sigma) \\ L_{xy}(x,\sigma) & L_{yy}(x,\sigma) \end{bmatrix} \tag{3.2}$$

$$\det H_{approx} = L_{xx}L_{yy} - L_{xy}^2 = D_{xx}D_{yy} - \omega^2 D_{xy}^2 \tag{3.3}$$

where ω^2 is a weight coefficient equal to 0.91 (Bay et al. 2006); the D_{xx}, D_{yy} and D_{xy} are computed respectively by the integral image and the box filters (see Figure 3.1(a)–(c)). Figure 3.1(a) is a box

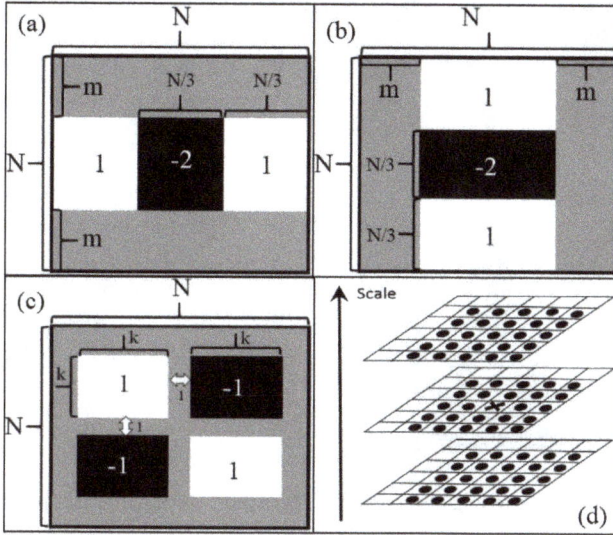

FIGURE 3.1   (**a**) Filter box in X direction; (**b**) filter box in Y direction; (**c**) filter box in 45° directions; (**d**) 3D non-maximal suppression.

filter in X direction, Figure 3.1(b) is a box filter in Y direction and Figure 3.1(c) is a box filter at XY direction. N can be selected at 9, 15, 21…; m can be selected at 2, 3, 4…; and k can be selected at 3, 5, 7…, respectively. In addition, the rectangles with gray mean their value with 0; the rectangles with white mean their values with 1; and the rectangles with black mean their value with −2.

3.2.1.1.3   *Using the 3D Non-Maximal Suppression*

A 3D non-maximal suppression is performed to find a set of candidate points. To keep the feature point with a strong robustness, a window with a size of 5 × 5 is used instead of a 3 × 3 window. To do this each pixel in the scale space is compared to its 74 neighbors (see black points in Figure 3.1(d)), comprised of the 24 points in the native scale and the 25 points in each of the scales above and below.

3.2.1.2   **BRIEF Descriptor**

A BRIEF descriptor of the feature point consists of a binary vector (Calonder et al. 2010). The length of the binary vector is generally defined as 128 bits, 256 bits and 512 bits. According to the comparative results analyzed by Calonder et al. (2010), the performance with the 256 bits was similar to that with 512 bits, while only the marginally worse in other cases. To save the FPGA resources, the 256 bits is suggested in this chapter. The following equation presents the definition of a BRIEF descriptor:

$$\tau\left(p; r_1, c_1, r_2, c_2\right) = \begin{cases} 1 : p(r_1, c_1) < p(r_2, c_2) \\ 0 : p(r_1, c_1) \geq p(r_2, c_2) \end{cases} \tag{3.4}$$

where (r_1, c_1) and (r_2, c_2) represent the rows and columns of one point-pair, respectively. $p(r_1, c_1)$ and $p(r_2, c_2)$ represent the intensity values at location (r_1, c_1) and (r_2, c_2). If the value at $p(r_1, c_1)$ is less than one at $p(r_2, c_2)$, then value of $\tau = 0$, otherwise, $\tau = 1$.

Because the BRIEF descriptor is sensitive to noise, the intensity value of patch-pair is computed by a smoothing filtering with a 5 × 5 sub-window centered on (r_i, c_i, i = 1,2…, and 512)

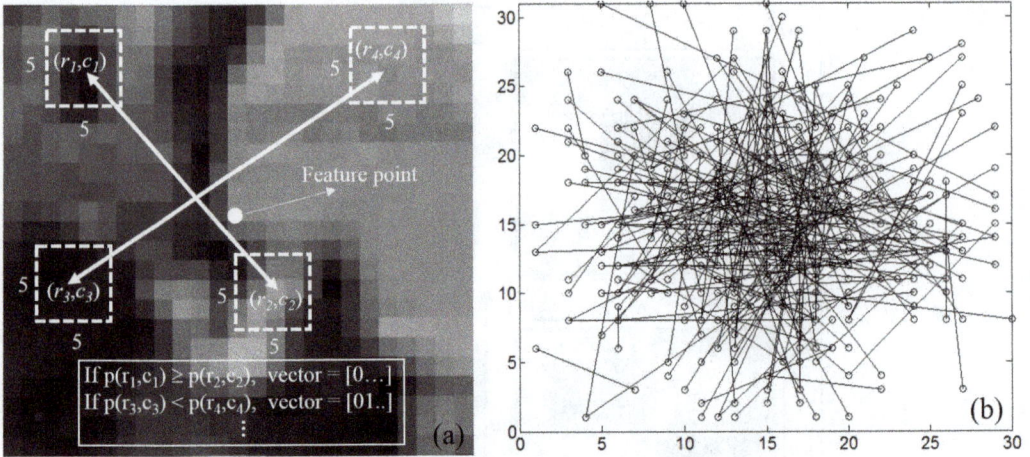

FIGURE 3.2 **(a)** Operation processing of the BRIEF descriptor; **(b)** its 256 pixel-pairs.

(see Figure 3.2(a)). Meanwhile, $\{(r_1,c_1),(r_2,c_2)\}$ is defined as a patch-pair, $\{(r_3,c_3),(r_4,c_4)\}$ is defined as another patch-pair, meanwhile, there are 256 patch-pairs in total (see Figure 3.2(b)). The locations (r_i,c_i) of the 256 point pairs are determined by the Gaussian distribution. (r_i,c_i)–i.i.d. Gaussian $(0, S^2/25)$. (r_i,c_i) are determined from an isotropic Gaussian distribution; S is the size of a patch (Calonder et al. 2010). For the details, it can be referenced to Calonder et al. (2010).

3.2.1.3 Hamming Distance Matching

In BRIEF descriptor, Hamming distance, which can be efficiently calculated by *XOR* operation, is used to match (Calonder et al. 2010; David et al. 2010). To match two feature points in one pair of images, the corresponding descriptors of the feature points have to first be created. Then a Hamming distance is computed by XOR operation (1 XOR 1 = 0 and 1 XOR 0 = 1). A threshold, *(t)* is setup to observe if the feature points are successfully matched when using the Hamming distance. The results and the matching process are listed in Table 3.1. As observed from Table 3.1, if the Hamming distance is less than 2, the two feature points are successfully matched. Otherwise, they are unmatched.

3.2.2 FPGA-BASED DETECTION AND MATCHING ALGORITHM

Implementation of the proposed algorithm on the basis of the FPGA can largely increase the speed of execution. If the proposed algorithm is implemented on FPGA directly, the speed and the clock frequency may be restricted largely, and the usage of FPGA resources might be increased significantly. Hence, to guarantee the accuracy and speed, the modifications of the algorithm are essential.

TABLE 3.1

Matching Process

Descriptor 1	Descriptor 2	XOR Operation	Hamming Distance	Result
110011	110011	000000	0	Matched
110011	110000	000011	2	Matched
110011	110100	000111	3	Unmatched
110011	001100	111111	6	Unmatched

Note: The length of a BRIEF descriptor is assumed 6, and the threshold *(t)* is given 2

In this chapter, the modifications on how to reduce the usage of memory, multipliers and dividers, are mainly focused. The modifications of BRIEF descriptor and matching are not necessary, since the performances of BRIEF descriptor and matching with FPGA implementation are excellent (Huang et al. 2018; Zhou 2020).

3.2.2.1 Modification of Integral Image

Although the integral image can speed up the calculation of summation of pixels, the large bit width of integral image may seriously impact the memory of FPGA where the whole integral image is stored on. For example, as an 8 bits format gray image with a size of 512×512 pixel^2, the bit width of the integral image is 28 bits. To store a frame of such integral image, $512 \times 512 \times 28$ bits (about 7.0 Mb) memory are required. That may be unacceptable in many cases, especially if the size of image is larger. To deal with this problem, several work focus on how to reduce the bit width on FPGA implementation, such as Ma et al. (2014, 2016, 2015) proposed a full-image evaluation methodology to reduce bit width when implementing the histogram of oriented gradients on FPGA (Ma et al. 2014; Ma et al. 2015; Ma et al. 2016), Sousa et al. (2013) presented a fixed-point algorithm when implemented a Harris corner detector on Tightly-Coupled Processor Arrays (Sousa et al. 2013). In this chapter, we use a computing through the overflow technique proposed by Belt (Belt 2008) to reduce the bit width of integral image and maintain the accuracy of the calculated results. If the width and the height of the maximal rectangular are W_{max} and H_{max}, respectively, the bit width L_{ii} of the final integral image is decided by

$$\left(2^{L_{ii}} - 1\right) \geq \left(2^{L_i} - 1\right) W_{max} H_{max} \tag{3.5}$$

where L_i is the bit width of input image. Although there are some intermediate overflowing results during the process of computing integral image, the final summation of pixels in any rectangular whose size is less than $W_{max} \times H_{max}$ is still correct. In the FPGA implementation described in this chapter, two octaves (see Table 3.2) are suggested in the FPGA-based implementation (Zhao et al. 2013; Weberruss et al. 2015), because of a higher resource consumption and less feature points in more scale space. As seen from Table 3.2 and Figures 3.1(a)–(c), the maximal size of box filter is 51×51 (namely, N = 51 and m = 7). The box filter with a size of 51 consists of gray rectangle, white rectangle and black rectangle. In these rectangles, the maximal size of these rectangles is 17×37 or 37×17 without considering the gray rectangle. Thus, the $W_{max} = 17$ (or 37), and the $H_{max} = 37$ (or 17) in this chapter, and the bit width of integral image is 18 bits if L_i is equal to 8 in accordance with Equation (3.5).

3.2.2.2 Modification of Hessian Matrix Responses

To reduce the use of multipliers or dividers, the value of ω^2 is changed from the original value 0.91 (Bay et al. 2006) to 0.875 (Schaeferling and Kiefer 2011; Sledevič and Serackis 2012), as the latter is more suitable to be calculated by FPGA. Then, Equation (3.3) is translated into

$$\det H_{approx} = D_{xx}D_{yy} - D_{xy}^2 + D_{xy}^2/8 = D_{xx}D_{yy} - D_{xy}^2 + D_{xy}^2 \gg 3 \tag{3.6}$$

where only two multipliers are needed and a right shift for division, while three multipliers are needed in Equation (3.3).

TABLE 3.2

The Relationship between Scale and the Size of Box Filter

Octave	1				2			
Size of box filter	9	15	21	27	15	27	39	51
Scale	1.2	2	2.8	3.6	2	3.6	5.2	6.8

3.3 FPGA-BASED IMPLEMENTATION

3.3.1 THE ARCHITECTURE OF FPGA-BASED DETECTION

To achieve the on-board detection and matching, a FPGA-based method is proposed in Figure 3.3. In the proposed method, a pipeline structure and a parallel processing model are proposed to ensure the real-time processing performance. The detailed descriptions of each module are given below (Huang 2018):

Firstly, a series of gray images are converts into the corresponding gray images by "RGB_Gray" module, and then the gray images are sent to "DDR3_Control" module to generate the write data and the write the address. The data and the address are sent to DDR3 SDRAM (DDR3), a memory with 512 M outside of the FPGA chip. The stored data are used for the feature point detection and the descriptor generation.

Secondly, an integral image of a gray image is generated by "Inte_Ima" module, and the Hessian matrixes are computed by "H_M Scale: i" modules (i = 9, 15, 21, 27, 35 and 51) in parallel. Then, the determinants of Hessian matrixes computed in parallel by "Det H_M Scale: i" modules (i = 9, 15, 21, 27, 35 and 51) are sent to "3D Non_Max_j" modules (j = 1, 2, 3 and 4). The locations (rows and columns) of feature points in different scales are determined by the 3D non-maximal suppression.

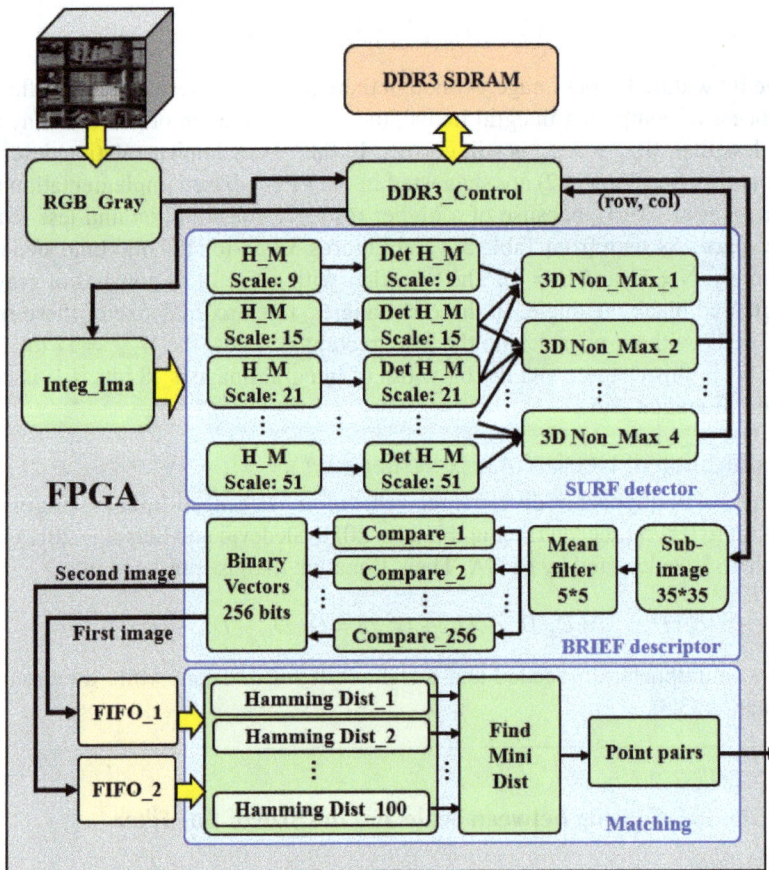

FIGURE 3.3   Hardware architecture of FPGA-based detection and matching.

Thirdly, a sub-image with the size of 35×35 pixels^2 centered on the feature point is reading out from the "DDR3_Control" module. To reduce the sensitiveness to noise, a mean filter with a size of 5×5 pixels^2 is used. The filtered results are sent to the "compare_k" module (k = 1, 2..., 256) to compute the binary vectors in parallel. The 256 binary vectors are used to create a BRIEF descriptor. The generated descriptors in the first image and the second image are sent to "FIFO_1"and "FIFO_2" IP cores, respectively.

Fourthly, when the computation of the descriptors in the first image is finished, the descriptors in the second image start to be computed. All descriptors in the first image and the first descriptor in the second image are sent to "Hamming Dist_g" modules (g = 1, 2..., 100), which are used to compute Hamming distances at the same clock period. The computed Hamming distances are sent to "Find Mini Dist" module, which is used to find out the minimum value of Hamming distances. The matching points can be outputted by the minimum value and a threshold (t). Similarly, the processes of the late descriptor in the second image are the same as that in the first descriptor.

3.3.2 IMPLEMENTATION OF DDR3 WRITE-READ CONTROL

To write the gray images into DDR3 and then read them out successfully, a write-read control module for six signals is redesigned below:

1. "app_cmd". When "app_cmd" = 3'b000, the write signal is active-high. When "app_cmd" = 3'b001, the read signal is active-high;
2. "app_addr". This input indicates the address for the current request;
3. "app_en". This is the active-high strobe for the "app_cmd", "add_addr" et al.;
4. "app_wdf_data". This provides the gray image data for write commands;
5. "app_wdf_wren". This is the active-high strobe for "app_wdf_data";
6. "app_wdf_end". This signal equals to "app_wdf_wren".

To achieve a DDR3 write-read control successfully, the six signals must meet the requirements of sequential relationships which are presented in Figure 3.4. As shown in Figure 3.4, the six signals are all tightly aligned with clock signal when writing data into DDR3 (Figure 3.4(a)), and the "app_cmd", "app_addr" and "app_en" are also tightly aligned with clock signal when reading data out of DDR3(Figure 3.4(b)). More details can be referenced to Xilinx (2011).

3.3.3 FPGA-BASED IMPLEMENTATION OF INTEGRAL IMAGE

The FPGA implementation of integral image computation is illustrated in Figure 3.5, in which the values of integral image in the first row are firstly computed, and then the results are sent into a line buffer in which its data width is 512 bits. The values of the integral image in later rows are computed by two parts. The first part is the sum of the gray values in native row, and the second part is the value of the integral image which is located at the last row and the same column. With Section 3.2.2.1, if the bit width of a gray image is 8 bits, the bit width of the integral image should be 18 bits (Huang 2018).

3.3.4 FPGA IMPLEMENTATION OF HESSIAN MATRIX RESPONSES

Feature extraction is regarded to be the most time-consuming step in SURF algorithm when implemented by a PC (Cornelis and Gool 2008). To accelerate the processing speed of the determinants of Fast-Hessian matrix at different scales, a parallel sliding window method is proposed.

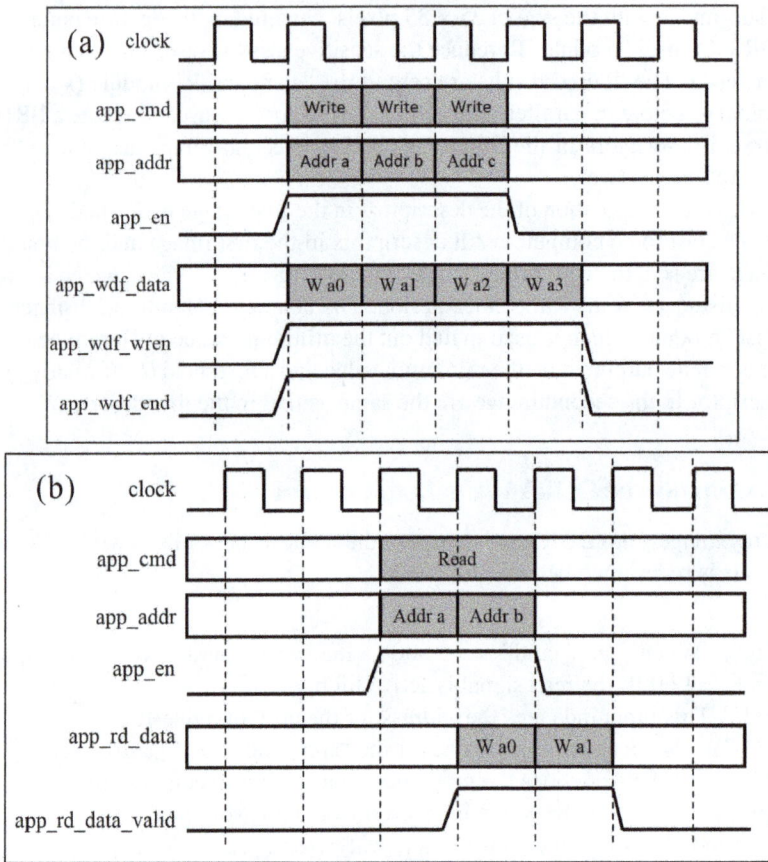

FIGURE 3.4   Sequential relationship of writing (**a**) and reading (**b**).

As depicted in Figure 3.6(a), the D_{xx}, D_{yy}, and D_{xy} are computed by the convolution of the integral image and the filter boxes (Figure 3.1(a)–(c)) in different scales in parallel. In addition, the FPGA implementation of Equation (3.6) is shown in Figure 3.6(b). As seen in Figure 3.6(b), the hessian matrix responses are computed by two multipliers, two subtracters and a right-shifting operation (Huang 2018).

3.3.5   FPGA-BASED IMPLEMENTATION OF 3D NON-MAXIMAL SUPPRESSION

After finishing the computation of Hessian matrix responses in parallel, a 3D non-maximal suppression method is adopted to determine the locations of the feature points. The FPGA-based implementation of 3D non-maximal suppression is presented in Figure 3.7, in which a candidate point (i.e., b_1) is compared with its 74 neighbors (Figure 3.1(d)) in parallel. An "AND" operation is

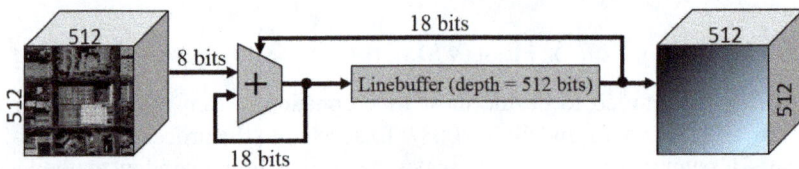

FIGURE 3.5   FPGA-based implementation of the integral image.

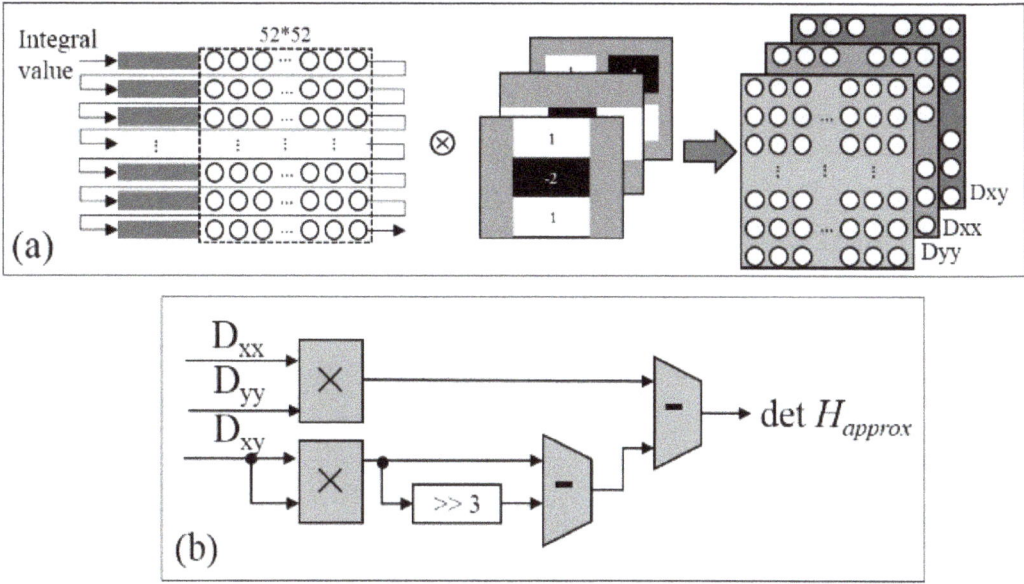

FIGURE 3.6 (a) Parallel computation of D_{xx}, D_{yy} and D_{xy} in different scales; (b) FPGA-based implementation of determinant of Hessian matrix.

adopted in the 74 comparative results. If the "AND" operation result is "1", then the candidate point is regarded as a feature point. Otherwise, the candidate point is not a feature point. In this operation implementation, 74 comparators and an "AND" operation are used to determine the feature points (Huang 2018).

3.3.6 FPGA-Based Implementation of BRIEF Descriptor

Once the location of feature point is determined, the BRIEF descriptor is generated. In BRIEF descriptor module, a sub-image (35×35) centered on the feature point are sent to 35-line buffers, and a mean filter with a size of 5×5 is implemented on the sub-image (see Figure 3.8(a)). Then the filter values of 256 patch-pairs are selected according to Figure 3.2(b). The FPGA-based implementation of Equation (3.4) is presented in Figure 3.8(b). As seen from Figure 3.8(b), the 256 patch-pairs are compared to generate a 256 bits' binary vector, and 256 comparators and a combination operation are adopted. The generated descriptors are stored into "FIFO" IP core waiting for matching.

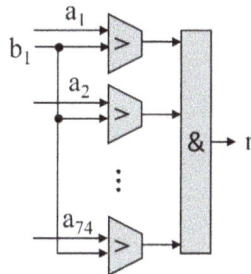

FIGURE 3.7 FPGA implementation non-maximal suppression.

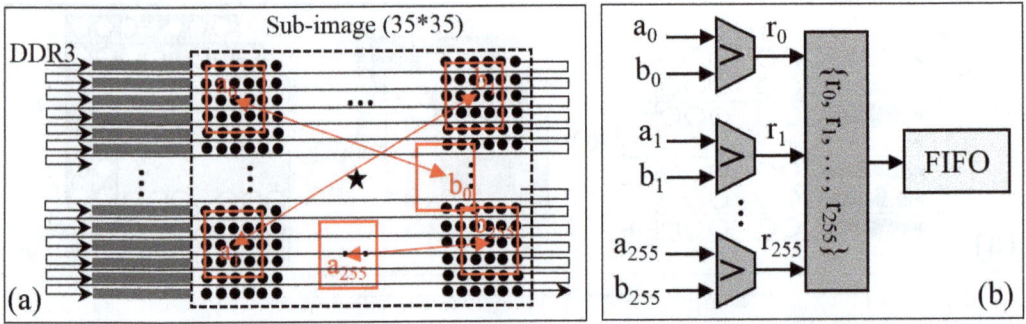

FIGURE 3.8 (**a**) A sub-image extraction by line-buffer; (**b**) generation of a BRIEF descriptor.

3.3.7 FPGA-Based Implementation of Matching

The FPGA-based implementation of matching is presented in Figure 3.9, with which the BRIEF descriptors with 256 bits in the first image and the second image are stored into "FIFO_1" IP core and "FIFO_2" IP core, respectively. To reduce running time and save hardware resources, the maximum number of descriptors is defined as 100. The first "XOR" operations of the first descriptor in the second image and the 100 descriptors in the first image are implemented in parallel, and the second "XOR" operations are implemented between the second descriptor in the second image and the 100 descriptors in the first image, the later "XOR" operation is in order. As seen from Figure 3.9(b), the Hamming distance of the two descriptors is computed by "+" operation. Meanwhile, the 100 Hamming distances are computed in parallel and multilevel. To determine a point-pair, a minimum value can be determined from the 100 hamming distances by compare operations (see Figure 3.9(c)) in parallel (Huang 2018).

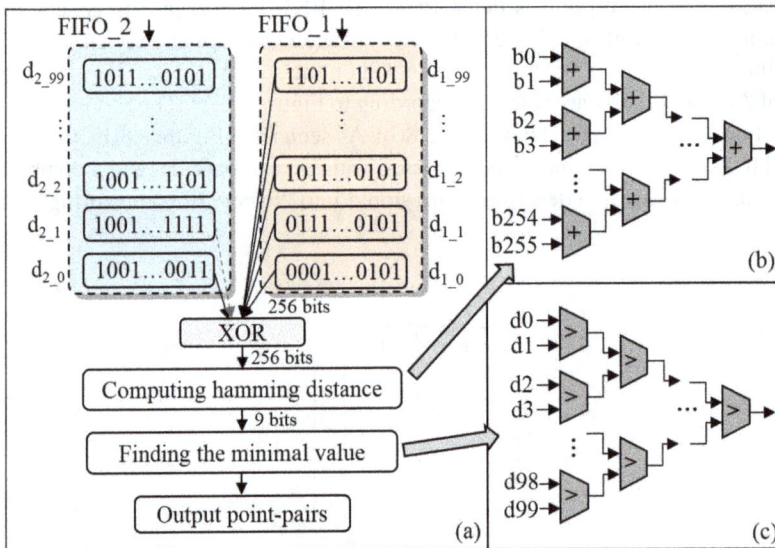

FIGURE 3.9 (**a**) Implementation process of matching; (**b**) Hamming distance computation; (**c**) Locating of the minimal value.

3.4 VERIFICATIONS AND ANALYSIS

3.4.1 HARDWARE ENVIRONMENT AND DATA SET

The proposed architecture on a custom-designed board which contains a Xilinx XC72K325T FPGA are implemented. The selected FPGA has 326,080 Logic Cells, 4000 kb Block RAMS and 840 DSP Slices. The resources of board are enough to implement the whole design (see Chapter 2). In addition, the design tool is Vivado (2014.2 version), the simulation tool is Modelsim-SE 10.4, and the hardware design language is Verilog HDL. Microsoft Visual Studio software (2015 version) and OpenCV library (2.4.9 version) are used for comparison on PC.

Three pairs of images with different land coverages (buildings, roads and woods) are downloaded from http://www.isprs.org/data/ikonos/default.aspx, which are captured by IKONOS sensor with 1-meter black-and-white (panchromatic) resolution. Homograhpies of each pair of images are computed using a "findHomography" function, which is called from OpenCV library (2.4.9 version) (http://opencv.org/). The computed results of the homographies for the three pairs of images are described as follows:

$$H_{Building} = \begin{bmatrix} 0.938443 & -0.029160 & 5.169672 \\ -0.008925 & 0.924411 & 9.490082 \\ -0.000121 & -0.000222 & 1 \end{bmatrix} \tag{3.7}$$

$$H_{Road} = \begin{bmatrix} 0.945168 & -0.020939 & 3.765825 \\ -0.055158 & 0.924799 & 12.459159 \\ -0.000151 & -0.000221 & 1 \end{bmatrix} \tag{3.8}$$

$$H_{Wood} = \begin{bmatrix} 1.064960 & 0.049669 & -3.542474 \\ -0.054367 & 1.120186 & -21.377591 \\ -1.031071 & 0.000489 & 1 \end{bmatrix} \tag{3.9}$$

3.4.2 VERIFICATIONS AND RESULTS

To evaluate the performance of the proposed FPGA-based implementation, the image pairs are sent to the proposed FPGA architecture as the origin input, and the locations of point pairs are outputted to the final results. To compare the proposed method, the experimental results are compared with the ones from MATLAB R2014a version software on PC. The result is depicted in Figure 3.10. As seen from Figure 3.10(a), when the image pair consists of the buildings, most of the point pairs are successfully matched except several mismatches. when the image pairs consist of roads, most of the point pairs are still successfully matched (Figure 3.10(b)), but the matching rate is low than that in the building area in Figure 3.10(a). In although lots of the point pairs are effectively matched, the number of mismatches is more (Figure 3.10(c)). To quantitatively analyze the performance of the FPGA-based implementation, a comprehensive analysis, which includes accuracy and speed, are presented in Section 3.4.3.

3.4.3 PERFORMANCE EVALUATION

3.4.3.1 Accuracy Analysis

A criterion based on the number of correct matches and the number of false matches is first proposed by Mikolajczyk and Schmid (2005). The criterion is widely used to assess the performance of detection and matching algorithms. The evaluation method is presented as a curve of *recall* vs. *1-precision* (Mikolajczyk and Schmid 2005). The curve is generated using a given threshold t which is used to determine if two descriptors are correctly matched. Giving one image pair covering the

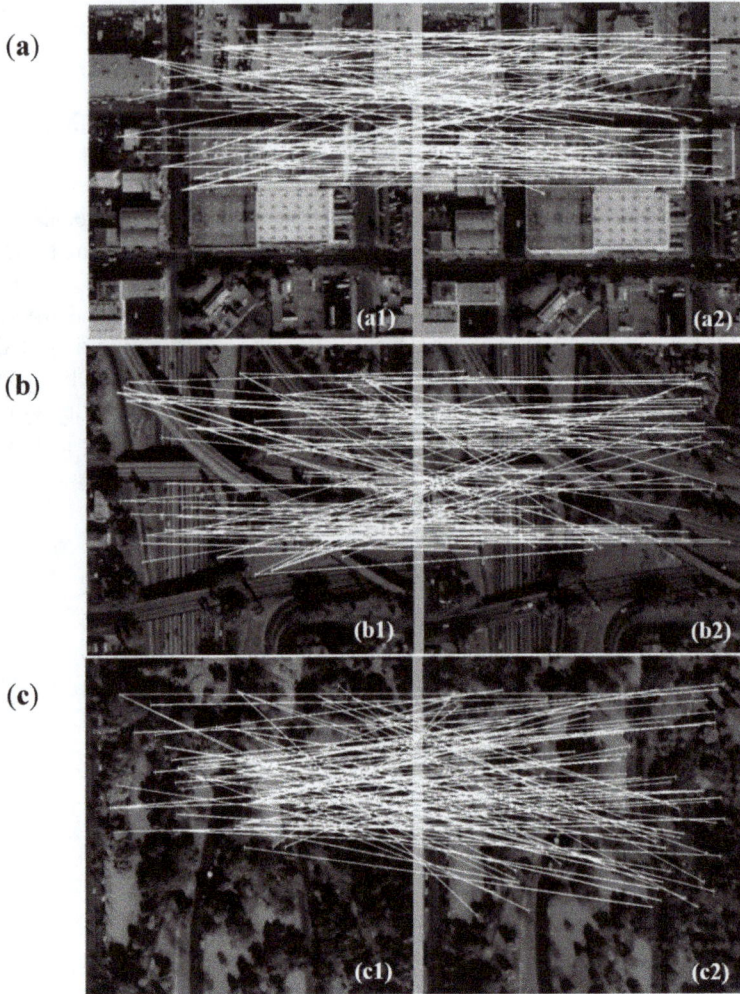

FIGURE 3.10 Detection and matching for three pairs of images with different land covers, including building (a1–a2), road (b1–b2) and wood (c1–c2), respectively.

same land coverages, the *recall* is the ratio of the number of the correctly matched points to the number of corresponding matched points, i.e.,

$$recall = \#correct\_matches / \#correspondence \qquad (3.10)$$

In practice, the number of corresponding points is determined by the overlapping of the points in the image pair. In addition, the *1-precision* is the ratio of the total number of the falsely matched points to the sum of the number of the correctly matched points and the number of the falsely matched points, i.e.,

$$1- precision = \#false\_matches / (\#correct\_matches + \#false\_matches) \qquad (3.11)$$

For the purpose of comparison, the corresponding OpenCV-based models of the SURF + BRIEF and SURF have also been implemented on PC with the Microsoft Visual Studio software (2015 version) and C++ programming language. The corresponding detection and matching functions are

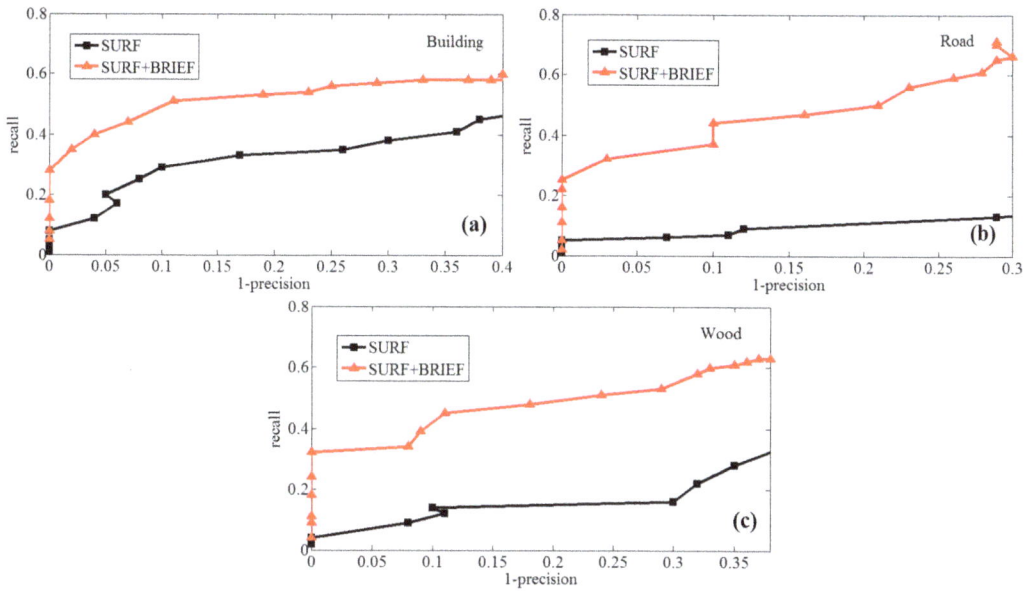

FIGURE 3.11   Accuracy comparison for SURF + BRIEF and SURF in three pairs of images with different land covers of building (**a**), road (**b**) and wood (**c**), respectively.

called from the OpenCV2.4.9 library (http://opencv.org/) which is added into the source library of Microsoft Visual Studio 2015. Three pairs of images processed by FPGA are also selected as the input data of PC-based implementation, and the number of detection and matching is about 100. The matching performance for the two methods are also obtained and presented with the curves of *recall* vs. *1-precision* in Figure 3.11. As seen from Figure 3.11, no matter what the image pair with buildings, roads, or woods, the red curve is always keep the similar level, which means that the SURF + BRIEF has a better matching performance in different land coverages. While the black curve is getting low, which means that the matching performance of SURF is worse than the performance for the image pair covering the natural features, such as woods. The reasons may be caused by the fact that the descriptor of SURF + BRIEF is more robust. Hence, the SURF + BRIEF is feasible to be implemented in FPGA.

After evaluating the performances of SURF + BRIEF and SURF implemented on the PC, on the FPGA, the results are depicted in Figure 3.11. The curves of *recall* vs. *1-precision* for both the FPGA-based and the PC-based implementations are presented in Figure 3.12. As observed from Figure 3.12(a), when the image pair consists of the buildings, the FPGA-based implementation can achieve a very close performance to the PC-based implementation. When the image pair consists of roads, the performance of FPGA-based implementation is slightly worse than that of the PC-based implementation (Figure 3.12(b)). When the image pair consists of woods, the performance of the FPGA-based implementation is worse than that of PC-based implementation (see Figure 3.12(c)). The reasons may be caused by:

1. Only two octaves are used to extract the feature points on FPGA-based implementation, which will inevitably lead to performance degradation.
2. Some divisions of the algorithm are implemented on FPGA by right shift operation which may cause errors. Additionally, the fix points are adopted in the whole system. A few calculation errors may propagate and accumulate which finally lead to false matching points.
3. Because of the different land coverages, the descriptors generated by these image pairs with artificial features (such as buildings and roads) are more robust than those generated by the image pairs with natural features (such as woods).

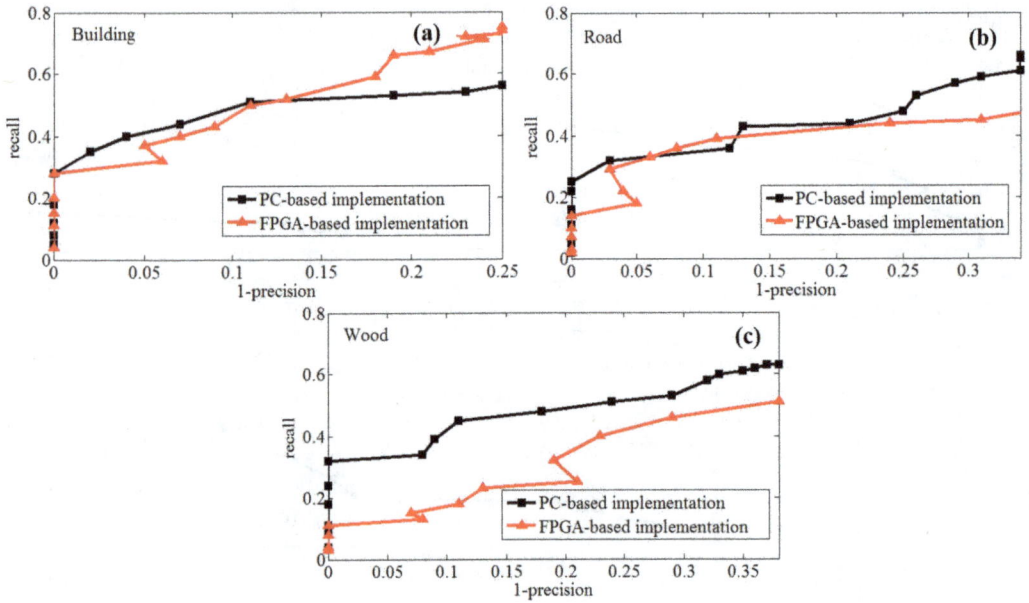

FIGURE 3.12 Comparison analysis of the FPGA-based and the PC-based implementation for SURF + BRIEF algorithm in three pairs of images with different land covers of building (**a**), road (**b**) and wood (**c**), respectively.

3.4.3.2 Comparison for Computational Speed

The computational speed is one of the most importance factors on on-board detection and matching. In this section, the comparison analysis consists of two aspects. The first aspect is a comparison of the FPGA-based and the PC-based implementations. The other aspect is the comparison of the FPGA-based implementation and the previous work.

The proposed method is running a PC with a Windows 7 (64-bit) operation system, which is equipped with an Intel(R) Core (TM) i7-4790 CPU @ 3.6GHz and a RAM 8 GB. To keep the same situations in comparison, the size of the image pair is same as the image pair processed by FPGA, and the number of point pairs is also about 100. The run time for the PC-based and the FPGA-based implementations are 90 ms (11 fps) and 3.29 ms (304 fps), respectively. The results demonstrate that the FPGA-based implementation can achieve 27 times faster than the PC-based does.

Considering both the SURF algorithm and FAST + BRIEF algorithm are successfully implemented on FPGA chips by the previous work, the proposed algorithm is compared with SURF algorithm and FAST + BRIEF algorithm. The comparison results are listed in Table 3.3. As observed in this table, the SURF algorithm has a highest fps, while the clock frequency is also highest, if

TABLE 3.3

Comparison with Previous Work

Algorithm	Row × Column (pixels²)	Clock (MHz)	Feature Points	fps	Speed-up (second)	FPGA Type
SURF (Fan et al. 2013)	640 × 480	156	100	356	6	XC6CSX475T, Xilinx
FAST + BRIEF (Fularz et al. 2015)	640 × 480	100	/	325	15–25	XC7Z020, Xilinx
Proposed Method	512 × 512	100	100	304	27	XC72K325T, Xilinx

the clock frequency is defined as 100 MHz, the fps can't achieve 356. The FAST + BRIEF reach 325 fps, while the clock frequency is only 100 MHz. The proposed method reaches 304 fps with the same clock frequency. In addition, the speed-up of the proposed method is most significant, but is lower than the method proposed by Fularz et al. (2015), which may be caused by

1. In detection phase, the SURF detector is more time-consuming than FAST detector is;
2. The image column is larger than the image column of Fularz et al. (2015). The larger column takes more time in operation on FPGA.

While the speed of the proposed method is still acceptable and satisfactory when implement on FPGA.

3.4.3.3 FPGA Resources Utilization Analysis

For the FPGA-based implementation with the proposed method, the utilization of FPGA resources is analyzed in Table 3.4. As seen from Table 3.4, the maximal utilization is look-up tables (LUTs), which is about 43%, while only 19% in Fularz et al. (2015). The minimal utilization is block random access memory (BRAM) which is about 0%, while increases to 27% and 28% in Fularz et al. (2015) and Fan et al. (2013), respectively. The utilization of digital signal processings (DSPs) is 59% in Fan et al. (2013), while 0% in Fularz et al. (2015) and this work.

3.4.4 Discussion

A FPGA-based implementation for SURF + BRIEF algorithm is presented and is firstly applied in three pairs of images with different land coverages to evaluate the FPGA-based implementation. The experimental results demonstrate that the performance of the FPGA-based implementation is satisfied. Especially when the image pairs with the artificial features, such as buildings and roads, the accuracy of the FPGA-based implementation is similar to that of the PC-based implementation. However, when the image pairs with woods, the accuracy of the FPGA-based implementation is lower than that of the PC-based implementation.

Furthermore, no matter which land coverages are covered in image pairs, the accuracy of the FPGA-based implementation still can't reach exactly the same as the PC-based implementation (Fularz et al. 2015). The main reasons are that (1) only two octaves are used in FPGA, while more octaves are adopted in PC; (2) the fix-point and shift operation are used, while floating point with 64 bits and multiplier/divider are adopted directly in PC.

Based on the hardware characteristic of FPGA, the FPGA architecture can achieve task parallel processing and pipeline processing. For instance, the task parallel means that each module is operated in parallel and independent modes, and the pipeline processing mode means that each module can deal with different parts of a same image. Meanwhile, the next module is not affected by the last module.

TABLE 3.4

The Resource Utilization by Entity (XC7325T FPGA)

Resources	SURF (Fan et al. 2013)	SURF + BRIEF (Huang et al. 2018)	The Proposed Method
FFs	/	17,412 (16%)	122,000 (30%)
LUTs	107,873 (36%)	9866 (19%)	88,462 (43%)
BRAMs	295 (28%)	38 (27%)	1 (~0%)
DSPs	1185 (59%)	/	/

3.5   CONCLUSIONS

This chapter describes a FPGA-based method for on-board detection and matching. A pipeline structure and a parallel computation are introduced. A model which combines the modified SURF detector, and a BRIEF descriptor is described. During the process of implementation, (1) a computation through the overflow technique is used to reduce bit width of integral image, and a right shift operation is used instead of a divider; (2) the responses of Hessian matrix in different scales are computed in parallel. The parallel processing also can be found in the 74 comparators in 3D non-maximal suppression module, the 256 comparators in BRIEF generator module and 100 comparators in matching module.

Three pairs of images with different land coverages are applied to evaluate the performance of FPGA-based implementation. The experimental results demonstrate that (1) when the image pairs consist of artificial features, such as buildings and roads, the performance from FPGA-based implementation is very close to that from the PC-based implementation. When the image pairs consist of natural features, such as woods, the performance from FPGA-based implementation is worse; (2) the FPGA-based implementation can achieve a 304 fps under a 100 MHz clock frequency, which is about 27 times fast than that from PC-based implementation.

REFERENCES

Ball, J., The NIOS II family of configurable soft-core processors. *In Proceedings of the 2005 IEEE Hot Chips XVII Symposium (HCS)*, Stanford, CA, USA, 14–16 August 2005; pp. 1–40.

Battezzati, N., Colazzo, S., Maffione, M., Senepa, L., SURF algorithm in FPGA: A novel architecture for high demanding industrial applications. *In Proceedings of the Conference on Design, Automation and Test in Europe*, Dresden, Germany, 12–16 March 2012; pp. 161–162.

Bay, H., Tuytelaars, T., Gool, L.V., SURF: Speeded up robust features. *In Computer Vision—ECCV 2010, Proceedings of the 9th European Conference on Computer Vision*, Graz, Austria, 7–13 May 2006; Springer Berlin/Heidelberg, Germany; pp. 404–417.

Belt, H.J., Word length reduction for the integral image. *In Proceedings of the 15th IEEE International Conference on image processing*, San Diego, CA, USA, 12–15 October 2008; pp. 805–808.

Calonder, M., Lepetit, V., Strecha, C., Fua, P., Brief: Binary robust independent elementary features. *In Computer Vision—ECCV 2010, Proceedings of the 11th European Conference on Computer Vision*, Crete, Greece, 5 September 2010; Springer: Berlin/Heidelberg, Germany, 2010; pp. 778–792.

Chen, C., Yong, H., Zhong, S., Yan, L., A real-time FPGA-based architecture for OpenSURF. *In Proceedings of the MIPPR 2015: Pattern Recognition and Computer Vision*, Enshi, China, 31 October 2015; pp. 1–8.

Cornelis, N., Gool, L.V., Fast scale invariant feature detection and matching on programmable graphics hardware. *In Proceedings of the IEEE Computer Society Conference on Computer Vision and Pattern Recognition Work-Shops*, Anchorage, AK, USA, 23–28 June 2008; pp. 1–8.

David, G., Decker, P., and Paulus, D., An evaluation of open source SURF implementations. In *RoboCup*. Springer, Berlin, Heidelberg, 2010, pp.169–179.

Dawood, A.S., Visser, S.J., Williams, J.A., Reconfigurable FPGAs for real time image processing in space. *In Proceedings of the 14th International Conference on Digital Signal Processing*, Santorini, Greece, 1–3 July 2002; Volume 2, pp. 845–848.

Evans, C., Notes on the opensurf library. *University of Bristol*, Anaheim, California, USA, 2009, January, CSTR-09-001.

Fan, X., Wu, C., Cao, W., Zhou, X., Wang, S., Wang, L., Implementation of high-performance hardware architecture of OpenSURF algorithm on FPGA. *In Proceedings of the 2013 International Conference on Field-Programmable Technology*, Kyoto, Japan, 9–11 December 2013; pp. 152–159.

Fularz, M., Kraft, M., Schmidt, A., Kasinski, A., A high-performance FPGA-based image feature detector and matcher based on the fast and brief algorithms. *International Journal of Advanced Robotic Systems*, 2015, 12(10): 1–15.

Gonzalez, C., Bernabe, S., Mozos, D., Plaza, A., FPGA implementation of an algorithm for automatically detecting targets in remotely sensed hyperspectral images. *IEEE Journal of Selected Topics in Applied Earth*, 2016, 9: 4334–4343.

González, D., Botella, G., García, C., Bäse, M., Bäse, U.M., Prieto-Matías, M. Customized NIOS II multi-cycle instructions to accelerate block-matching techniques. *In Proceedings of the SPIE/IS&T Electronic Imaging. International Society for Optics and Photonics*, San Francisco, CA, USA, 8–12 February 2015; pp. 940002-1–940002-14.

González, D., Botella, G., García, C., Prieto, M., Tirado, F., Acceleration of block-matching algorithms using a custom instruction-based paradigm on a NIOS II microprocessor. *EURASIP Journal on Advances in Signal Processing*, 2013, 2013: 1–20.

González, D., Botella, G., Meyer-Baese, U., García, C., Sanz, C., Prieto-Matías, M., Tirado, F., A low-cost matching motion estimation sensor based on the NIOS II microprocessor. *Sensors* 2012, 12: 13126–13149.

Huang, J.J., FPGA-Based Optimization and Hardware Implementation of P-H Method for Satellite Relative Attitude and Absolute Attitude Solution, Dissertation, Tianjin University, September 2018.

Huang, J.J., Zhou, G., Zhou, X., Zhang, R.T., An FPGA-based implementation of satellite image registration with outlier rejection. *International of Journal of Remote Sensing*, February 2018, 39(23): 8905–8933

Huang, J.J., Zhou, G., Zhou, X., Zhang, R.T., FPGA architecture of FAST and BRIEF algorithm for on-board corner detection and matching, *Sensors*, 2018, 18: 1014. doi:10.3390/s18041014.

Krajnik, T., Svab, J., Pedre, S., Cizek, P., Preucil, L., FPGA-based module for SURF extraction. *Machine Vision and Applications*, 2014, 25: 787–800.

Lentaris, G., Stamoulias, I., Diamantopoulos, D., Siozios, K., Soudris, D., An FPGA implementation of the SURF algorithm for the ExoMars programme. *In Proceecdings of the 7th HiPEAC Workshop on Reconfigurable Computing (WRC2013)*, Berlin, Germany, 21–23 January 2013.

Ma, X., Borbon, J.R., Najjar, W., Roy-Chowdhury, A.K., Optimizing hardware design for Human Action Recognition. *In Proceedings of the 2016 26th International Conference on Field Programmable Logic and Applications (FPL)*, Lausanne, Switzerland, 29 August–2 September 2016; pp. 1–11.

Ma, X., Najjar, A., Roy-Chowdhury, A.K., Evaluation and acceleration of high-throughput fixed-point object detection on FPGAs. *IEEE Transactions on Circuits and Systems for Video Technology*, 2015, 25: 1051–1062.

Ma, X., Najjar, W., Chowdhury, A.R., High-throughput fixed-point object detection on FPGAs. *In Proceedings of the 2014 IEEE 22nd Annual International Symposium on Field-Programmable Custom Computing Machines (FCCM)*, Boston, MA, USA, 11–13 May 2014; pp. 107–107.

Mair, E., Hager, G.D., Burschka, D., Suppa, M., Hirzinger, G., Adaptive and generic corner detection based on the accelerated segment test. *In Computer Vision—ECCV 2010, Proceedings of the 11th European Conference on Computer Vision*, Crete, Greece, 5 September 2010; Springer: Berlin/Heidelberg, Germany, 2010; pp. 183–196.

Meyer-Bäse, U., Botella, G., Castillo, E., García, A., NIOS II hardware acceleration of the epsilon quadratic sieve algorithm. *In Proceedings of the SPIE Defense, Security, and Sensing*, Orlando, Florida, USA, 5–9 April 2010; pp. 77030M-1–77030M-10.

Mikolajczyk, K., Schmid, C., A performance evaluation of local descriptors. *IEEE Transactions on Pattern Analysis and Machine Intelligence*, 2005, 27: 1615–1630.

Mueller, R., Teubner, J., Alonso, G., Data processing on FPGAs. *Proceedings of The VLDB Endowment*, 2009, 2: 910–921.

Pingree, P., Chapter 5: Advancing NASA's on-board processing capabilities with reconfigurable FPGA technologies. *In Aerospace Technologies Advancements*; InTech: Rijeka, Croatia, 2010; pp. 69–86.

Rublee, E., Rabaud, V., Konolige, K., Bradski, G., ORB: An efficient alternative to SIFT or SURF. *In Proceedings of the 2011 IEEE International Conference on Computer Vision (ICCV)*, Barcelona, Spain, 6–13 November 2011; pp. 2564–2571.

Schaeferling, M., Kiefer, G., Flex-SURF: A flexible architecture for FPGA-based robust feature extraction for optical tracking systems. *In Proceedings of the 2010 International Conference on Reconfigurable Computing and FPGAs*, Cancun, Mexico, 13–15 December 2010, pp. 458–463.

Schaeferling, M., Kiefer, G., Object recognition on a chip: A complete SURF-based system on a single FPGA. *In Proceedings of the 2011 International Conference on Reconfigurable Computing and FPGAs*, Cancun, Mexico, 30 November–2 December 2011, pp. 49–54.

Sledevič, T., Serackis, A., SURF algorithm implementation on FPGA. *In Proceedings of the 2012 13th Biennial Baltic Electronics Conference*, Tallinn, Estonia, 3–5 October 2012; pp. 291–294.

Sousa, R., Tanase, A., Hannig, F., Teich, J., Accuracy and performance analysis of Harris Corner computation on tightly-coupled processor arrays. *In Proceedings of the 2013 Conference on Design and Architectures for Signal and Image Processing (DASIP)*, Cagliari, Italy, 8–10 October 2013; pp. 88–95.

Svab, J., Krajnik, T., Faigl, J., Preucil, L., FPGA based speeded up robust features. *In Proceedings of the IEEE International conference on Technologies for Practical Robot Applications*, Woburn, MA, USA, 9–10 November 2009, pp. 35–41.

Weberruss, J., Kleeman, L., Drummond, T., ORB feature extraction and matching in hardware. *In Proceedings of the Australasian Conference on Robotics and Automation, the Australian National University*, Canberra, Australia, 2–4 December 2015; pp. 1–10.

Xilinx., 7 Series FPGAs Memory Interface Solutions. Available online: *http://www.xilinx.com/support/ documentation/ip_documentation/ug586_7Series_MIS.pdf* (accessed on March 1, 2011).

Yao, L., Feng, H., Zhu, Y., Jiang, Z., Zhao, D., An architecture of optimised SIFT feature detection for an FPGA implementation of an image matcher. *In Proceedings of the International Conference on Field-Programmable Technology*, Sydney, NSW, Australia, 9–11 December 2009; pp. 30–37.

Zhang, B., Intelligent remote sensing satellite system. *Remote Sensing.* 2011, 15: 415–431.

Zhao, J., Zhu, S., Huang, X., Real-time traffic sign detection using SURF features on FPGA. *In Proceedings of the 2013 IEEE High Performance Extreme Computing Conference (HPEC)*, Waltham, MA, USA, 10–12 September 2013; pp. 1–6.

Zhou, G., Baysal, O., Kaye, J., Habib, S., Wang, C., Concept design of future intelligent earth observing satellites. *International Journal of Remote Sensing*, 2004, 25: 2667–2685.

Zhou, G., *Urban High-Resolution Remote Sensing: Algorithms and Modelling*, Taylor & Francis/CRC Press, 2020. ISBN: 978-03-67-857509, 465 pp.

4 On-Board Detection of Ground Control Points

4.1 INTRODUCTION

Traditionally, the ground control points (GCPs) are acquired by a special equipment or manually measured from a referenced image. Such a traditional method cannot meet the performance requirement for real-time response to time-critical events in practice (Zhou et al. 2017; Leng et al. 2018, Deng et al. 2008). Therefore, this chapter aims to automatically extract GCPs from the remotely sensed image in real-time (Liu 2022).

The literatures on local feature point detection are wide and can be back to 1950s, which was first observed by Attneave (Attneave 1954) who though that shape information concentrated at dominant points has a high curvature (Tuytelaars and Mikolajczyk 2008). In 1980, Moravec (Moravec 1981) proposed a detector that was repeatable under small variation and near edges and was applied for stereo image matching. Harris and Stephens (Harris and Stephens 1988) improved the prior Moravec detector in 1988. Harris corner detector includes the gradient information and the eigenvalues of symmetric positive which are defined as 2×2 matrix to make it more repeatable. Harris corner detector is one of widely used feature detection techniques that is combined corner and edge detector based on the local auto-correlation function, but it is not scale-invariant and sensitive to noise (Idris et al. 2019; Zhou 2020). Smith and Brady (1997) developed smallest univalue segment assimilating nucleus (SUSAN) detector in 1997. SUSAN is not sensitive to local noise and high anti-interference ability (Deng et al. 2008). To obtain the scale invariant feature, Lindenberg (Lindeberg 1994; 1998; 1996) had studied the scale invariant theory and presented a framework to select local appropriate scales that can be used for automatic scales selection. Lowe (2004) proposed the scale-invariant feature transform (SIFT) algorithm, but since each layer relied on the previous, and images need to be resized, it was not computationally efficient (Christopher 2009). Mikolajczyk and Schmid (2004) presented Harris-Laplace and Harris-Affine detectors and gradient location and orientation histogram (GLOH) detector (Mikolajczyk and Schmid 2005). There are many SIFT variants, such as PCA-SIFT (Ke and Sukthankar 2004), GSIFT (Mortensen et al. 2005), CSIFT (Abdel-Hakim and Farag 2006) and ASIFT (Morel and Yu 2009) are relatively highly efficient. The speeded-up robust features (SURF) is a local descriptor inspired by SIFT. It was first introduced in Bay et al. (2006) and later more fully explained in Bay et al. (2008).

In addition, binary descriptor was developed to image processing. Leutenegger et al. (2011) proposed a binary robust invariant scalable keypoints (BRISK) detector, which is a novel method for keypoint detection, description and matching. BRISK is constructed by pixel comparisons whose distribution forms a concentric circle surrounding the feature point. Calonder et al. (2012) studied a method to extract feature very efficiently, which was named as binary robust independent elementary features (BRIEF) often suffice to obtain very good matching results, the descriptor vector is 512, 256 or 128 bits. The size of the feature descriptor largely decides the time consumed when matching the features, and accuracy (Wang et al. 2013). Rublee et al. (2011) proposed a very fast binary descriptor based on BRIEF, called ORB, which was rotation invariant and resistant to noise. the main contributions in this work lied in adding an orientation component to the feature from accelerated segment test (FAST) (Rosten and Drummond 2005) and proposing a learning method for choosing pairwise tests with good discrimination power and low correlation response among them (Lima et al. 2015). Alahi et al. (2012) suggested a fast retina keypoint (FREAK) as a fast compact and robust keypoint descriptor.

The feature-based matching algorithm has been widely used for seeking for the conjugate points from one image to another image. Many researchers have been applying the matching algorithm to real-time track the conjugate points for GCPs extraction (Zhou and Li 2000; Zhao et al. 2013; Cai et al. 2014; Cai et al. 2017). These studies can be grouped into two main categories:

The first category aims to reduce the complexity of matching algorithm without performance loss. Principal component analysis (PCA) (Mikolajczyk and Schmid 2005) and linear discriminant analysis (LDA) (Hua et al. 2007) are dimensionality reduction techniques that reduce the size of original descriptor, such as SIFT or SURF (Calonder et al. 2012; Rani et al. 2019). Calonder et al. (2009) proposed a concept that used a shorten descriptor to quantize its floating-point coordinate into integers codes on fewer bits, the same result in (Je´gou et al. 2011; Huang et al. 2018a). It is also a very effective way to replace the original complex detector or descriptor with speeded-up detector or binary description, such as FAST (Rosten and Drummond 2005), BRISK (Leutenegger et al. 2011), BRIEF (Calonder et al. 2012), ORB (Rublee et al. 2011) and FREAK (Alahi et al. 2012). Lowe (Lowe 2004) approximated the Laplacian of Gaussian (LoG) through the difference of Gaussians (DoG) filter. Wang et al. (2013) proposed approximation to the LoG by using box filter representations of the respective kernels based on SIFT.

The second category is focused on improving the processing speed using dedicated hardware such as multi-core central processing units (CPUs), graphic processing units (GPUs), application-specific integrated circuit (ASIC) and field programmable gate arrays (FPGAs). Čížek et al. (2016) proposed a processor-centric FPGA-based architecture for a latency reduction in the vision-based robotic navigation. Krajník et al. (2014) presented a complete hardware and software solution of an FPGA-based computer vision embedded module capable of carrying out SURF image feature extraction algorithm. Yao et al. (2009) proposed an architecture of the optimized SIFT feature detection for an FPGA implementation of an image matching, the total dimension of the feature descriptor had been reduced to 72 from 128 of the original SIFT. Kim et al. (2014) proposed a parallelization and optimization method to effectively accelerate SURF, which can achieve a maximum of 83.80 fps in a real-machine experiment, enabling real-time processing. Cheon et al. (2016) analyzed the SURF algorithm and presented a fast descriptor extraction method that eliminated redundant operations in the Haar wavelet response step without additional resources. Kim et al. (2009) presented a parallel processing technique for real-time feature extraction for object recognition by autonomous mobile robots, which utilized both CPU and GPU by combining OpenMP, SSE (Streaming SIMD Extension) and CUDA programming. Schaeferling et al. (2012) described two embedded systems (ARM-based microcontroller and intelligent FPGA) for object detection and pose estimation using sophisticated point features, the feature detection step of the SURF algorithm was accelerated by a special IP core. Huang et al. (2018b) designed an architecture which was combined FAST detector and BRIEF descriptor for detection and matching with sub-pixel precision. Lima et al. (2015) proposed hardware architecture based on SURF detector and BRIEF descriptor. This approach contributed to reduce the number of memory accesses required to obtain the descriptor while maintaining its discrimination quality. Zhao et al. (2014) presented an efficient real-time FPGA implementation for object detection. The system was employed SURF algorithm to detect keypoint on every video frame and applied FREAK method to describe the keypoint. Huang and Zhou (2017) and Wang et al. (2013) proposed a FPGA-based modified SURF detector and BRIEF descriptor matching algorithm. To accelerate the SURF algorithm, an improved FAST feature point combined with SURF descriptor matching algorithm were proposed in Li et al. (2017) and Liu et al. (2020) and Liu et al. (2019), which realized the real-time matching of target.

Inspired by the previous research, this chapter proposes a hardware architecture for automatically extract GCPs on the basis of FPGA.

4.2   FEATURE DETECTOR AND DESCRIPTOR

The process of feature-based matching has three steps: feature detection, feature description and feature matching. In this chapter, an efficient real-time FPGA-based algorithm is presented which consists of SURF feature detection, BRIEF descriptor and BRIEF matching on a single chip.

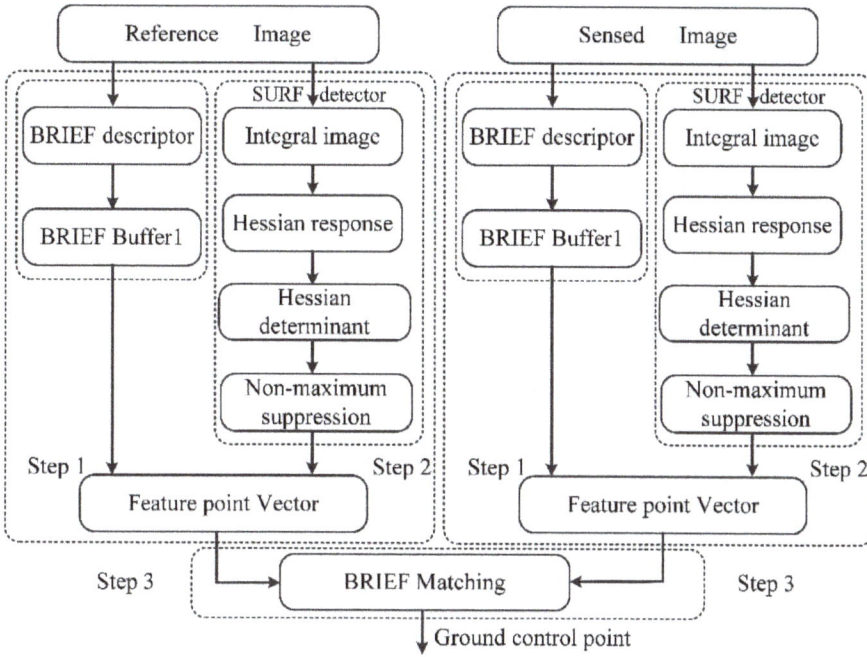

FIGURE 4.1 The flowchart of the ground control point (GCPs) extraction on remote sensing image.

The diagram of a proposed process is presented in Figure 4.1. The feature detection and feature description module are to extract GCPs automatically. The feature matching module is to match GCPs automatically (Deng et al. 2008).

4.2.1 SURF FEATURE DETECTOR

The SURF, which takes a grayscale image as an input, is a scale- and rotation-invariant feature point detector (Bay et al. 2006; 2008). As shown in Figure 4.1, SURF detector consists of four steps, Integral image, Hessian response, Hessian determinant and non-maximum suppression, which can be briefly summarized below (Liu et al. 2020; Liu 2022).

4.2.1.1 Integral Image

Integral image is a novel method to improve the performance of the subsequent steps for SURF detector (Lima et al. 2015). The integral image is used as a rapid and effective way to calculate summations over image sub regions (Bay et al. 2008). Given the pixel value $i(x,y)$ for the coordinate (x,y) of an image with W-width and H-height, the value of coordinate (x,y) in the integral image $ii(x,y)$ can be defined as (Bay et al. 2006; Kasezawa et al. 2016).

$$ii(x,y) = \sum_{x'=0}^{x}\sum_{y'=0}^{y}i(x',y') \quad 0 \leq x \leq W, 0 \leq y \leq H \tag{4.1}$$

Using the integral image in Figure 4.2, the total pixel value of the rectangular region with coordinate (x,y) of the top-left pixel, w width and h height in the source image is (Fan et al. 2013)

$$S_{w,h}(x,y) = \sum_{x'=x}^{x+w-1}\sum_{y'=y}^{y+h-1}i(x',y')$$

$$= ii(x-1,y-1) + ii(x+w-1,y+h-1) - ii(x-1,y+h-1) - ii(x+w-1,y+h-1) \tag{4.2}$$

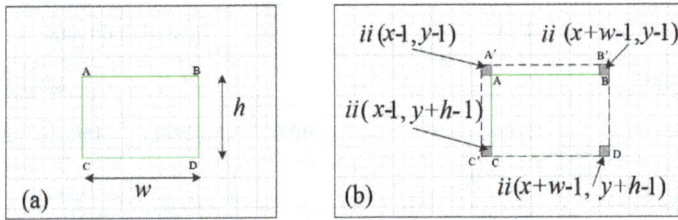

FIGURE 4.2   Calculation of the total pixel value of the rectangular region by Equation (4.2). With an integral image, the sum of pixel values inside the green rectangular region bounded by A, B, C and D is calculated by $ii(A') + ii(D) - ii(C') - ii(B')$

The initial condition for Equation (4.2) is

$$ii(-1, y) = ii(x, -1) = ii(-1, -1) = 0 \qquad (4.3)$$

From Equation (4.2), Integral image provides a fast way to get the histogram of an arbitrary-sized rectangle, requiring only four adders and calculating near a constant time.

4.2.1.2   Extraction of Feature Points

SURF algorithm (Bay et al. 2006) detects feature point based on the scale-space analysis for scale invariance. The scale-space can be divided into $o(o \geq 1)$ octaves, and each octave is further divided into $v(v \geq 3)$ intervals. Totally, there are $o \times v$ scale levels. Each interval represents the determinant response of Hessian matrix that can be approximately defined as Equation (4.4), where $w = 0.912$ is a weight coefficient, which is used to correct the error caused by approximation. The D_{xx}, D_{yy} and D_{xy} are respectively computed by the corresponding box filter as demonstrated in Figure 4.3 (Pohl et al. 2014), where white, gray and black pixels refer to the weight values of $\{1, 0, -2\}$ for D_{xx} and D_{yy} box filters and $\{1, 0, -1\}$ for D_{xy} box filter, respectively (Schaeferling et al. 2012). Through the concept of integral image, the calculation of D_{xx} or D_{yy} takes 8 memory accesses, while the calculation of D_{xy} needs 16 memory accesses. The 32 memories are marked with a dot in Figure 4.3. The scale σ is defined in SURF method by analogy to the linear scale space. With the increasing of σ, the distances between the marked points are increasing, but the number of points to be accessed remains the same (Terriberry et al. 2008).

$$\det(H) = D_{xx}D_{yy} - w^2 D_{xy}^2 \qquad (4.4)$$

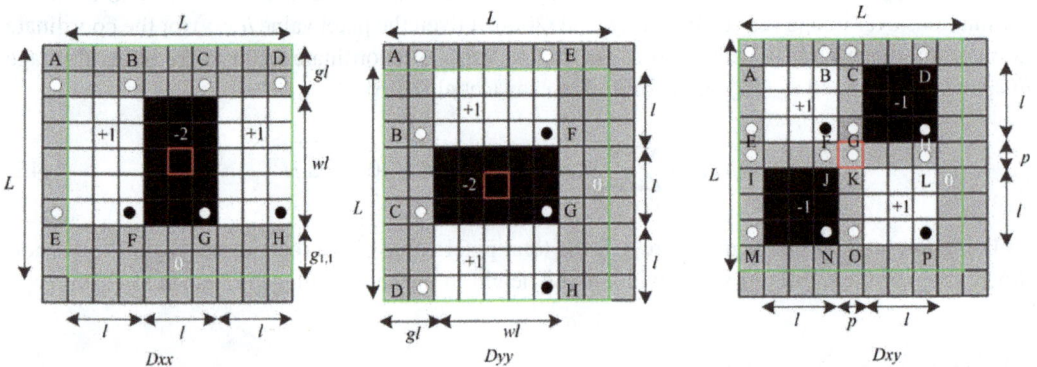

FIGURE 4.3   The 9×9 box filter approximating the second order Gaussian derivative in the x, y and xy direction to get D_{xx}, D_{yy} and D_{xy}.

In the O-SURF method, 9×9 box filter is defined by Bay et al. (2008) at the first scale. However, they did not specify exact values at the remaining scales (Zhang 2010; Oyallon and Rabin 2015), hence, a number of parameters for a particular scale level (i, j) $i \in [1, o]$ and $j \in [1, v]$ need to be defined. The lobe l, being one-third of the size (L) of the box filters D_{xx}, D_{yy} and D_{xy} at the scale level (i, j), is $2^o \times v + 1$. The scale σ is $\frac{1.2}{3}(2^o \times v + 1) = 0.4l$. wl, the length of the white area in the D_{xx} and D_{yy} box filters, is $2l + 1$ in (Zhang 2010) or $2l\text{-}1$ in (David et al. 2010) or $\frac{(3l+1)}{2}$ in MATLAB with OpenSURF by Chris Evans https://ww2.mathworks.cn/matlabcentral/fileexchange/28300-opensurf-including-image-warp. gl, the length of the side of the gray area in the D_{xx} and D_{yy} box filters, is $\frac{l-1}{2}$ in (Zhang 2010) or $\frac{l+1}{2}$ in (David et al. 2010) or $\frac{(3l+1)}{4}$ in https://ww2.mathworks.cn/matlabcentral/fileexchange/28300-opensurf-including-image-warp. Table 4.1 shows the value of box-filters. The size of the octave increases by $6 \times 2^{o-1}$ pixels per interval. p is a constant of one pixel in the D_{xy} box filter.

Based on Equation (4.2), the separable convolution response of D_{xx}, D_{xy} and D_{yy} can be computed by Equation (4.4) thru (4.6).

$$D_{xx} = (A + F - B - E) - 2(B + G - C - F) + (C + H - D - G) \tag{4.4}$$

$$D_{yy} = (A + F - B - E) - 2(B + G - C - F) + (C + H - D - G) \tag{4.5}$$

$$D_{xy} = (A + F - B - E) - (C + H - G - D) - (I + N - J - M) + (K + P - L - O) \tag{4.6}$$

4.2.1.3 Non-Maximum Suppression

To localize the interest point, the non-maximum suppression (NMS) is applied in the three adjacent scales. The NMS compares a determinant with its 8 direction neighbors in its native scale interval, and 9 direction neighbors in each of the interval above and below, totally 26. Moreover, a threshold is set to determine only the most distinctive image point as a candidate point (Bay et al. 2008).

TABLE 4.1

Box-Spaced Sampling Values

o	V	l	L	σ	2l+1	(l−1)/2	2l−1	(l+1)/2	(3l+1)/2	(3l−1)/2
1	1	3	9	1.2	7	1	5	2	5	4
	2	5	15	2.0	11	2	9	3	8	7
	3	7	21	2.8	15	3	13	4	11	10
	4	9	27	3.6	19	4	17	5	14	13
2	1	5	15	2.0	11	2	9	3	8	7
	2	9	27	3.6	19	4	17	5	14	13
	3	13	39	5.2	27	6	25	7	20	19
	4	17	51	6.8	35	8	33	9	26	25
3	1	9	27	3.6	19	4	17	5	14	13
	2	17	51	6.8	35	8	33	9	26	25
	3	25	75	10	51	12	49	13	38	37
	4	33	99	13.2	67	16	65	17	50	49
4	1	17	51	6.8	35	8	33	9	26	25
	2	33	99	13.2	67	16	65	17	50	49
	3	49	147	19.6	99	24	97	25	74	73
	4	65	195	26.0	131	32	129	33	98	97

4.2.2 BRIEF DESCRIPTOR

BRIEF is a descriptor that uses binary tests between pixels in a smoothed image patch. More specifically, if p is a smoothed image patch, corresponding binary test τ is defined by (Calonder et al. 2012).

$$\tau(p;x,y) = \begin{cases} 1 & if\ \ I(p,x) < I(p,y) \\ 0 & otherwise \end{cases} \tag{4.7}$$

where $p(x)$ is the intensity of p at a point x. The descriptor is defined as a vector of n_d binary tests:

$$f_{nd}(p) = \sum_{1 < i \leq n_d} 2^{i-1} \tau(p;x_i,y_i) \tag{4.8}$$

The dimension of the n_d is generally defined as 128, 256 and 512 bits. The result of the experiment in Calonder et al. (2012) demonstrated that the performance of 256 bits was similar to that of 512 bits, while only the marginally worse in other case (Huang et al. 2018b). Due to the limited hardware resources, 256 bits is used as an example for explanation in this chapter.

4.2.3 HAMMING DISTANCE MATCHING

In BRIEF descriptor (Calonder et al. 2012), Hamming distance is used to match. Hamming distance can be calculated by *XOR* operation. The number of "1" in the result of *XOR* operation is Hamming distance. A threshold is used to check whether the points truly correspond to each other or not. If the Hamming distance is less than a threshold, the feature points-pair are corresponding points, otherwise, they are nonmatching points (David et al. 2010).

4.3 OPTIMIZATION OF SURF DETECTOR

To make SURF detector is efficient to implement in a signal FPGA, five technical approaches are applied to optimize the SURF detector. The first approach, word length reduction (WLR) is applied to reduce the word length of integral image without loss of accuracy. The second approach, a memory-efficient parallel architecture (MEPA) is applied to parallel compute the output of First In First Out (FIFO). The third approach, shift and subtraction strategies (SAS) are adopted to simplify the determinant response of Hessian matrix, which transform the floating-point operation into a shift and subtraction operation. The fourth approach is called sliding widow, which is applied to parallel compute the D_{xx}, D_{yy} and D_{xy} of the box filter. The fifth approach, called optimization space-scale (Liu et al. 2020; Liu 2022). All of the five approaches are introduced in this section.

4.3.1 WORD LENGTH REDUCTION (WLR)

SURF detector is to use integral images for image convolutions to produce a comparable or even a better result for feature point detection (David et al. 2010). However, the large word length of integral image may seriously impact on the performance of designed hardware, especially for those implementations that need to store the whole integral image on FPGA (Schaeferling and Kiefer 2010). To solve this problem, Hsu and Chien (2011) presented a row-based stream processing method can be further shared to increase the memory access efficiently, which only need 34-rows memory space to process filter of size 9, 5, 21, 27, 39, to 51 simultaneously. The overflow based on two's complement-coded arithmetic and rounding with error diffusion techniques were proposed by Belt (2008), which can work on face detector for a VGA resolution with 16 bits. However, this approach exposes a few drawbacks, including rounding errors and an additional constraint of fixed size of box filter. Lee and Jeong (2014) proposed a new structure for memory size reduction which including an integral image, a row integral image, a column integral and an input image four types image information. Using this method, integral memory can be reduced by 42.6% on a 640 × 480 8-bit gray-scale image. Ehsan et al. (Ehsan and McDonald-Maier 2009; Ehsan et al. 2009; 2015) proposed a parallel recursive equation to computer integral image, this method not only substantial decrease in the number of operation and memory requirement (at least 44.44%) but also maintain the accuracy of the calculated results. This section mainly describes how to reduce the size of memory and parallelization computation.

Ehsan and McDonald-Maier (2009) proposed the maximum value of integral image in the worst-case (WS) by

$$ii_{\max} = (2^{L_i} - 1) \times W \times H \tag{4.9}$$

where ii is the integral image, $ii_{\max}$ is the WS value, W and H are the width and height of input image i, respectively. From Belt (2008), the number of bits L_{ii} required for representing WS integral image value is $(2^{L_{ii}} - 1) \ge (2^{L_i} - 1) \times W \times H$. All of the memory in byte is $(W \times H) \times L_{ii} / 8$. In general, the maximum width and height of the box filter are known, the word length for the integral image using the exact method with complement-coded arithmetic needs to satisfy (Ehsan et al. 2015):

$$(2^{L_{ii}} - 1) \ge (2^{L_i} - 1) \times W_{\max} \times H_{\max} \tag{4.10}$$

where $W_{\max}$ and $H_{\max}$ are the maximum width and height of box filter, i.e., $l_{\max} \times wl_{\max}$ or $wl_{\max} \times l_{\max}$ in Figure 4.3. For an example, the input image is 8-bits gray data with the size of 512*512 pixel^2, Table 4.2 lists the WLR value of the different size of box filter. All of the memories in byte are shown in Figure 4.4. As observed from Table 4.2 and Figure 4.4, the box filters in Zhang (2010) and David et al. (2010) have the same memory. The OpenSURF (https://ww2.mathworks.cn/matlabcentral/fileexchange/28300-opensurf-including-image-warp) has smaller memory than that in Zhang (2010) and David et al. (2010). However, the value of $\frac{1}{2}$ is not an integer and unsuitable for FPGA. The WLR method significantly reduces the space of memory.

4.3.2 PARALLEL COMPUTATION FOR INTEGRAL IMAGE

Equation (4.1) can be transformed into the pipeline recursive equation that was presented by Viola and Jones (2001):

$$S(x, y) = i(x, y) + S(x, y - 1) \tag{4.11}$$

$$ii(x, y) = ii(x - 1, y) + S(x, y) \tag{4.12}$$

where $S(x, y)$ is the cumulative row sum value at image location (x, y).

TABLE 4.2

The Word Length (Bit) of Integral Image with the Maximum Width and Height of the Box Filter

Octave	WS L_{ii}	$l_{\max} \times wl_{\max}$	L_{ii} (Zhang 2010)	$l_{\max} \times wl_{\max}$	L_{ii} (David et al. 2010)	$l_{\max} \times wl_{\max}$	L_{ii} (Chris E., Available online)
1	26	9×19	16	9×17	16	9×14	15
2		17×35	18	17×33	18	17×26	17
3		33×67	20	33×65	20	33×50	19
4		65×131	22	65×129	22	65×98	21

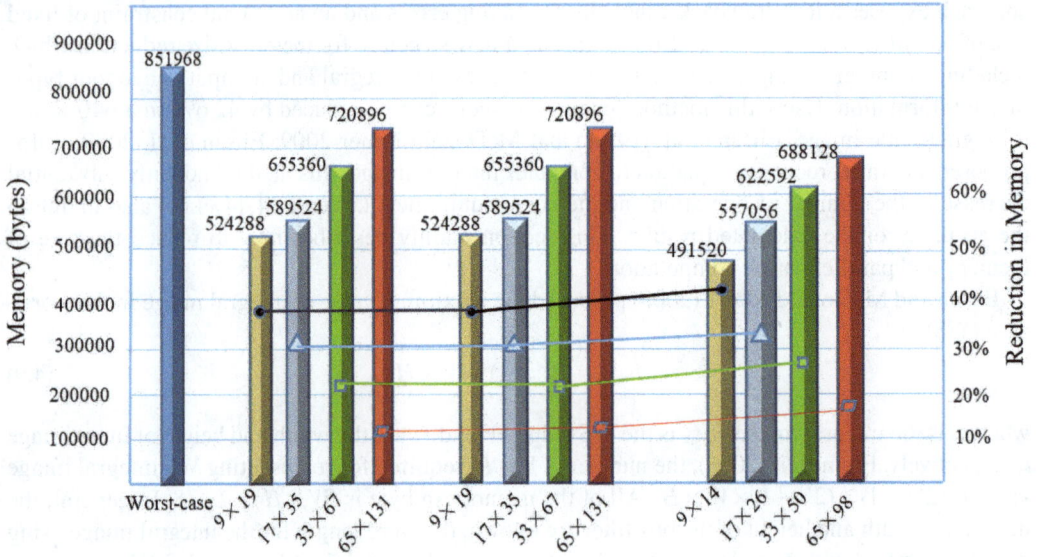

FIGURE 4.4 The comparison for the occupied memories between the worst-case and the maximum width and the height of box filter.

In Equation (4.1), $\frac{1}{4}M^2N^2$ adder are used to compute integral image for an input of size $M \times N$ pixel^2 (Kisacanin 2008). Apparently, Equation (4.1) is not suitable for a medium or high-resolution image. In Viola-Jones parallel recursive Equation (4.11) thru (4.12), the number of additions is $2MN$. However, the Viola-Jones method has time delay drawbacks. To speeded-up the processing of integral image, Ehsan et al. (2015) proposed a n stage, pipelined system that processes n rows if an input image in parallel, providing n integral image value per clock cycles without delay when the pipeline is full. The method can be defined mathematically by:

$$S(x+j,y) = ii(x+j,y) + S(x+j,y-1) \qquad (4.13)$$

For odd rows:

$$ii(x+2k,y) = ii(x+2k-1,y) + S(x+2k,y) \qquad (4.14)$$

For even rows:

$$ii(x+2m+1,y) = ii(x+2m-1,y) + S(x+2m,y) + S(x+2m+1,y) \qquad (4.15)$$

where n is the number of row to be calculated (always a multiple of 2), $j = 0,....,n-1, k = 0,....,\left(\frac{n}{2}-1\right)$ and $m = 0,....,\left(\frac{n}{2}-1\right)$. This set of equations requires $2MN + \frac{MN}{2}$ addition operation for an input image size $M \times N$ pixels (Ehsan et al. 2015), which is not significant increase, when compared to Viola-Jones equation.

To trade-off computation time and consumption memory, the 4-rows parallel method are adopted in the integral image.

4.3.3 Shift and Subtraction (SAS)

In Equation (4.4), the weight coefficient $w^2 = 0.831744$ ($w = 0.912$) is derived to minimize the approximation error caused by box filters in the O-SURF. Hence, an architecture with floating-point format is required to compute det(H). OpenSURF does not use this value, but 0.81 ($w = 0.9$) is replaced. However, the floating-point operation is more complex than a fixed-point format does. To overcome this problem, a value of $w^2 = 0.875$ is firstly used that may be disregarded for common tracking tasks in Flex-SURF (Schaeferling and Kiefer 2010). The same value is used to simplify the processing in Cai et al. (2014), Huang et al. (2018b), Zhao et al. (2014), Sledevič and Serackis (2012) and Schaeferling and Kiefer (2011). Equation (4.4) can be rewritten with

$$\det(H_{approx}) = D_{xx} \times D_{xy} - 0.875(D_{xy})^2 = D_{xx} \times D_{xy} - (D_{xy}^2 - D_{xy}^2/8)$$
$$= D_{xx} \times D_{xy} - D_{xy}^2 + (D_{xy}^2 >> 3) \tag{4.16}$$

Here, Equation (4.4) can be replaced by Equation (4.16) with shift and subtraction operation. Two multipliers, two adders/subtractors and one right shifting operation are used in Equation (4.16).

Figure 4.5 shows the difference of calculating integral image between SAS and O-SURF. As observed from Figure 4.5, it indicated that the SAS highly requires shift operations than O-SURF does. However, the shift operation consumes one clock period in the FPGA architecture. Comparing with O-SURF, addition/subtraction (abbreviate Add-sub), multiplication and division are reduced by 4.44%, 13.33% and 33.33%, respectively.

4.3.4 SLIDING WINDOW

It has been demonstrated that a sliding window technique is a good choice for parallel computer data (Schmidt et al. 2012; Zhao et al. 2014; Lima et al. 2015; Ni et al. 2019), which includes four portions: the stream of input image, buffer, slice registers and a function module. The described fabric is depicted in Figure 4.6. The input data stream is buffered in a custom pipeline structure, organized in pixel rows. The buffer is implemented by a combination of block random-access memory first in first outs (Block-RAM FIFOs) and slice registers (SRs). The SR sections allow to

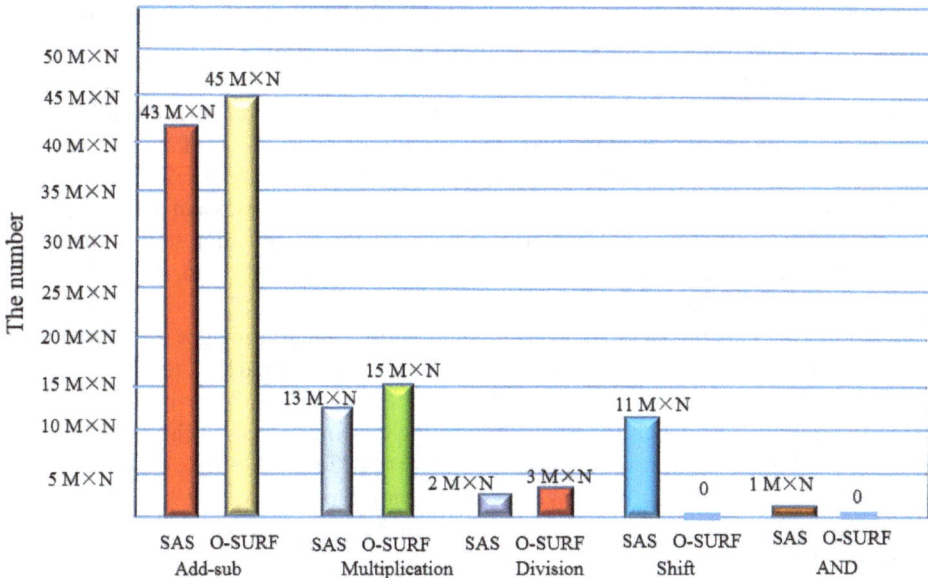

FIGURE 4.5   The comparison analysis between SAS and O-SURF with an image size of $M \times N$ pixels².

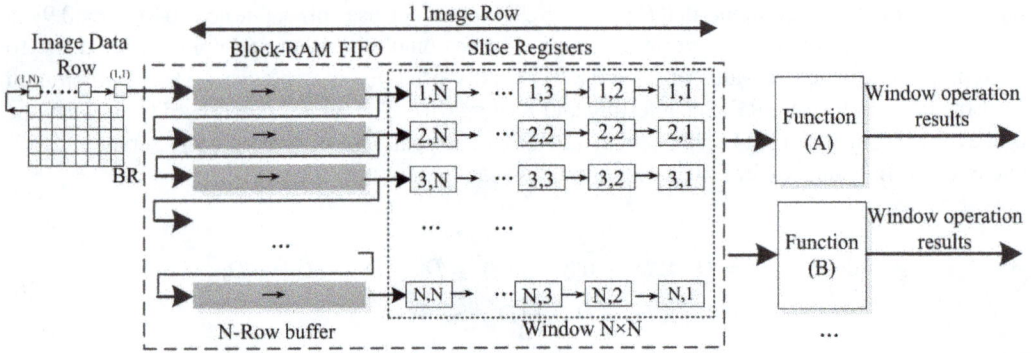

FIGURE 4.6   The architecture of a sliding window.

access all pixels in the respective pipeline elements, comparable to a window. With each incoming pixel, data is shifted by one pixel, causing the window to virtually slide forward. Hence, while the input image is streamed into the pipeline, the window traverses the entire input image from the top left (TL) to the bottom right (BR) corner and each line buffer can hold exactly one line of the image data. A function module, determined for a certain window operation, interconnects to a subset of pixel registers in order to fetch all data required for one calculation simultaneously. With each pixel shifted into the structure, the function module calculates a new result, producing an output data stream (Pohl et al. 2014). It has three types of sliding Window ($N \times N$) in overall system, with N equals to 5, 25 and 35 in Hessian response, non-max suppression and BRIEF descriptor, respectively.

4.3.5   PARALLEL MULTI-SCALE-SPACE HESSIAN DETERMINANT

The more octaves are used; the more hardware resources be consumed (Chen et al. 2015). Avoiding consume more resource on the FPGA, a parallel multi-scale space architecture is designed to simultaneously compute the Hessian determinant. The recommended two octaves and six scales are implemented to extract feature point, corresponding the box filters sizes {9, 15, 21, 27} and {15, 27, 39, 51}(Bay et al. 2008).

The interpolation step in Hessian determinant of the O-SURF is computationally expensive, because it requires the calculation of the first- and second-order derivatives of the Hessian matrices and their inverses. Two box filters with the sizes of 33 ($l = 11$, L = 33, $2l - 1 = 21$, $(2l+1)/2 = 7$) and 45 ($l = 15$, L = 45, $2l - 1 = 29$, $(2l + 1)/2 = 8$) are added to calculate the Hessian determinants which are to create a scale-space of higher granularity and remove the interpolation step, without sacrificing accuracy (Wilson et al. 2014). Finally, eight box filters with the sizes of {9, 15, 21, 27, 33, 39, 45, 51} are used to compute the Hessian determinants.

To efficiently calculate the Hessian determinant, multiple integral images need to be accessed by row-based stream processing (RBSP) (Hsu and Chien 2011) in parallel to perform the separable convolution of the integral image with all 24 box filters (8 × 3). The r-line buffer cores are visualized in Figure 4.7, in which Figure 4.7(a) lists the ten-line buffers of the box filter with the size 15 × 15, and Figure 4.7(b) reveals the memory foots of 32 (8 + 8 + 16) points of box filter in Figure 4.7(a), where Wx, Wy and W show the corresponding points in the Dxx,Dyy and Dxy box filters, and Figure 4.7(c) shows the all memory foots of 24 box filters with the size of {9, 15, 21, 27, 33, 39, 45, 51}.

The sampled 32-points are separable convolution for a 15 × 15 box filter as follow. L_0 thru L_{15} data are input to the r-line buffer cores in parallel mode. Sixteen registers R_0 thru R_{15} are used to store

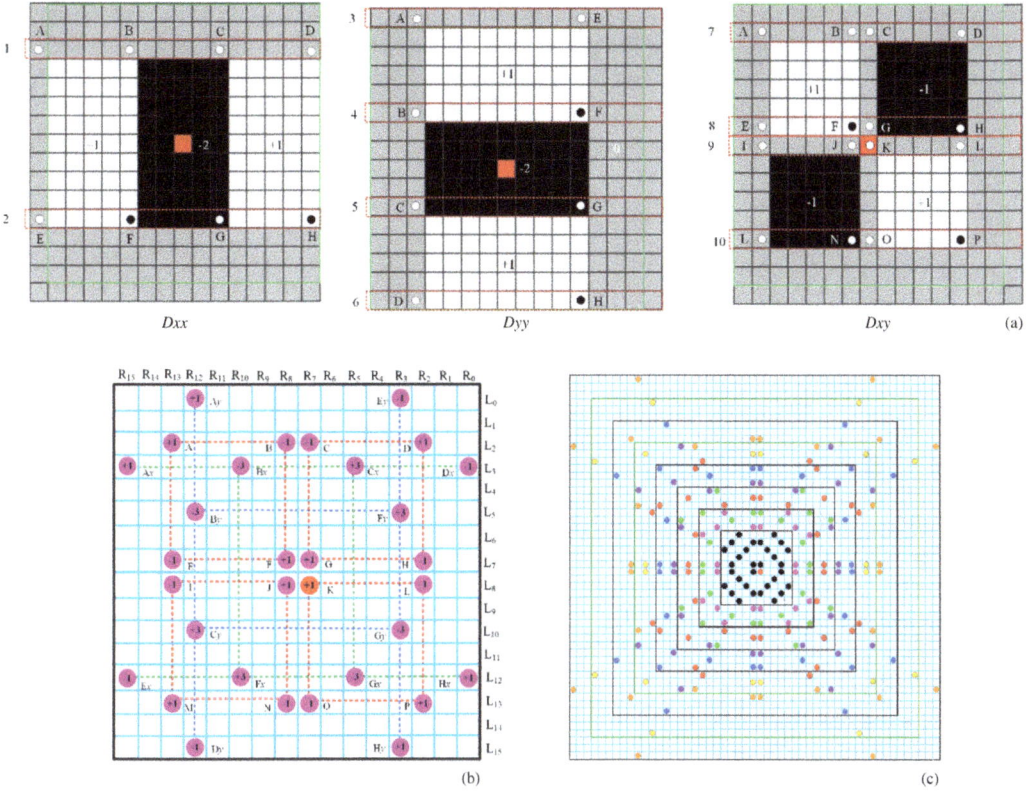

FIGURE 4.7 Parallel Multi-Scale Space Hessian Detector. (a) ten rows box filters with the size of 15; (b) the sampled points for 15×15 box filter; (c) the pixel-access window required for parallel convolution between 24 box filters.

the data of each line. 32 points can be selected to calculate the determinant of Fast-Hessian. Then, the box filter response is given by:

$$
\begin{aligned}
Dxx &= (L_3{}_{-}R_{15} + L_{12}{}_{-}R_{10} - L_3{}_{-}R_{10} - L_{12}{}_{-}R_{15}) - 2(L_3{}_{-}R_{10} + L_{12}{}_{-}R_5 - L_3{}_{-}R_5 - L_{12}{}_{-}R_{10}) \\
&\quad + (L_3{}_{-}R_5 + L_{12}{}_{-}R_0 - L_3{}_{-}R_0 - L_{12}{}_{-}R_5) \\
&= (L_3{}_{-}R_{15} - L_3{}_{-}R_{10}) - (L_{12}{}_{-}R_{15} - L_{12}{}_{-}R_{10}) - 2(L_3{}_{-}R_{10} - L_3{}_{-}R_5) + 2(L_{12}{}_{-}R_{10} - L_{12}{}_{-}R_5) \\
&\quad + (L_3{}_{-}R_5 - L_3{}_{-}R_0) - (L_{12}{}_{-}R_5 - L_{12}{}_{-}R_0)
\end{aligned}
\tag{4.17}
$$

$$
\begin{aligned}
Dyy &= (L_0{}_{-}R_{12} + L_5{}_{-}R_3 - L_0{}_{-}R_3 - L_5{}_{-}R_{12}) - 2(L_5{}_{-}R_{12} + L_{10}{}_{-}R_3 - L_5{}_{-}R_3 - L_{10}{}_{-}R_{12}) \\
&\quad + (L_{10}{}_{-}R_{12} + L_{15}{}_{-}R_3 - L_{10}{}_{-}R_3 - L_{15}{}_{-}R_{12}) \\
&= (L_0{}_{-}R_{12} - L_0{}_{-}R_3) - (L_5{}_{-}R_{12} - L_5{}_{-}R_3) - 2(L_5{}_{-}R_{12} - L_5{}_{-}R_3) + 2(L_{10}{}_{-}R_{12} - L_{10}{}_{-}R_3) \\
&\quad + (L_{10}{}_{-}R_{12} - L_{10}{}_{-}R_3) - (L_{15}{}_{-}R_{12} - L_{15}{}_{-}R_3)
\end{aligned}
\tag{4.18}
$$

$$
\begin{aligned}
Dxy &= (L_2{}_{-}R_{13} - L_2{}_{-}R_2) - (L_7{}_{-}R_{13} - L_7{}_{-}R_8) - (L_2{}_{-}R_6 - L_2{}_{-}R_2) + (L_7{}_{-}R_7 - L_7{}_{-}R_2) \\
&\quad - (L_8{}_{-}R_{13} - L_8{}_{-}R_8) + (L_{13}{}_{-}R_{13} - L_{13}{}_{-}R_8)
\end{aligned}
\tag{4.19}
$$

Due to the values of R_0 thru R_{15} are integral, then the multiply operation can be transformed into shift operation. Equations (4.17) through (4.19) are decomposed into the vertical and horizontal convolution, which are implemented by addition/subtraction and shift operation. The proposed separable convolution is simpler than proposed by Čížek and Faigl (2017). The proposed separable convolution utilization is only $O(n)$ and not $O(n^2)$ for parallel implementation of the full sliding window (Čížek and Faigl 2017).

In Figure 4.7(c), the color dots are the selected positions for the convolution operation. The sliding window shifts from the left to the right and the top to the bottom of the image. After scanning through the whole image, Hessian determinants are simultaneously computed at the same period clock.

4.4 HARDWARE IMPLEMENTATION

4.4.1 ARCHITECTURE OF HARDWARE IMPLEMENTATION

Considering the space, power and real-time constraints of an embedded system, FPGA is selected to ensure the system in real-time. The proposed architecture for hardware implementation is shown in Figure 4.8, which contains memory controller mode, integral image generation (IIG), SURF feature detection, BRIEF descriptor and BRIEF matching (Liu et al. 2019; 2020).

1. **Memory controller module**
   To drive on-board DDR3, Xilinx IP Memory Interface Generator (MIG) is chosen to create a logic connection with DDR3.
2. **IIG module**
   Integral image is a novel method to improve the performance of the subsequent steps for SURF detector. The word length reduction algorithm and 4-rows parallel method are adopted to optimize Integral image. IIG module transforms the integral image to SURF

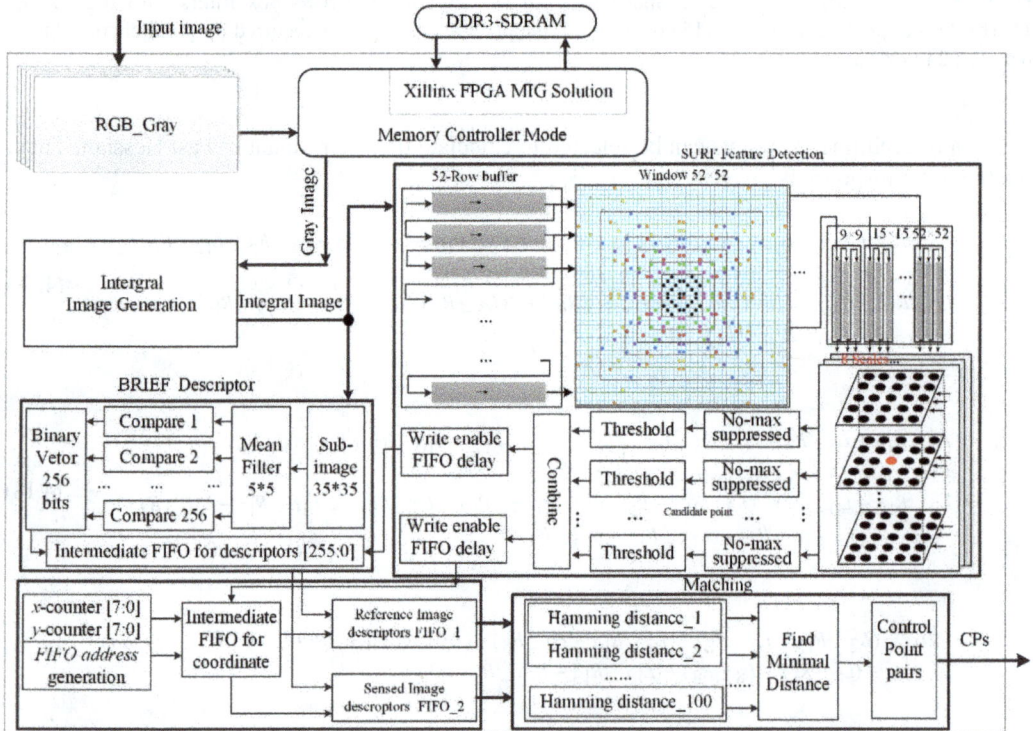

FIGURE 4.8 The proposed architecture for hardware implementation.

feature detector, memory controller module and BRIEF descriptor module, Therefore, the IIG module is considered as a separate part from the SURF feature detector Module.

3. **SURF feature detection module**

   SURF feature detection finds all the local maxima Fast-Hessian determinant as candidate points in multi-scale and then locates the corresponding index and scale. The FPGA-based SURF feature detection architecture is divided into 2 sub-modules: Fast-Hessian response generation and location of interest point. Location of interest point is divided into three steps: Non-Maximal suppression, threshold and interpolation. To solve these problems, a parallel architecture for the modified SURF algorithm is proposed. A sliding window buffer is used to store the shifted pixels of integral image at each clock. The buffer is shared with Hessian determinant. SAS algorithm and parallel multi-scale space are used to implement Hessian determinant. Additional 33 and 45 scales are used to replace interpolation without sacrificing accuracy (Čížek and Faigl 2017).

4. **BRIEF descriptor and matching module**

   256 bits' BRIEF descriptor and matching require a low hardware cost. To reduce its complexity, an optimized parallel adder tree and parallel comparators are employed to perform BRIEF descriptor and matching.

4.4.2 INTEGRAL IMAGE GENERATOR (IIG)

The integral image generator (IIG) module generates the integral image from incoming 8-bits grayscale image via memory interface generator (MIG) (Wang et al. 2013) to communicate with off-chip on-board DDR3 SDRAM (see Figure 4.9(a)), which is stored by the row sequence.

The hardware architecture of integral image (see Figure 4.9(b)) consists of address generator, row accumulator, multiplier and adder. Address generator module is designed to generate a read address (rd_addr) and a write address (wr_addr) though column counter and row counter. The value of the row counter generates a selector (sel) signal of the multiplexer (MUX).

4.4.3 SURF DETECTOR IMPLEMENTATION

4.4.3.1 Fast-Hessian Responses Implementation

After calculating integral image in the IIG module, integral image is sent to 52 lines buffer (Svab et al. 2009) and 52 silence registers in the sliding window for a separable convolution. The detailed

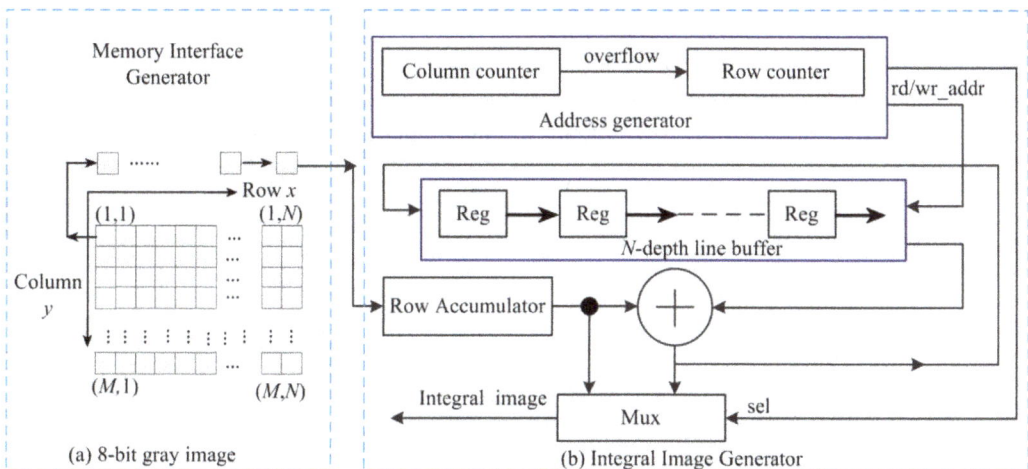

(a) 8-bit gray image

(b) Integral Image Generator

FIGURE 4.9 Hardware architecture for integral image generator.

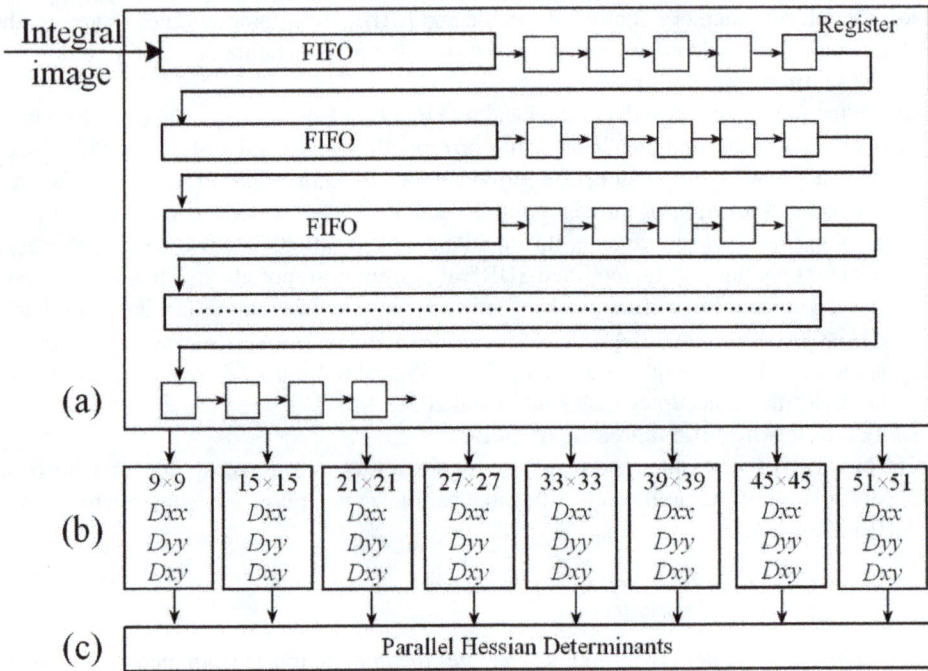

FIGURE 4.10 Parallel multi-scale space Hessian detector. (a) sliding window. (b) parallel implementation D_{xx}, D_{yy} and D_{xy}. (c) parallel calculation Hessian determinant.

design of the multi-scale-space Hessian detector is depicted in Figure 4.10. This architecture includes sliding window, parallel implementation of the second-order Gaussian derivatives and parallel calculation Hessian determinants. To construct a sliding window for the Hessian response, a FIFO architecture including 52 lines buffer and 52×52 silence registers are used. The sliding window provides a parallel access to the integral image required for all second-order Gaussian derivatives D_{xx}, D_{yy} and D_{xy} using Equations (4.17)–(4.19). The 256 pixels (due to overlap, in fact, there are 220 pixels) access for the response of the 8 box filters.

After D_{xx}, D_{yy} and D_{xy} are obtained, the Hessian response can be calculated in the three pipelines by Equation (4.16) and is shown in Figure 4.11. The eight Hessian determinants are concurrently calculated and then parallel output to the Non-Maximal suppression module.

4.4.3.2 Non-Maximal Suppression Implementation

Non-Maximal Suppression (NMS) module selects the local maximal Hessian determinant as a candidate point. A 5×5 sliding window is chosen in NMS module (Figure 4.12). Eight multi-scales can be divided into 6 multi-space-scales to concurrently find the local maximal interest point.

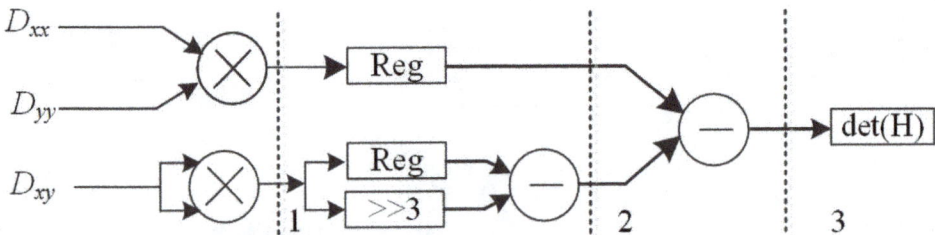

FIGURE 4.11 Architecture of the pipelined Hessian determinant calculation.

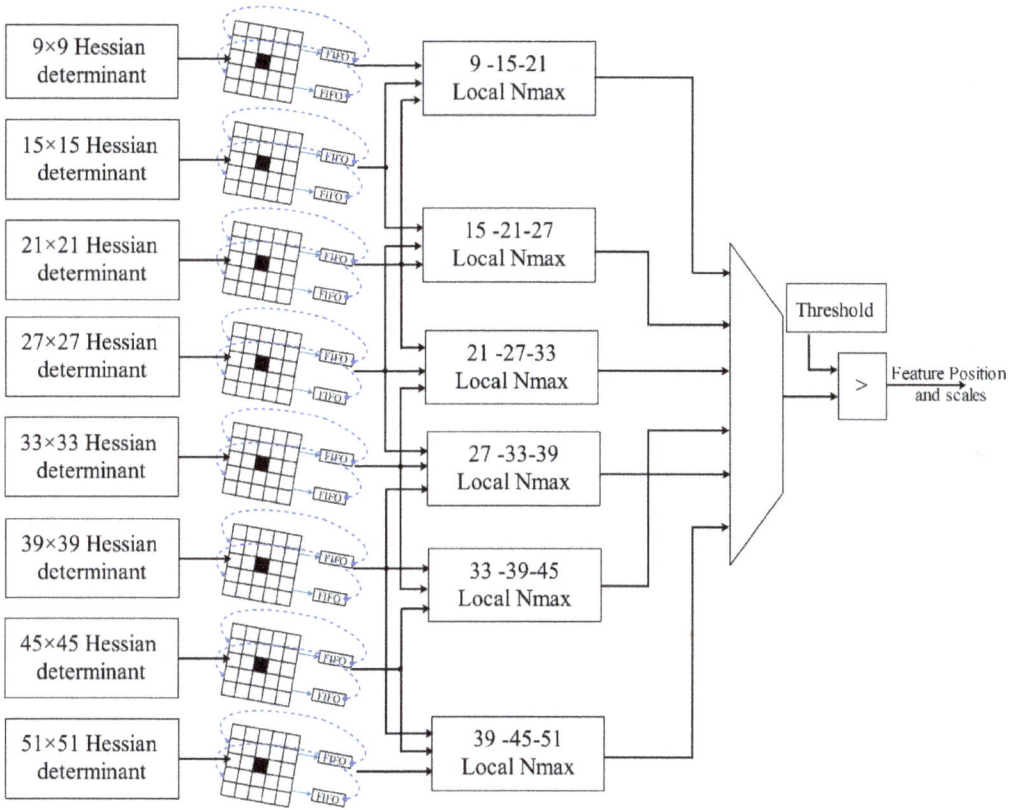

FIGURE 4.12 The architecture of Non-Maximal suppression module.

The 5×5 points of each matrix can be presented as $Top\_m_{i,j}$, $Middle\_m_{i,j}$ and $Bottom\_m_{i,j}$ ($i = 1,2,3,4,5; j = 1,2,3,4,5$). The $Middle\_m_{3,3}$ is compared with its 74 neighbor points (24 neighbors in the same scale and 25 neighbors in the consecutive scales above and below it, respectively) in parallel. If the result of *AND* operation is *true*, the center point is regarded as a candidate point (Huang and Zhou 2017) (Figure 4.13). Then, the candidate point is fed into user-defined

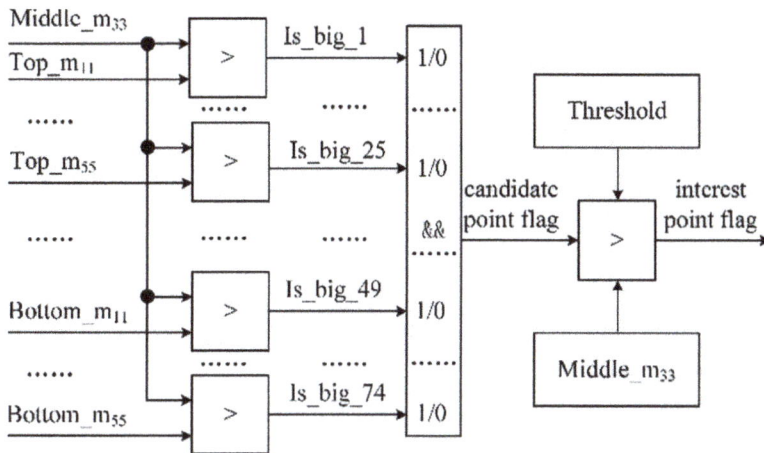

FIGURE 4.13 The architecture of location maximum across the 74 neighbor points.

threshold module. Only when candidate point is more than the threshold, then can be thought as a feature point, and 1-bit interest point flag is set to *true* (Wilson et al. 2014; Chen et al. 2015; Ni et al. 2019).

Since the Hessian determinant computations are parallelized to compute, all of 8 scale determinants are synchronously available to compare. The search for a local maximum in the scale-space domain is also pipelined and parallelized using register FIFOs. Only the determinants exceeding a certain threshold are saved while the others are set to zeros points that are not local maxima in the three-dimensional scale-space domain are suppressed (=0). This method ensures that feature don't overlap and are evenly spread over the input image (Čížek and Faigl 2017). Finally, every local maximum is compared to a user-defined threshold value. The threshold controls the overall sensitivity of the detector by fine tuning the number of interest point that populate the image (Zhang 2010). Considering that the features in the same octave can generate the same descriptor if they have the same coordinate even though in different scale, we only store one of them if there is more than one feature at the same time (Fularz et al. 2015).

4.4.4 IMPLEMENTATION FOR BRIEF DESCRIPTOR

BRIEF-32 algorithm is adopted to generate descriptors (Calonder et al. 2012). The BRIEF-32 structure includes two modules: image buffer, point pair comparator (Figure 4.14). In image buffer module, a 35×35 sub-window was proposed by Huang (Huang et al. 2018b). Calonder et al. (2012) provided consume time when using Gaussian smoothing, simple box filtering and box filtering using integral image, the latter was much faster. Furthermore, it did not result in any matching performance loss. For this reason, a 5×5 box filtering using integral images is used to smooth the original image. 256 binary (32 byte) tests-patch in the point comparator module, the tests-patch are related to blue lines which are sampled from an isotropic $\left(0, \dfrac{S^2}{25}\right)$ Gaussian distribution. Then, 256 patch-points $p(r_i, c_j)$ ($i = 1, 2, ..., 256$; $j = 1, 2, ..., 256$) are parallel compared to corresponding point using Equation (4.7) at the same cycle, and one bit result of comparison 1 or 0 is stored into 256-bit descriptor register in sequence.

4.4.5 IMPLEMENTATION FOR BRIEF MATCHING

Hamming distance is used to match the 256-bit descriptor in the BRIEF matching. BRIEF descriptor is robust to illumination changes and small rotations (Calonder et al. 2012), which makes it a good candidate point for the georeferencing. The BRIEF matching module consists of the computing

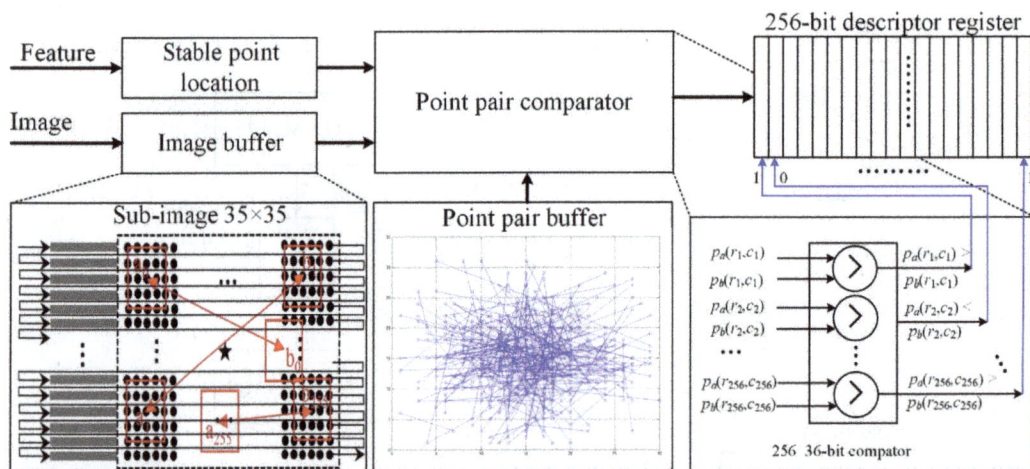

FIGURE 4.14 Exemplary 256 patch-pairs for BRIEF descriptor.

Hamming distance module and finding the minimal distance module. The reference image descriptors are stored into the FIFO1, and the sensed image descriptors are stored into the FIFO2. To trade-off the speed time and resource. The number of descriptor point determines the accuracy of the match, however, the more descriptor points, the more resources are consumed. For example, 100 pair-descriptors are used to implement the Hamming distance in Figure 4.15 at parallel mode (Zhao et al. 2014). In the Hamming distance module (see Figure 4.15(a)), the Hamming distance is computed using 256 XOR gates and the results are stored into 256-bit register. Then improving parallel pipelined adder trees (Chen et al. 2016) are used to compute the number of '1' in the 256-bits register. The 256-bit result of XOR is divided into eight 32-bit registers which are paralleled to calculate the Hamming distance by 5-level pipeline adder trees (see Figure 4.15(c)). However, 9-level pipeline adder trees were used in (Zhao et al. 2014).

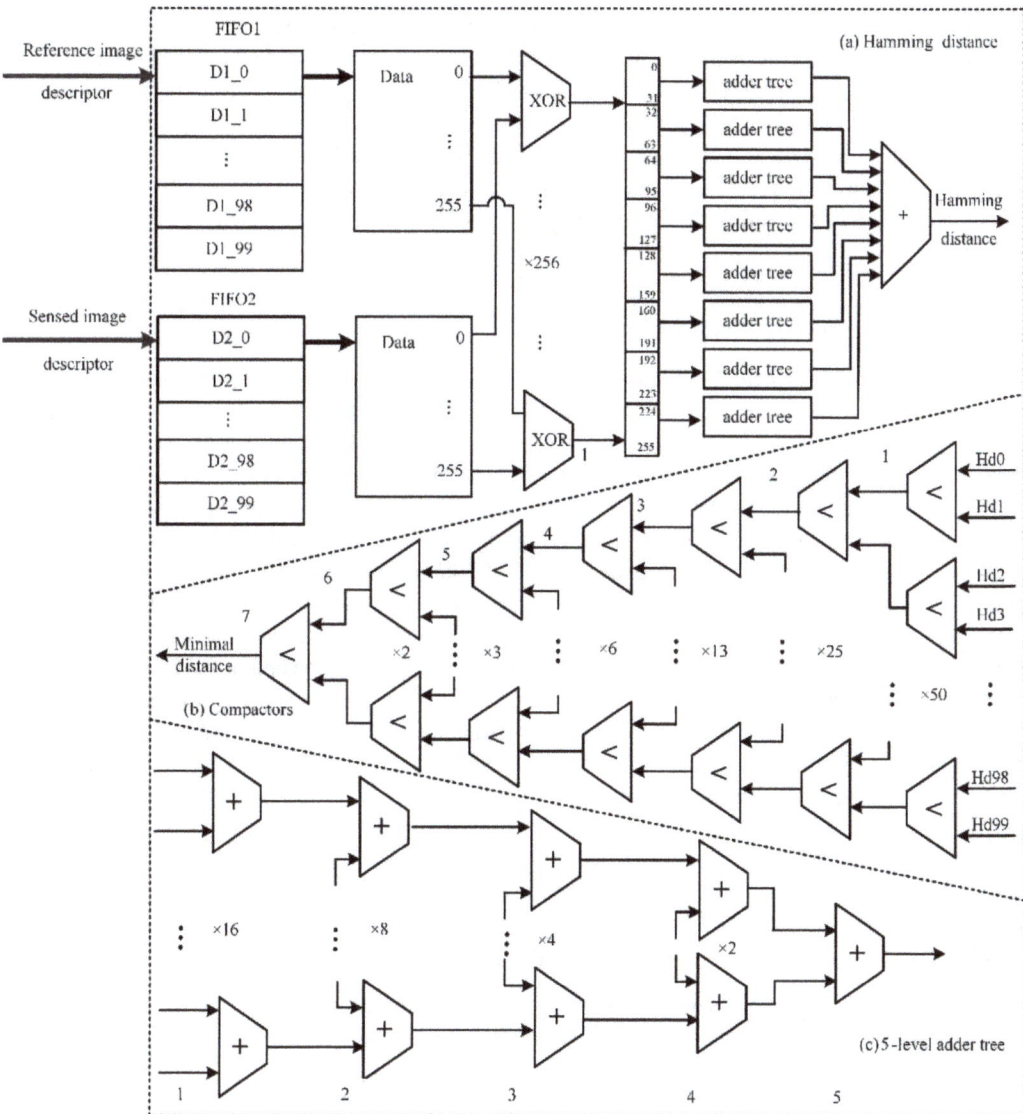

FIGURE 4.15 The detailed structure of the matching coprocessor. (a) Computation Hamming distance, (b) Locating minimal Hamming distance and (c) 5-level adder trees, respectively.

The finding minimal distance module begins with receiving the 100 Hamming distances, the minimum Hamming distance is defined as the marching feature, and the corresponding feature point is control point. The finding minimal distance module is shown in Figure 4.15(b), an improving compactor module is adopted to compare the Hamming distance with 7 levels pipeline. Three compactor modules are reduced in comparison with Zhao et al. (2014).

The coordinates that are obtained in BRIEF matching core are scanning coordinates. However, in georeferencing method, the geodetic coordinates are used in the projection transformation equation. So, the scanning coordinates must be transformed into geodetic coordinates (Liu et al. 2019).

4.5 VERIFICATION AND DISCUSSION

4.5.1 HARDWARE ENVIRONMENT AND DATA SET

The proposed system is implemented in a signal Xilinx XC7VX980T FPGA that has 612,000 logic cells, 1,224,000 Flip-Flops, 1500 kB Block RAM and 3600 digital signal processing (DSP) slices. The development kit Vivado (14.2 version) is used to design the hardware of the system in Verilog HDL, the simulation tool is Vivado simulator. The first data sets are from Huang et al. (2018a), the second data sets are downloaded from BIGMAP software (Figure 4.17(c)–(e)). The image dimension is 512 × 512. In addition, the results that are generated by the implemented FPGA-based are compared with that of OpenCV library which is written by Chris Evans in the MATLAB.

4.5.2 INTEREST POINTS EXTRACTION

The number of interest point is affected by some parameters such as octave, scale, resampling, the size of Non-Maximal matrix, threshold, etc (Ehsan et al. 2010). Table 4.3 shows the number of interest points with the different threshold. As seen from Table 4.3, the variation of blob response threshold has a significant effect on interest point distribution.

Figure 4.16 shows the result of different pair-points matching. The result indicated that the uniform distribution of matching point is affected by the number of matching points and the textures of object. As seen from Figure 4.16, when the 33 and 45 scales are added, the number of interest points increase as well, which make the interest points more evenly distributed. But it meanwhile increases the consumption of hardware resource. Furthermore, the extra scales create a scale-space of higher granularity, allowing us to remove the interpolation step at the end of the Hessian determinant calculation (Wang et al. 2013). In addition, there are some error matched point pairs, but these points can be eliminated by using robust fitting methods, such as RANSAC (Ehsan et al. 2010) or a combined algorithm of the slope-based rejection (SR) and the correlation-coefficient-based rejection (CCR) (Huang et al. 2018b).

TABLE 4.3
The Number of Feature Points with the Different Threshold

Image pairs	No Threshold	Threshold = 1000	Threshold = 8000	Threshold = 15,400
Bungalows 1/2	6666/6526	1467/1452	125/118	68/63
High-rise building 1/2	10,110/10,548	2309/2277	1169/1125	703/576
Flyover pairs 1/2	3898/3928	2931/2918	1270/1220	718/683
Bare soil 1/2	3349/3349	491/491	32/32	26/26
Farmland 1/2	1803/2327	1238/1235	1057/1067	728/760

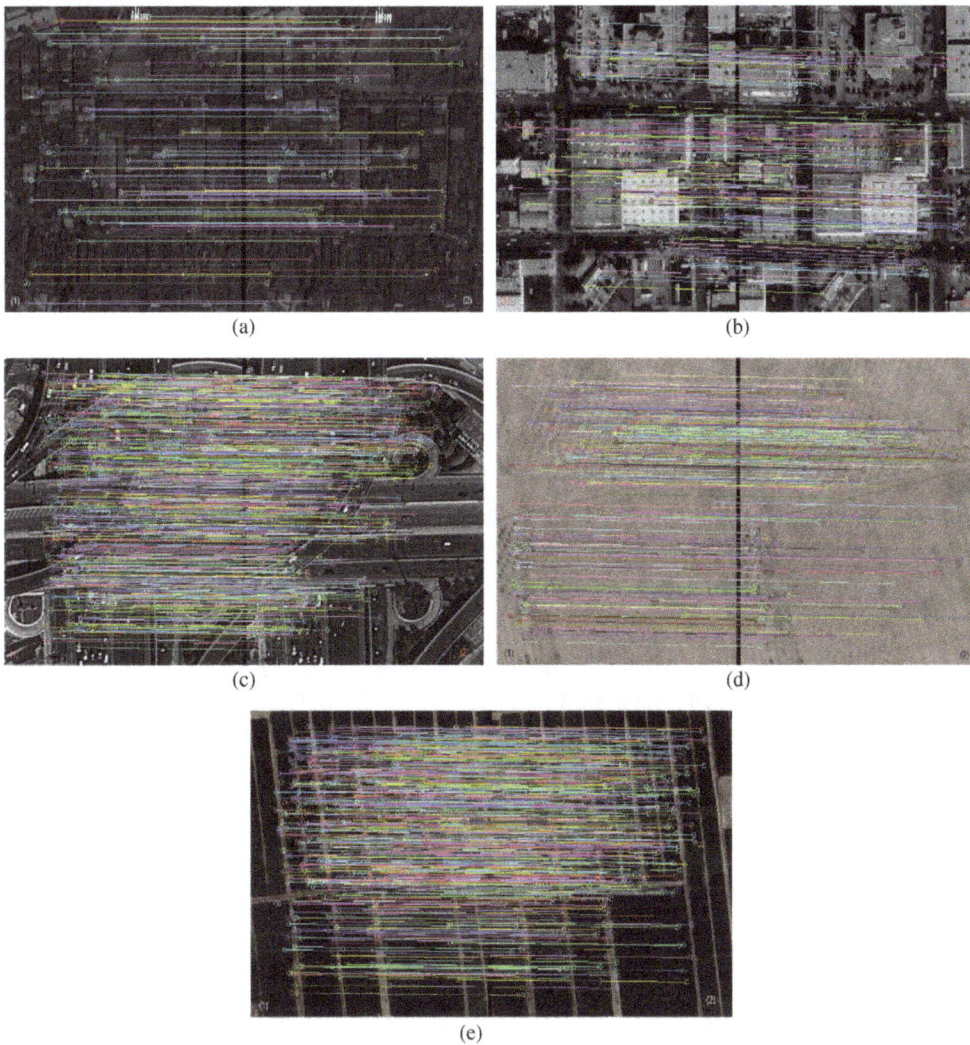

FIGURE 4.16 Matching of five pairs of remote sensing images. (a) 100 pair-points matching with bunga-lows; (b) 100 pair-points matching with high buildings; (c) 800 pair-points matching with flyover; (d) 220 pair-point matching with bare soils; (e) 700 pair-point matching with farmlands.

4.5.3 ERROR ANALYSIS

Error is inevitable when the fixed-point data is replaced by a floating-point data format. The error analysis is conducted using called *recall* versus 1-*precision* curve, which were defined as (Mikolajczyk and Schmid 2004):

$$recall = \frac{\#\,correct\ matches}{\#\,correspondences} \tag{4.21}$$

$$1 - precision = \frac{\#\,false\ matches}{\#\,correct\ matches + \#\,false\ matches} \tag{4.22}$$

where the *recall* is the ratio of the number correctly matched points relative to the number of corresponding matched points. And the 1-*precision* is defined as the radio of the number of false matches to the total number of matches which including the false matches and the correct matches.

It has been demonstrated that the SURF+BRIEF has a better matching performance in different land coverages than that from SURF when using software with OpenCV2.4.9+ Microsoft Visual Studio 2015 (Huang et al. 2018a). The relationship between two images can be described by the Homography matrix (Huang et al. 2018a). The Homography matrix of the five image pairs are calculated by MATLAB on the basis of a PC. The results are

$$H_{village} = \begin{bmatrix} 1.024682517497177 & 0.018725015152393 & -5.404435483561459 \\ 0.015206213894699 & 1.014273691140137 & -2.552171349203126 \\ 0.000024222830514 & 0.000023299587958 & 1.000000000000000 \end{bmatrix} \quad (4.23)$$

$$H_{high-rise\ building} = \begin{bmatrix} 1.099260397052538 & 0.002259608451860 & -7.957861186118295 \\ 0.090326101809412 & 0.955409181508876 & -4.421034816054379 \\ 0.000271661969942 & -0.000085315081328 & 1.000000000000000 \end{bmatrix} \quad (4.24)$$

$$H_{bare\ soil} = \begin{bmatrix} 1.061661003420602 & -0.064774241169707 & -22.139327368389228 \\ 0.030845167584049 & 0.861666704860821 & 6.043495802072064 \\ 0.000065697594479 & -0.000382940095883 & 1.000000000000000 \end{bmatrix} \quad (4.25)$$

$$H_{flyover} = \begin{bmatrix} 1.000297662301777 & 0.025695798331601 & -66.220856181085963 \\ -0.032698271963169 & 1.023565673743905 & 8.462326102920549 \\ -0.000064630077908 & 0.000065894730718 & 1.000000000000000 \end{bmatrix} \quad (4.26)$$

$$H_{framland} = \begin{bmatrix} 1.000297662301777 & 0.025695798331601 & -66.220856181085963 \\ -0.032698271963169 & 1.023565673743905 & 8.462326102920549 \\ -0.000064630077908 & 0.000065894730718 & 1.000000000000000 \end{bmatrix} \quad (4.27)$$

FIGURE 4.17 shows the comparison between the FPGA-based and the PC-based implementation for SURF+BRIEF algorithm. A high recall and a low 1-precision mean a good matching performance (Huang et al. 2018a). As seen from Figure 4.17, the performance from the FPGA-based implementation is slightly worse than that from the PC-based implementation. The reasons may be caused by:

1. Several fixed-point approximations are implemented to save the resource and fit the entire architecture on a single Xilinx chip. The Hessian determinant is approximated using only shifting and adding operation.
2. Only two octaves are implemented to detect the feature points, which inevitably lead to a performance degradation.

4.5.4   PERFORMANCE ANALYSIS OF FPGA

4.5.4.1   FPGA Resources Utilization Analysis

249,811 (40.82%) slice look-up tables (LUTs), 244,567 (19.98%) slice flip flops (FFs) and 11 memories are used to implement the proposed architecture. Two octaves and 8 scales (6 O-SURF scales, two extra scales (33 and 45) are used to build Hessian response. SAS method is adopted, which makes floating-point transform into shift and subtraction, and other multiplication are also constructed using slice logic only. The DSP resource is significantly reduced when compared with other methods (Wilson et al. 2014), but slice logical resource increase significantly (Svab et al. 2009; Wang et al. 2013; Zhao et al. 2013; Cai et al. 2017; Ni et al. 2019). SAS, sliding window, parallel

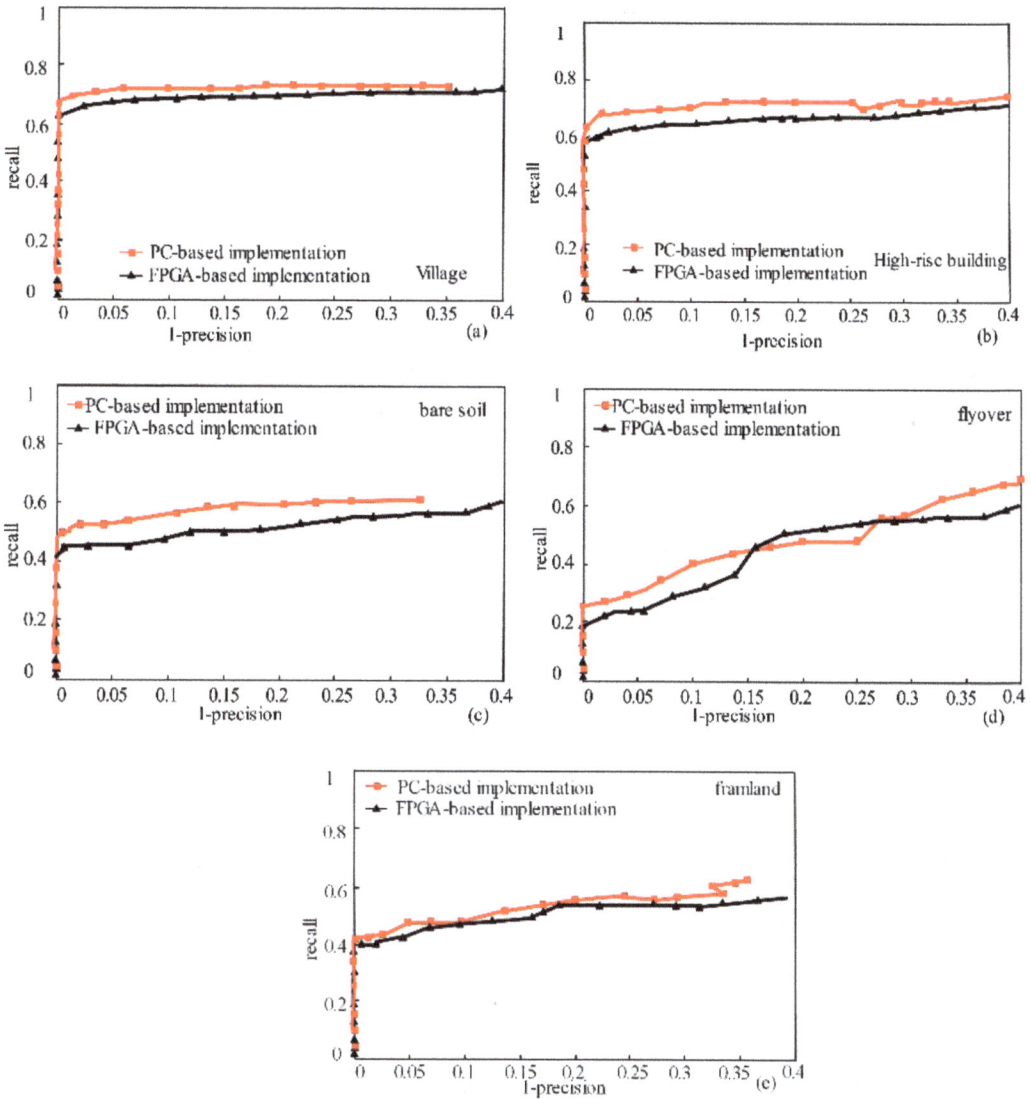

FIGURE 4.17 The 1-precision versus the recall of five pairs of RS images with land cover in (a) village; (b) rise-high building; (c) flyover; (d) bare soil; and (e) farmland, respectively.

multi-scale space and parallel pained add-trees are used to optimize the SURF detector and BRIEF descriptor, block random access memory (BRAM) resources are greatly reduce when compared with other methods (Svab et al. 2009; Wang et al. 2013). Table 4.4 lists the compared between proposed method and have been published using hardware.

4.5.4.2 Computational Speed Comparison

The speed is one of the most import factors for on-board detection and matching. The run time of integral image, SURF detection, BRIEF descriptor and BRIEF matching is 2.62 μs, 2.9 ms, 2.81μs and 32.66 μs, respectively. The total run time is 2.9 ms, guaranteeing a frame rate of 340 fps (frame per second) for a stream of 512×512 pixels^2 at the 100 MHz. An efficient image matching system based on a single FPGA chip was proposed by Fularz et al. (2015), which can implement SIFT feature, BRIEF descriptor and BRIEF matching for image of 1280×720 pixels^2 within 33 ms, at 30 fps

TABLE 4.4

Comparison of FPGA Utilization

Method	FFs	LUTs	BRAMs	DSPs
SURF+BRIEF (Huang et al. 2018b)	122,000	88,462	1	0
OpenSURF (Ni et al. 2019)	n.a.	107,873	295	1185
SURF+BRIEF (Chen et al. 2015)	29,841	25,463	116	160
SURF (Zhao et al. 2013)	29,165	37,592	68	178
SURF (Alahi et al. 2012)	108,581	179,559	80	244
SURF (Huang et al. 2018a)	35,804	37,662	80	105
SIFT+BRIEF (Chen et al. 2016)	30,002	21,119	4672	80
SURF (Wang et al. 2013)	1273	6541	1,544,102	24
SURF (Fularz et al. 2015)	6542	12,732	985,566	24
Proposed	244,567	249,811	11	0

for image sequences with a dimension of 640×480 pixels^2, the hardware-accelerated version using the FPGA obtained an execution times of 0.047 seconds (Bouris et al. 2010). The same architecture was proposed, in which a frame rate was 132.36 fps for a video stream at VGA resolution of 40.355 MHz (Wilson et al. 2014). The comparisons of the proposed method with other methods are shown in Tables 4.5 and 4.6.

TABLE 4.5

Comparison of Performance Using fps (SW, Software; HW, Hardware)

Method	Clock (MHz)	Resolution	fps	SW/HW	Descriptor
SURF (Cai et al. 2017)	100	640×480	270	SW+HW	Yes
SURF (Wang et al. 2013)	100	1024×480	~10	SW+HW	No
SURF (Svab et al. 2009)	136.3	640×480	420	SW+HW	Yes
SURF+BRIEF (Huang et al. 2018b)	100	512×512	304	HW	Yes
SURF+ FREAK (Zhao et al. 2014)	100	800×600	60	HW	Yes
SURF (Fularz et al. 2015)	66.7	640×480	50	HW	Yes
SURF (Ni et al. 2019)	66	800×600	60	HW	Yes
SURF (Wilson et al. 2014)	40.355	640×480	131.36	HW	Yes
SURF (Fularz et al. 2015)	200	640×480	56	HW	Yes
Proposed	100	512×512	340	HW	Yes

TABLE 4.6

Comparison of Performance for Speed Time (s = Second; ms = Micro Second)

Method	Clock (MHz)	Resolution	SW/HW	Descriptor	Matching	Speed
SURF+ BRIEF (Zhao et al. 2014)	100	512×512	HW(FPGA)	Yes	Yes	3.29 ms
SURF (Yao et al. 2009)	200	1024×768	HW(FPGA)	No	No	0.100 s
SIFT+BRIEF (Wang et al. 2013)	359	1280×720	HW(FPGA)	Yes	Yes	33 ms
SURF (Chen et al. 2016)	66.7	640×480	HW(FPGA)	Yes	No	0.247 s
SURF (Chen et al. 2016)	3400	640×480	SW+HW	Yes	No	11.72 s
SURF (Chen et al. 2016)	200	640×480	HW(ARM)	Yes	No	1.8 s
SURF (Bouris et al. 2010)	200	640×480	HW	Yes	No	0.047 s
Proposed	100	512×512	HW	Yes	Yes	2.94 ms

4.6 CONCLUSIONS

An optimized FPGA-based architecture is proposed for performing robust, real-time automatic extraction GCPs on-chip at a frame rate of 340 fps under 100 MHz. The performance of combined SURF diction, BRIEF descriptor and BRIEF feature matching system is similar to that of OpenSURF in the MATLAB. Real-time performance is guaranteed by an optimized parallel and pipelined architecture. WLR, MEPA, SAS, sliding window and optimized space-scale are adopted to optimize SURF detection. The parallel pipelined adder trees are used to accelerate the BRIEF matching. The whole system consumes approximately 8 Watts according to Xilinx Power Estimator Tool. Hence, the combined SURF feature detection, BRIEF descriptor and BRIEF matching system has advantages in real-time processing, low-power consumption.

REFERENCES

Abdel-Hakim, A.E., Farag, A.A., CSIFT: A SIFT descriptor with color invariant characteristics. *In Computer Vision and Pattern Recognition (CVPR 2006)*, New York, USA, 17–22 June 2006, 2: 1978–1983.

Alahi, A., Ortiz, R., Vandergheynst, P., Freak: Fast retina keypoint. *2012 IEEE Conference on Computer Vision and Pattern Recognition. Providence*, RI, USA 510–517, 16–21 June 2012, pp. 510–517.

Attneave, F., Some informational aspects of visual perception. *Psychological Review*, 1954, 61(3): 183–193.

Bay, H., Ess, A., Tuytelaars, T., Gool, L.V., Speeded-up robust features (SURF), *Computer Vision and Image Understanding*. 110(3): 346–359, 2008.

Bay, H., Tuytelaars, T., Goo, L.V., SURF: Speeded up robust features. *European Conference on Computer Vision*. Berlin, Heidelberg, 7–13 May 2006, Part II, pp. 404–417.

Belt, H.J.W., Word length reduction for the integral image. *In Proceedings of the 15th IEEE International Conference on Image Processing*, San Diego, CA, USA, 12–15 October 2008; pp. 805–808.

Bouris, D., Nikitakis, A., Papaefstathiou, I., Fast and efficient FPGA-based feature detection employing the SURF algorithm. *In 2010 18th IEEE Annual International Symposium on Field-Programmable Custom Computing Machines*. Charlotte, NC, USA, 2–4 May 2010, pp. 3–10.

Cai, S.S., Liu, L.B., Yin, S.Y., Zhou, R.Y., Zhang, W.L., Wei, S.J., Optimization of speeded-up robust feature algorithm for hardware implementation. *Science China Information Sciences*, 2014, 57(4): 1–15.

Cai, W., Xu, Z., Li, Z., A high performance SURF image feature detecting system based on ZYNQ. *DEStech Transactions on Computer Science and Engineering*, 2017, 12: 256–261.

Calonder, M., Lepetit, V., Fua, P., Konolige, K., Bowman, J., Mihelich, P., Compact signatures for high-speed interest point description and matching. *In 2009 IEEE 12th International Conference on Computer Vision*. Kyoto, Japan, 29 September–2 October 2009, pp. 357–364.

Calonder, M., Lepetit, V., Oezuysal, M., Trzcinski, T., Strecha, C., Fua, P., BRIEF: Computing a local binary descriptor very fast. *IEEE Transactions on Pattern Analysis and Machine Intelligence*, 2012, 34(7): 1281–1298.

Chen, C., Yong, H., Zhong, S., Yan, L., A real-time FPGA-based architecture for OpenSURF. *Proc. SPIE 9813, MIPPR 2015: Pattern Recognition and Computer Vision*, 98130K-1, 2015.

Chen, W., Ding, S., Chai, Z., He, D., Zhang, W., Zhang, G., Luo, W., FPGA-based parallel implementation of SURF algorithm. *In 2016 IEEE 22nd International Conference on Parallel and Distributed Systems (ICPADS)*, Wuhan, China, 13–16 December 2016. pp. 308–315.

Cheon, S.H., Eom, I.K., Moon, Y.H., Fast descriptor extraction method for a SURF-based interest point. *Electronics Letters*, 2016, 52(4): 274–275.

Chris, E., OpenSURF (including Image Warp), Available online. https://ww2.mathworks.cn/matlabcentral/fileexchange/28300-opensurf-including-image-warp.

Christopher, E., Notes on the Opensurf library. *University of Bristol*, Tech. Rep. CSTR-09-001, 2009, January.

Čížek, P., Faigl, J., Real-time FPGA-based detection of speeded-up robust features using separable convolution. *IEEE Transactions on Industrial Informatics*, 2017, 14(3): 1155–1163.

Čížek, P., Jan, F., Masri, D., Low-latency image processing for vision-based navigation systems. *In 2016 IEEE International Conference on Robotics and Automation (ICRA), IEEE*. Stockholm, Sweden, 16–21 May 2016, pp. 781–786.

David, G., Decker, P., Paulus, D., An evaluation of open-source SURF implementations. *Robot Soccer World Cup*. Springer, Berlin, Heidelberg, 2010, pp. 169–179.

Deng, X., Huang, Y., Feng, S.C., Wand. Ground control point extraction algorithm for remote sensing image based on adaptive curvature threshold. *In 2008 International Workshop on Education Technology and Training & 2008 International Workshop on Geoscience and Remote Sensing.* Shanghai, China, 21–22 December 2008, 2, pp. 137–140.

Ehsan, S., Clark, A.F., McDonald-Maier, K.D., Novel hardware algorithms for row-parallel integral image calculation. *In 2009 Digital Image Computing: Techniques and Applications, IEEE.* Melbourne, VIC, Australia, 1–3 December 2009. pp. 61–65.

Ehsan, S., Clark, A.F., Rehman, N., McDonald-Maier, K., Integral images: Efficient algorithms for their computation and storage in Resource-constrained embedded vision systems. *Sensors,* 2015, 15(7): 16804–16830.

Ehsan, S., Kanwal, N., Bostanci, E., Clark, A.F., McDonald-Maier, K.D., Analysis of interest point distribution in SURF octaves. *c,* Hong Kong, China, April 2010, pp. 411–415.

Ehsan, S., McDonald-Maier, K.D., Exploring integral image word length reduction techniques for SURF detector. *In 2009 Second International Conference on Computer and Electrical Engineering, IEEE. Dubai, United Arab Emirates,* Dubai, United Arab Emirates, 28–30 December 2009, pp. 635–639.

Fan, X., Wu, C., Cao, W., Zhou, X., Wang, S., Wang, L., Implementation of high-performance hardware architecture of OpenSURF algorithm on FPGA. *In 2013 International Conference on Field-Programmable Technology (FPT).* Kyoto Japan, 9–11 December 2013, pp. 152–159.

Fularz, M., Kraft, M., Schmidt, A., Kasiński, A., A high-performance FPGA-based image feature detector and matcher based on the FAST and BRIEF algorithms. *International Journal of Advanced Robotic Systems,* 2015, 12(10): 141.

Harris, C., Stephens, M., A combined corner and edge detector. *Proceedings of the Alvey Vision Conference,* 1988, 15(50): 147–151.

Hsu, P. H., Chien, S.Y., Reconfigurable cache memory architecture for integral image and integral histogram applications, Beirut, Lebanon, 4–7 October 2011, Beirut, Lebanon, 151–156.

Hua, G., Brown, M., Winder, S., Discriminant embedding for local image descriptors. *In 2007 IEEE 11th International Conference on Computer Vision,* Rio de Janeiro, Brazil, 14–21 October 2007, pp. 1–8.

Huang, J., Zhou, G., On-board detection and matching of feature points. *Remote Sensing,* 2017, 9(6): 601.

Huang, J., Zhou, G., Zhang, D., Zhang, G., Zhang, R., Baysal, O., An FPGA-based implementation of corner detection and matching with outlier rejection. *International Journal of Remote Sensing,* 2018b, 39(23): 8905–8933.

Huang, J., Zhou, G., Zhou, X., Zhang, R., A new FPGA architecture of FAST and BRIEF algorithm for on-board corner detection and matching. *Sensors,* 2018a, 18(4): 1014.

Idris, M.Y.I., Warif, N.B.A., Arof, N.M., Wahab, A.W.A., Razak, Z., Acceleration FPGA-SURF feature detection module by memory access reduction. *Malaysian Journal of Computer Science,* 2019, 32(1): 47–61.

Je´gou, H., Douze, M., Schmid, C., Product quantization for nearest neighbor search, *IEEE Transaction Pattern Analysis and Machine Intelligence,* 2011, 33(11): 117–128.

Kasezawa, T., Tanaka, H., Ito, H., Integral image word length reduction using overlapping rectangular regions. *2016 IEEE International Conference on Industrial Technology (ICIT),* Taipei, Taiwan, 14–17 March 2016, pp. 763–768. doi: 10.1109/ICIT.2016.7474847.

Ke, Y., Sukthankar, R., PCA-SIFT: A more distinctive representation for local image descriptors. *Proceedings of the 2004 IEEE Computer Society Conference on Computer Vision and Pattern Recognition. IEEE Computer Society,* 2004, 2, pp. 506–513. doi:10.1109/CVPR.2004.1315206.

Kim, D., Kim, M., Kim, K., Sung, M., Ro, W.W., Dynamic load balancing of parallel SURF with vertical partitioning. *IEEE Transactions on Parallel and Distributed Systems,* 2014, 26(12): 3358–3370. doi: 10.1109/TPDS.2014.2372763.

Kim, J., Park, E., Cui, X., Kim, H., A fast feature extraction in object recognition using parallel processing on CPU and GPU, San Antonio, TX, USA, 11–14 October 2009, 3842–3847.

Kisacanin, B., Integral image optimizations for embedded vision applications. *In Proceedings of the IEEE Southwest Symposium on Image Analysis and Interpretation,* Santa Fe, NM, USA, 24–26 March 2008; pp. 181–184. doi: 10.1109/SSIAI.2008.4512315.

Krajník, T., Šváb, J., Pedre, S., Čížek, P., Přeučil, L., FPGA-based module for SURF extraction. *Machine Vision and Applications,* 2014, 25(3): 787–800.

Lee, S.H., Jeong, Y.J., A new integral image structure for memory size reduction. *IEICE TRANSACTIONS on Information and Systems,* 2014, 97(4): 98–1000. doi: 10.1587/transinf. E97. D.998.

Leng, C.C., Zhang, H., Li, B., Cai, G.R., Pei, Z., He, L., Local feature descriptor for image matching: A survey, *IEEE Access,* 2018, 7: 6424–6434. doi:10.1109/ACCESS.2018.2888856.

Leutenegger, S., Chli, M., Siegwart, R.Y., BRISK: Binary robust invariant scalable keypoints, *In 2011 IEEE International Conference on Computer Vision*. Barcelona, Spain, 6–13 November. 2011, pp. 2548–2555. doi: 10.1109/ICCV.2011.6126542.

Li, A., Jiang, W., Yuan, W., Dai, D., Zhang, S., Wei, Z., An improved FAST+SURF fast matching algorithm, *Procedia Computer Science*, 2017, 107: 306–312.

Lima, R.D., Martinez-Carranza, J., Morales-Reyes, A., Cumplido, R. Accelerating the construction of BRIRF descriptors using an FPGA-based architecture, *2015 International Conference on Reconfigurable Computing and FPGAs*, Mexico City, Mexico, 7–9 December, 2015, pp. 1–6, doi: 10.1109/ReConFigure2015.7393285.

Lindeberg, T., Feature detection with automatic scale selection, *International Journal of Computer Vision*, 1998, 30(2): 79–116.

Lindeberg, T., Scale-space theory: A basic tool for analyzing structures at different scales, *Journal of Applied Statistics*, 1994, 21(1–2): 225–270. doi.org/10.1080/757582976.

Lindeberg, T., Scale-space: A framework for handling image structures at multiple scales, *Proc. CERN School of Computing*, 8–21 Sep. 1996, Egmond aan Zee, The Netherlands 1996, pp. 1–12.

Liu, D., Research on Parallel Algorithms Using FPGA for Geometric Correction and Registration of Remotely Sensed Imagery. Dissertation, *Tianjin University*, Tianjin China May 2022.

Liu, D., Zhou, G., Huang, J., Zhang, R., Shu, L., Zhou, X., Xin, C.S., On-board georeferencing using FPGA-based optimized second-order polynomial equation, *Remote Sensing*, 2019, 11(2): 124. doi: 10.3390/rs11020124.

Liu, D., Zhou, G., Zhang, Dianjun, et al. Ground control point automatic extraction for spaceborne georeferencing based on FPGA, *IEEE Journal of Selected Topics in Applied Earth Observations and Remote Sensing*, 2020, 13: 3350–3366. doi:10.1109/JSTARS.2020.2998838.

Lowe, D.G., Distinctive image features from scale-invariant keypoints, *International Journal of Computer Vision*, 2004, 60(2): 91–110.

Mikolajczyk, K., Schmid, C., A performance evaluation of local descriptors. IEEE transactions on pattern analysis and machine intelligence, *Institute of Electrical and Electronics Engineers*, 2005, 27(10): 1615–1630. doi: 10.1109/TPAMI.2005.188.

Mikolajczyk, K., Schmid, C., Scale & affine invariant interest point detectors, *International Journal of Computer Vision*, 2004, 60(1): 63–86.

Moravec, H.P., Rover visual obstacle avoidance, *Proceedings of the 7th International Joint Conference on Artificial Intelligence (IJCAI '81)*, Vancouver, BC, Canada, Morgan Kaufmann Publishers Inc, August 1981, pp. 785–790.

Morel, J.M., Yu, G., ASIFT: A new framework for fully affine invariant image comparison, *SIAM Journal on Imaging Sciences*, 2009, 2(2): 438–469.

Mortensen, E.N., Deng, H., Shapiro, L., A SIFT descriptor with global context, *2005 IEEE Computer Society Conference on Computer Vision and Pattern Recognition*, San Diego, CA, USA, 20–25 June 2005, 1, pp. 184–190. doi: 10.1109/CVPR.2005.45.

Ni, Q., Wang, F., Zhao, Z., Gao, P., *FPGA-based Binocular Image Feature Extraction and Matching System*. arXiv preprint arXiv:1905.04890, 2019.

Oyallon, E., Rabin, J., An analysis of the SURF method, *Image Processing Online*, 2015, 5: 176–218.

Pohl, M., Schaeferling, M., Kiefer, G., An efficient FPGA-based hardware framework for natural feature extraction and related computer vision tasks, Munich, Germany, 2–4 September 2014, 1–8.

Rani, R., Singh, A.P., Kumar, R., Impact of reduction in descriptor size on object detection and classification, *Multimedia Tools and Applications*, 2019, 78(7): 8965–8979.

Rosten, E., Drummond, T., Fusing points and lines for high performance tracking, *10th IEEE International Conference on Computer Vision*, 17–21 Beijing, China, October 2005, 2, pp. 1508–1515.

Rublee, E., Rabaud, V., Konolige, K., Bradski, G., ORB: An efficient alternative to SIFT or SURF, *IEEE International Conference on Computer Vision*, Barcelona, Spain, November. 2011, 11(1): 2564–2571.

Schaeferling, M., Hornung, U., Kiefer, G., Object recognition and pose estimation on embedded hardware: SURF-based system designs accelerated by FPGA logic, *International Journal of Reconfigurable Computing*, 2012, 6: 1–16.

Schaeferling, M., Kiefer, G., Flex-SURF: A flexible architecture for FPGA-based robust feature extraction for optical tracking systems, *In: IEEE International Conference on Reconfigurable Computing and FPGAs (ReConFig)*, Quintana Roo, Mexico, 13–15 December 2010. pp. 458–463. doi: 10.1109/ReConFigure2010.11.

Schaeferling, M., Kiefer, G., Object recognition on a chip: A complete SURF-based system on a single FPGA, *International Conference on Reconfigurable Computing and FPGAs*, Cancun, Mexico, 30 November–2 December 2011, pp. 49–54. doi: 10.1109/ReConFigure2011.65.

Schmidt, M., Reichenbach, M., Fey, D., A generic VHDL template for 2d stencil code applications on FPGAs, *In 2012 IEEE 15th International Symposium on Object/Component/Service-Oriented Real-Time Distributed Computing Workshops*, Shenzhen, Guangdong, China, 11 April 2012; pp. 180–187. doi: 10.1109/ISORCW.2012.39.

Sledeviˇc, T., Serackis, A., SURF algorithm implementation on FPGA, *In Proceedings of the 2012, 13th Biennial Baltic Electronics Conference*, Tallinn, Estonia; 3–5 October 2012; pp. 291–294. doi: 10.1109/BEC.2012.6376874.

Smith, S.M., Brady, J.M., SUSANA new approach to low level image processing, *International Journal of Computer Vision*, 1997, 23(1): 45–78.

Svab, J., Krajnik, T., Faigl, J., Preucil, L., FPGA based speeded up robust features, *In 2009 IEEE International Conference on Technologies for Practical Robot Applications*, 9–10 November 2009, Woburn, MA, USA. pp. 35–41.

Terriberry, T.B., French, L.M., Helmsen, J., GPU accelerating speeded-up robust features, *In Proceedings of 3DPVT*, 2008, 8: 355–362. doi: 10.1016/j.cviu.2007.09.014.

Tuytelaars, T., Mikolajczyk, K., Local invariant feature detectors: A survey, *Foundations and Trends® in Computer Graphics and Vision*, 2008, 3(3): 177–280.

Viola, P., Jones, M., Rapid object detection using a boosted cascade of simple features, *In Proceedings of the IEEE computer society conference on computer vision and pattern recognition*, Kauai, HI, USA, 8–14 December 2001, pp. 511–518. doi: 10.1109/CVPR.2001.990517.

Wang, J.H., Zhong, S., Xu, W.H., Zhang, W.J., Cao, Z.G., A FPGA-based architecture for real-time image matching. MIPPR 2013: Parallel processing of images and optimization and medical imaging processing, *International Society for Optics and Photonics*, 2013, 8920: 892003, doi: 10.1117/12.2031050.

Wilson, C., Zicari, P., Craciun, S., Gauvin, P., Carlisle, E., George, A., Lam, H., A power-efficient real-time architecture for SURF feature extraction, *In 2014 International Conference on Reconfigurable Computing and FPGAs*, 8–10 December 2014, Cancun, Mexico, pp. 1–8.

Yao, L., Feng, H., Zhu, Y., Jiang, Z., Zhao, D., Feng, W., An architecture of optimised SIFT feature detection for an FPGA implementation of an image matcher, Sydney, NSW, Australia, Sydney, NSW, Australia, 9–11 December 2009, pp. 30–37.

Zhang, H., Hu, Q., Fast image matching based-on improved SURF algorithm, *2011 International Conference on Electronics, Communications and Control*, IEEE, Ningbo, China, 9–11 September 2011, pp. 1460–1463. doi:10.1109/ICECC.2011.6066546.

Zhang, N., Computing optimised parallel speeded-up robust features (p-surf) on multi-core processors, *International Journal of Parallel Programming*, 2010, 38(2): 138–158. doi 10.1007/s10766-009-0122-9.

Zhao, J., Huang, X., Massoud, Y., An efficient real-time FPGA implementation for object detection, *2014 IEEE 12th International New Circuits and Systems Conference (NEWCAS)*, Trois-Rivieres, QC, Canada, 22–25 June 2014, pp. 313–316. doi: 10.1109/NEWCAS.2014.6934045.

Zhao, J., Zhu, S., Huang, X., Real-time traffic sign detection using SURF features on FPGA, *In 2013 IEEE High Performance Extreme Computing Conference (HPEC)*, IEEE, Waltham, MA, USA, 10–12 September 2013, pp. 1–6, doi: 10.1109/HPEC.2013.6670350.

Zhou, G., Li, R., Accuracy evaluation of ground points from IKONOS high-resolution satellite imagery, *Photogrammetry Engineering & Remote Sensing*, 2000, 66(9): 1103–1112. doi:10.1016/S0924-2716(00)00020-4.

Zhou, G., *Urban High-Resolution Remote Sensing: Algorithms and Modelling*, Taylor & Francis/CRC Press, 2020, ISBN: 978-03-67-857509, 465 pp.

Zhou, G., Zhang, R., Liu, N., Huang, J., Zhou, X., On-board ortho-rectification for images based on an FPGA, *Remote Sensing*, 2017, 9: 874. doi: 10.3390/rs9090874.

5 On-Board Geometric Calibration for Frame Camera

5.1 INTRODUCTION

Traditional on-orbit geometric calibration can effectively improve the geometric accuracy of satellite remote sensing images, which lies in a good imaging physical model (Wang et al. 2012). Many scholars have conducted many investigations and developed many mature on-orbit geometric calibration technologies and methods. The on-orbit geometric calibration is usually implemented under desk computers. The calibrated parameters are used to correct the systematic geometric error of the other image obtained from the same satellite. However, due to the confidentiality of satellite imaging system, the systematic geometric errors of images are usually corrected by satellite ground stations or satellite vendor. However, with the increase of images acquired by satellites every day, various image processing cannot meet the needs of some users with high timeliness requirements for images, such as military deployments, government emergency agencies, etc. In view of the mature on-orbit geometric calibration technology, the timeliness of satellite image application can be greatly improved in real time and automatically, so as to eliminate or weaken the system geometric error. For example, BIRD (Bispectral Infrared Detection) satellite launched by Germany in 2001 has certain on-board real-time image processing capability (Halle et al. 2002; Brieß et al. 2005; Jiang 2016)), including on-board geometric correction of sensor radiation, on-board geometric correction of systematic position error and on-board geometric correction of spacecraft attitude (DLR, Mission objectives of BIRD) (URL: http://www.dlr.de/os/en/desktopdefault.aspx/tabid-3508/5440_read-7886/). BIRD satellite can more accurately and automatically identify the geographical location of fire points or fire areas on the ground. However, the specific implementation strategies are not given. Zhou et al. (2004) have put forward the idea of the IEOSNS, in which the on-board real-time data processor includes the on-board real-time correction of geometric errors of imaging sensor system. However, so far, few literatures have conducted research on on-board geometric calibration in-depth. Therefore, how to implement the real-time and automatic on-board geometric calibration for images acquired by a frame camera is to be discussed in this chapter.

5.2 THE MATHEMATICAL MODEL OF GEOMETRIC CALIBRATION FOR FRAME CAMERA

Traditional image geometric calibration uses the observed image point coordinates and the coordinates of the corresponding ground control points (GCPs) on a single image to obtain the error equations according to the collinear equation, and then the equation is solved with the least square adjustment. Finally, the six exterior orientation parameters (EOPs) of a frame image are obtained (Zhang et al. 2009). The initial values of the six EOPs of a frame image can be obtained by the on-board positioning and navigation sensors. The collinear equation is

$$
\begin{cases}
x - x_0 = -f \dfrac{r_{11}(X - X_S) + r_{12}(Y - Y_S) + r_{13}(Z - Z_S)}{r_{31}(X - X_S) + r_{32}(Y - Y_S) + r_{33}(Z - Z_S)} \\[3mm]
y - y_0 = -f \dfrac{r_{21}(X - X_S) + r_{22}(Y - Y_S) + r_{23}(Z - Z_S)}{r_{31}(X - X_S) + r_{32}(Y - Y_S) + r_{33}(Z - Z_S)}
\end{cases}
\tag{5.1}
$$

DOI: 10.1201/9781003319634-5

where (x, y) are the coordinates of the image point (the coordinates in the frame coordinate system); $(x-x_0)$ and $(y-y_0)$ are the coordinates in the image plane coordinate system; (x_0, y_0, f) are the interior orientation parameters (IOPs) of a frame camera; (X_S, Y_S, Z_S) is the coordinate of the center of photography at the moment of imaging; (X, Y, Z) is the geodetic coordinate of the ground control points (GCPs); and r_{ij} $(i = 1, 2, 3; j = 1, 2, 3)$ are 9 elements of cosines functions calculated by using the 3 rotation angles $(\varphi, \omega, \kappa)$ of the image, namely the rotation matrix R (R is an orthogonal matrix, so $R^{-1} = R^{\mathrm{T}}$).

If φ-ω-κ rotation angle system is adopted, the rotation matrix R is:

$$R = R(\varphi)R(\omega)R(\kappa) = \begin{bmatrix} \cos\varphi & 0 & -\sin\varphi \\ 0 & 1 & 0 \\ \sin\varphi & 0 & \cos\varphi \end{bmatrix} \begin{bmatrix} 1 & 0 & 0 \\ 0 & \cos\omega & -\sin\omega \\ 0 & \sin\omega & \cos\omega \end{bmatrix} \begin{bmatrix} \cos\kappa & -\sin\kappa & 0 \\ \sin\kappa & \cos\kappa & 0 \\ 0 & 0 & 1 \end{bmatrix}$$

$$= \begin{pmatrix} \cos\varphi\cos\kappa - \sin\varphi\sin\omega\sin\kappa & -\cos\varphi\sin\kappa - \sin\varphi\sin\omega\cos\kappa & -\sin\varphi\cos\omega \\ \cos\omega\sin\kappa & \cos\omega\cos\kappa & -\sin\omega \\ \sin\varphi\cos\kappa + \cos\varphi\sin\omega\sin\kappa & -\sin\varphi\sin\kappa + \cos\varphi\sin\omega\cos\kappa & \cos\varphi\cos\omega \end{pmatrix}$$

$$(5.2)$$

where

$$\begin{cases} r_{11} = \cos\varphi\cos\kappa - \sin\varphi\sin\omega\sin\kappa \\ r_{12} = \cos\omega\sin\kappa \\ r_{13} = \sin\varphi\cos\kappa + \cos\varphi\sin\omega\sin\kappa \\ r_{21} = -\cos\varphi\sin\kappa - \sin\varphi\sin\omega\cos\kappa \\ r_{22} = \cos\omega\cos\kappa \\ r_{23} = -\sin\varphi\sin\kappa + \cos\varphi\sin\omega\cos\kappa \\ r_{31} = -\sin\varphi\cos\omega \\ r_{32} = -\sin\omega \\ r_{33} = \cos\varphi\cos\omega \end{cases} \qquad (5.3)$$

The coordinate (x, y) of an image point is observed, the coordinates (X, Y, Z) of the corresponding GCP are true values, and the IOPs (x_0, y_0, f) are known. The collinear equation of Equation (5.1) is linearized with Taylor series, and the following error equation can be obtained

$$\begin{cases} v_x = a_{11}\Delta X + a_{12}\Delta Y + a_{13}\Delta Z + a_{14}\Delta\varphi + a_{15}\Delta\omega + a_{16}\Delta\kappa - l_x \\ v_y = a_{21}\Delta X + a_{22}\Delta Y + a_{23}\Delta Z + a_{24}\Delta\varphi + a_{25}\Delta\omega + a_{26}\Delta k - l_y \end{cases} \qquad (5.4)$$

Rewrite (5.4) by a matrix form as

$$V = AX - l, \quad E \qquad (5.5)$$

where $V = \begin{bmatrix} v_x & v_y \end{bmatrix}^{\mathrm{T}}$ is the residual matrix of the error observation;

$A = \begin{bmatrix} a_{11} & a_{12} & a_{13} & a_{14} & a_{15} & a_{16} \\ a_{21} & a_{22} & a_{23} & a_{24} & a_{25} & a_{26} \end{bmatrix}$ is the coefficient matrix of the unknown parameters;

$X = \begin{bmatrix} \Delta X & \Delta Y & \Delta Z & \Delta\varphi & \Delta\omega & \Delta\kappa \end{bmatrix}^{\mathrm{T}}$ is the correction matrix of EOPs;

$l = \begin{bmatrix} l_x & l_y \end{bmatrix}^{\mathrm{T}} = \begin{bmatrix} x-(x) & y-(y) \end{bmatrix}^{\mathrm{T}}$ is the constant matrix of EOPs;

$\big((x), (y)\big)$ is the approximate value of the image coordinates calculated by using the initial value of the EOPs and IOPs according to Equation (5.1);

E is the weight matrix of the observed value of the image point coordinate, and the unit matrix is taken when the image point coordinate is regarded as the unrelated observation value of equal precision.

Each coefficient in matrix A is

$$a_{11} = \frac{\partial x}{\partial X_S} = \frac{1}{\bar{Z}}\left[r_{11}f + r_{31}(x - x_0)\right] \quad a_{21} = \frac{\partial y}{\partial X_S} = \frac{1}{\bar{Z}}\left[r_{21}f + r_{31}(y - y_0)\right]$$

$$a_{12} = \frac{\partial x}{\partial Y_S} = \frac{1}{\bar{Z}}\left[r_{12}f + r_{32}(x - x_0)\right] \quad a_{22} = \frac{\partial y}{\partial Y_S} = \frac{1}{\bar{Z}}\left[r_{22}f + r_{32}(y - y_0)\right]$$

$$a_{13} = \frac{\partial x}{\partial Z_S} = \frac{1}{\bar{Z}}\left[r_{13}f + r_{33}(x - x_0)\right] \quad a_{23} = \frac{\partial y}{\partial Z_S} = \frac{1}{\bar{Z}}\left[r_{23}f + r_{33}(y - y_0)\right]$$

$$a_{14} = \frac{\partial x}{\partial \varphi} = (y - y_0)\sin\omega - \left\{\frac{x - x_0}{f}\left[(x - x_0)\cos\kappa - (y - y_0)\sin\kappa + f\cos\kappa\right]\right\}\cos\omega$$

$$a_{15} = \frac{\partial x}{\partial \omega} = -f\sin\kappa - \frac{x - x_0}{f}\left[(x - x_0)\sin\kappa + (y - y_0)\cos\kappa\right] \qquad (5.6)$$

$$a_{16} = \frac{\partial x}{\partial \kappa} = y - y_0$$

$$a_{24} = \frac{\partial y}{\partial \varphi} = -(x - x_0)\sin\omega - \left\{\frac{y - y_0}{f}\left[(x - x_0)\cos\kappa - (y - y_0)\sin\kappa\right] - f\sin\kappa\right\}\cos\omega$$

$$a_{25} = \frac{\partial y}{\partial \omega} = -f\cos\kappa - \frac{y - y_0}{f}\left[(x - x_0)\sin\kappa + (y - y_0)\cos\kappa\right]$$

$$a_{26} = \frac{\partial y}{\partial \kappa} = -(x - x_0)$$

where $\bar{Z} = r_{31}(X - X_S) + r_{32}(Y - Y_S) + r_{33}(Z - Z_S)$.

For the solution of Equation (5.5), at least three GCPs are needed. In fact, more than three GCPs uniformly distributed on the image are selected. If n GCPs are selected, $2n$ error equations can be observed according to Equation (5.4), and the total error equations are also obtained by Equation (5.5). With the least squares adjustment, the following equations can be obtained

$$A^\mathsf{T}AX = A^\mathsf{T}l \qquad (5.7)$$

The solution of this normal equation is

$$X = (A^\mathsf{T}A)^{-1}(A^\mathsf{T}l) \qquad (5.8)$$

The corrections of six EOPs are obtained through iterative calculation for a single image until the correction of the unknowns is less than a given threshold (Zhang et al. 2009). The final corrections of the six EOPs is the sum of the correction from each iteration, i.e.,

$$\begin{cases} \Delta X = \Delta X_S^1 + \Delta X_S^2 + \cdots \quad \Delta\varphi = \Delta\varphi^1 + \Delta\varphi^2 + \cdots \\ \Delta Y = \Delta Y_S^1 + \Delta Y_S^2 + \cdots \quad \Delta\omega = \Delta\omega^1 + \Delta\omega^2 + \cdots \\ \Delta Z = \Delta Z_S^1 + \Delta Z_S^2 + \cdots \quad \Delta\kappa = \Delta\kappa^1 + \Delta\kappa^2 + \cdots \end{cases} \qquad (5.9)$$

A single image space resection algorithm requires an iterative adjustment calculation, which involves a series of floating-point mathematical operations, mainly including the solution of normal equation, matrix multiplication and matrix inversion and other complex operations (Jacobsen 2003; 2005; Zhou et al. 2018, Zhou et al. 2004; Zhou 2015a, 2015b, 2015c, 2015d).

5.3 FPGA-BASED DESIGN STRUCTURE FOR ON-BOARD GEOMETRIC CALIBRATION

An architecture for a single image space resection for field programmable gate array (FPGA)-based frame camera calibration is depicted in Figure 5.1. With the top-down FPGA design and development method, the architecture first divides the functional modules, and then connects each module to get its hardware implementation structure. The advantages of FPGA lie in its parallel hardware structure and the flexibility of arbitrary programming. Therefore, designing parallel computational structures can fully leverage the advantages of FPGA. However, the single image space resection algorithm is an iterative process, which means that parallel structure cannot be used for FPGA-based implementation of a single image space resection globally, and only can be used locally (Zhou et al. 2018).

According to the characteristics of the single image space algorithm, the top-level implementation structure of the algorithm based on FPGA is designed and shown in Figure 5.1. The structure is divided into four modules:

- Data input,
- Parameter calculation,
- Adjustment calculation,
- Threshold comparison, and
- Result output.

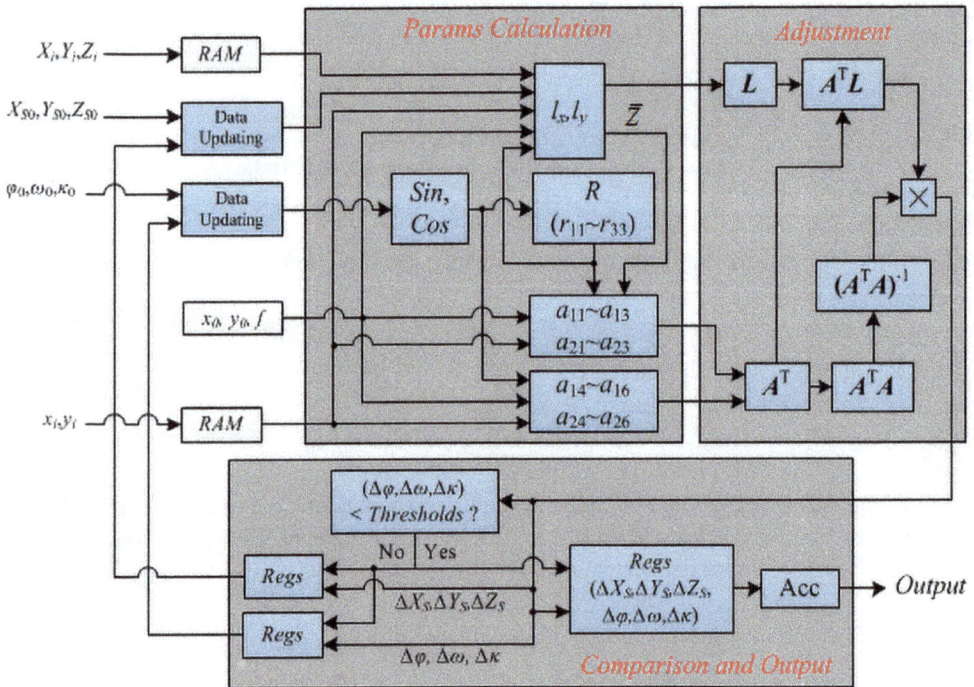

FIGURE 5.1 FPGA-based top-level implementation structure for a single frame image space resection.

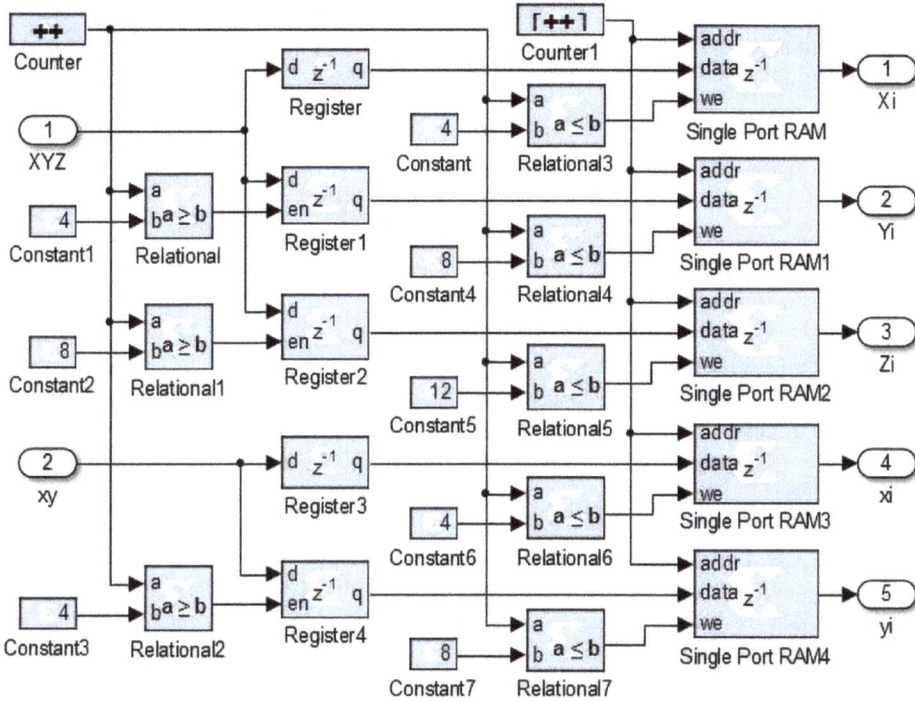

FIGURE 5.2   Implementation structure of ground control point (GCP) gateway.

Iterative adjustment calculation is achieved through the reuse of each module, which can greatly reduce the use of hardware resources. The gateway in module section is used to input the initial value of the EOPs and the image point coordinates of the ground control points (GCPs) (see Figure 5.2). Random access memory (RAM) implements the cache of control point data to facilitate subsequent modules. The data update module is used to implement updates of the input data required for the iterative calculation. The parameter calculation module is used to calculate the elements of the coefficient matrix A and the constant term matrix L. Among them, the FPGA implementation of sine and cosine calculation of three rotation angles is complex and time-consuming, so it is used as a separate module; the rotation matrix R, which is constructed by 9 elements, is also shared by subsequent sub-modules and also used as a separate module. Three submodules for calculating elements of matrices A and L can be executed in parallel. The matrix adjustment module carries out the solution of the normal equation, including the construction of matrix multiplication and matrix inversion. The matrix multiplier modules A^TA and A^TL can be executed in parallel. The threshold comparison and adjustment result output module are used to compare the correction values of three rotation angles with the preset threshold. When the increments of EOPs meet the requirement of the given threshold, and the iterative calculation is stop, and accumulate the increments and output the results (ACC module in Figure 5.1). In the entire computing structure, in addition to the parallel execution of local sub-modules, the internal implementation of sub-modules can also design parallel computing structure. The specific implementation will be discussed in detail below.

5.3.1   IMPLEMENTATION OF GATEWAY IN MODULE

What FPGA processes in each clock cycle is one piece of data in the data stream, so the input multiple data should also be converted into the form of the data stream. This can be done by building a matrix of n rows, and 2 columns using the "*reshape function*" in MATLAB. The elements in the first column are integers from 0 to $(n-1)$, indicating the sequence of input data, and the elements in the second column are the data to be processed in order. This implies that the data stream is inputted into FPGA. In addition, two aspects should be considered when inputting data.

- First, there is a limit to the number of I/O pins that a user can freely allocate on an experimental GA chip (there are 400 such pins), so the number of pins used for input and output data cannot exceed this limit. The number of the pins used is equal to the total binary digits of the input and output data, and the data accuracy requirements decide to the number of binary digits of the data. The higher the data accuracy, the more binary bits. This represents that the many pins are used. When the number of pins is limited, and data accuracy is required to be high, it is appropriate to consider setting up several data input ports.
- Second, the processing of input data streams (such as the input order of different types of data items, the input mode of sharing a port for different types of data, and the caching of data) also affects the structure of subsequent data processing modules somewhat, which in fact affect the use of hardware resources and time-consuming of the entire computing system (Zhou et al. 2018).

For example, when using four GCPs for camera calibration, the designed structure is shown in Figure 5.2, which requires input of geodetic and image coordinates of four GCPs and the initial values of six EOPs. In order to obtain high-accuracy of the calculation results, the input data are in the format of double-precision floating-point. The geodetic and image coordinates of the GCPs are input from two ports. The implementation structure of the corresponding module is shown in Figure 5.2. Due to many data items for four GCPs, the data is input in the order of X, Y, Z and x, y to facilitate the data shunting. Finally, the data is cached using single-port RAM, and the storage address of RAM is controlled by the counter so that it can output corresponding data of the four GCPs repeatedly to facilitate the use of subsequent modules. The first counter implements the control of a two-tier comparator so that the specific data is entered into RAM. Considering the limited number of pins, six EOPs are input from one port, and the corresponding module implementation structure is similar to Figure 5.3, but no RAM is required. In the subsequent calculations, the three rotation angles are first used.

5.3.2 Implementation of Parameter Calculation Module

With the design structure in the parameter calculation module in Figure 5.3, the conversion from the frame coordinate of the image point to the image plane coordinate is determined by xy_x0y0 module implementation. Two subtractors are used inside the module, and only the frame coordinates of the image points are subtracted from the principal point coordinates. It should be noted that the naming of any unit or submodule in the entire calculation module can only be a combination of letters, numbers and underscores. The design and implementation of the remaining sub-modules are complicated, and will be explained one by one below.

1. **Calculating model for sine and cosine functions of three rotational angles**
   CORDIC (Coordinate Rotational Digital Computer) algorithm is usually used in hardware implementation for sine and cosine function calculation. In the System Generator tool, the encapsulated CORDIC cores can be used directly to calculate the sine and cosine values of the angles (calculating the sine and cosine values is only one function of the CORDIC cores). For Artix-7 series FPGA chips, the CORDIC 5.0 IP core can be used. The input data is the radian value of the angle, the dout_tdata_image signal in the output signal represents the sine value, and the dout_tdata_real signal represents the cosine value. Three CORDIC cores are used to achieve the parallel solution of the sine and cosine values of the three angles. The calculation speed is three times that of

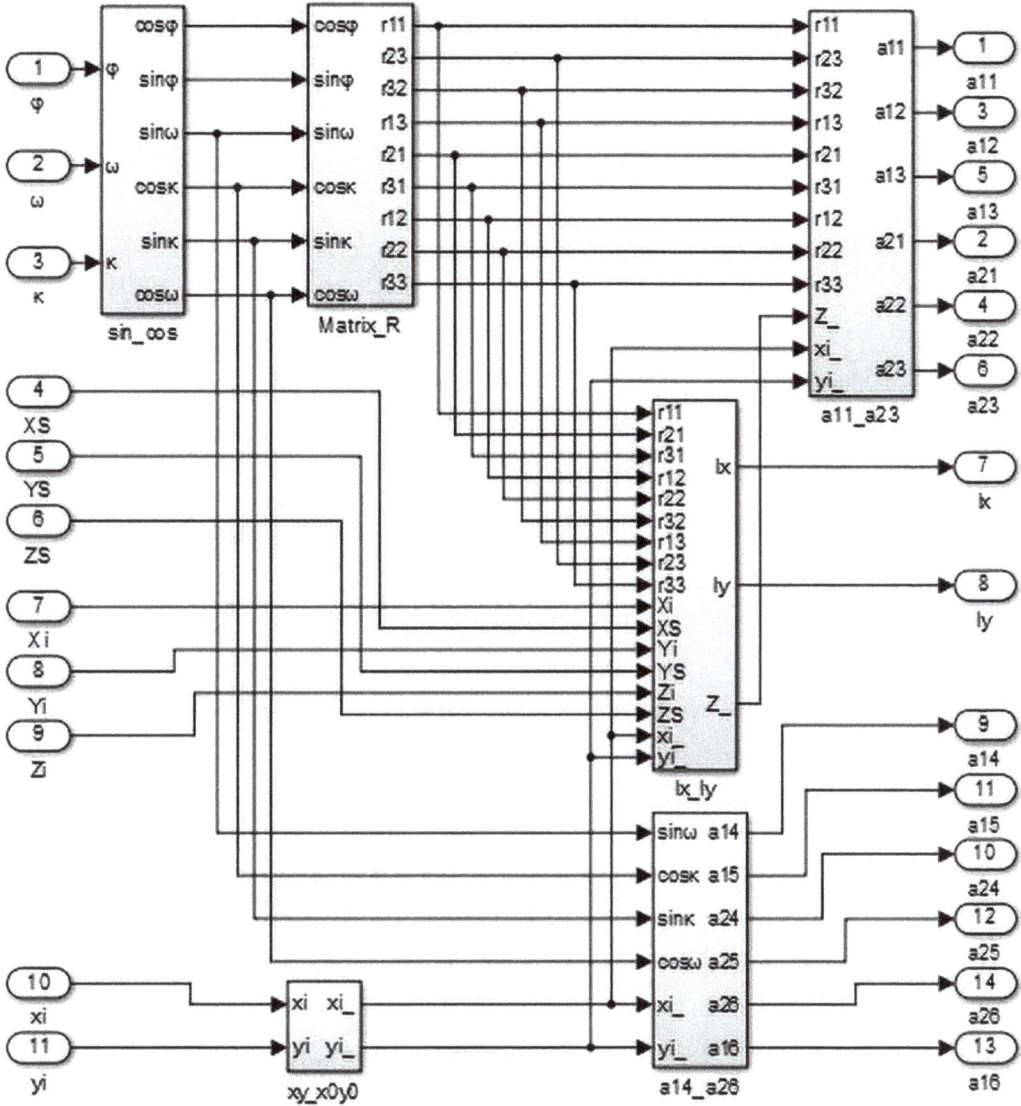

FIGURE 5.3 Structure of top module for parameter calculation.

using a CORDIC core, that is, the circuit area is exchanged for an increase in calculation speed, as shown in Figure 5.4. The input data of CORDIC core is specified as fixed-point number, so the arriving double-precision floating-point number needs to be converted to fixed-point number through data convert unit, and then to double-precision floating-point number after calculation is completed. In order to obtain high-precision results, the fixed-point bit width of the input and output data of the CORDIC core is set to the maximum allowable bit width of 48 bits, which requires 52 clock cycles to complete the calculation.

2. **Directional cosine calculation module**

   With Equation (5.3), it is known that the elements of the rotational matrix are the sine and cosine functions of, or their combination of three rotation angles. The designed parallel calculation module for them is shown in Figure 5.5, where r_{11} and r_{23} have common items,

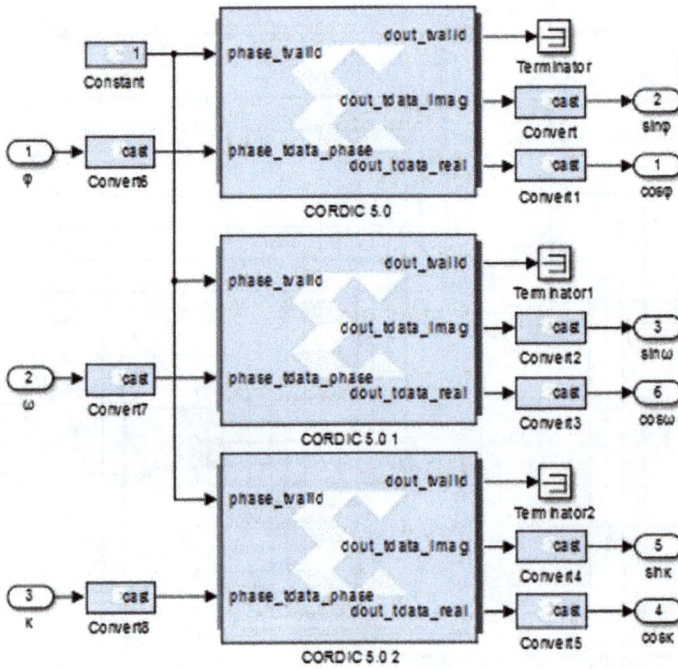

FIGURE 5.4   Calculating model for sine and cosine functions of rotational angles.

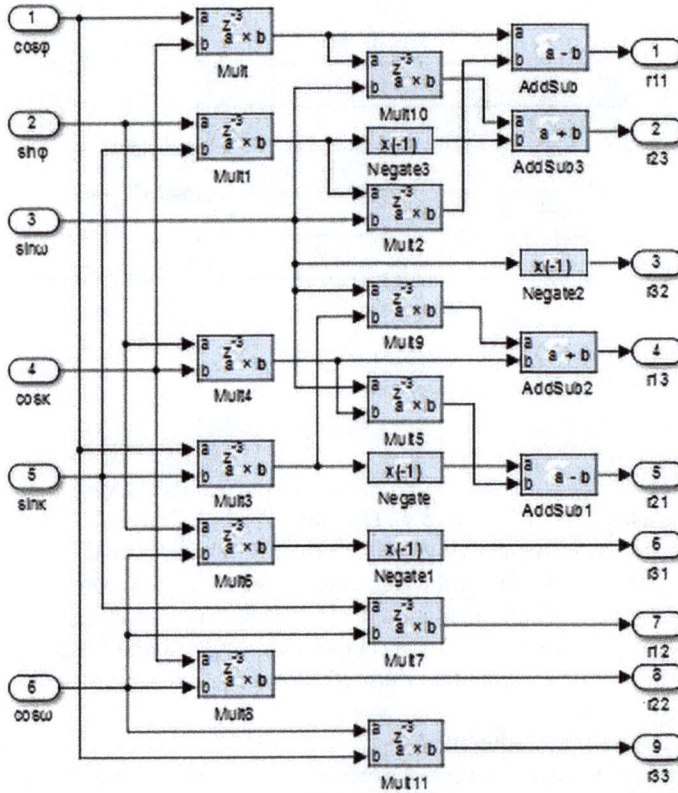

FIGURE 5.5   Cosine function calculation module.

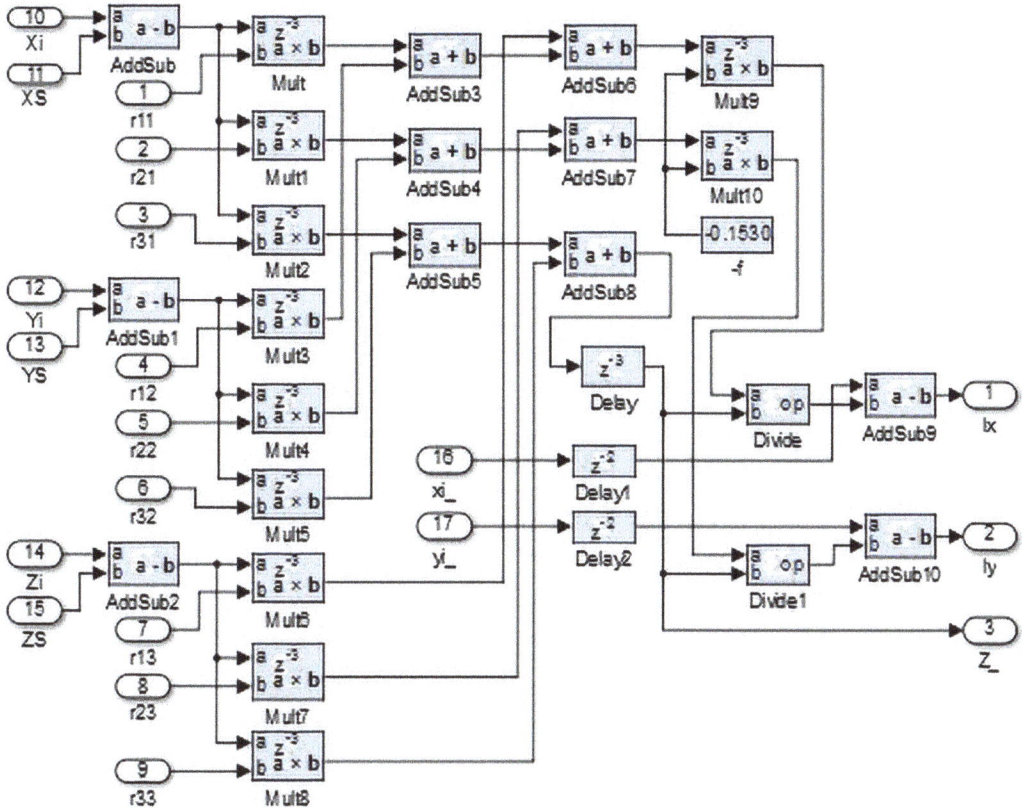

FIGURE 5.6 Constant matrix calculation module.

and r_{13} and r_{21} have common items. For these common items, only one calculation is needed in the design module and shared by the related calculation processes, to reduce the use of hardware resources. The Negate cell is used to calculate the inverse of a number.

3. **Calculation module for elements of constant matrix and coefficient matrix**

The elements of the constant matrix L are calculated, whose module is shown in Figure 5.6. In the diagram, xi_ represents x_i-x_0, yi_ represents y_i-y_0, Z_ represents $\bar{Z}$ and the Delay unit is used to align the time (clock cycle) at which the data signal arrives. The formula for calculating the elements of the coefficient matrix of the observational equations in Equation (5.6) is exactly the same for the elements of $(a_{11}, a_{12}, a_{13}, a_{21}, a_{22}, a_{23})$, and $(a_{14}, a_{15}, a_{16}, a_{24}, a_{25}, a_{26})$, which are suitable for implementation separately into two sub-modules. According to the characteristics of the calculation formula, the calculation modules for $(a_{11}, a_{12}, a_{13}, a_{21}, a_{22}, a_{23})$ and $(a_{14}, a_{15}, a_{16}, a_{24}, a_{25}, a_{26})$ are constructed in Figures 5.7 and 5.8, respectively.

5.3.3 PARALLEL COMPUTATION OF MATRIX MULTIPLICATION

Matrix multiplication is an operation of the row vectors of one matrix multiplying by the corresponding column vectors of another matrix, and then accumulating them to obtain the corresponding elements of the resulting matrix. According to the characteristics of matrix multiplication A^TA and A^TL, six-row vectors (denoted as A1, A2, A3, A4, A5, A6) of matrix A^T and

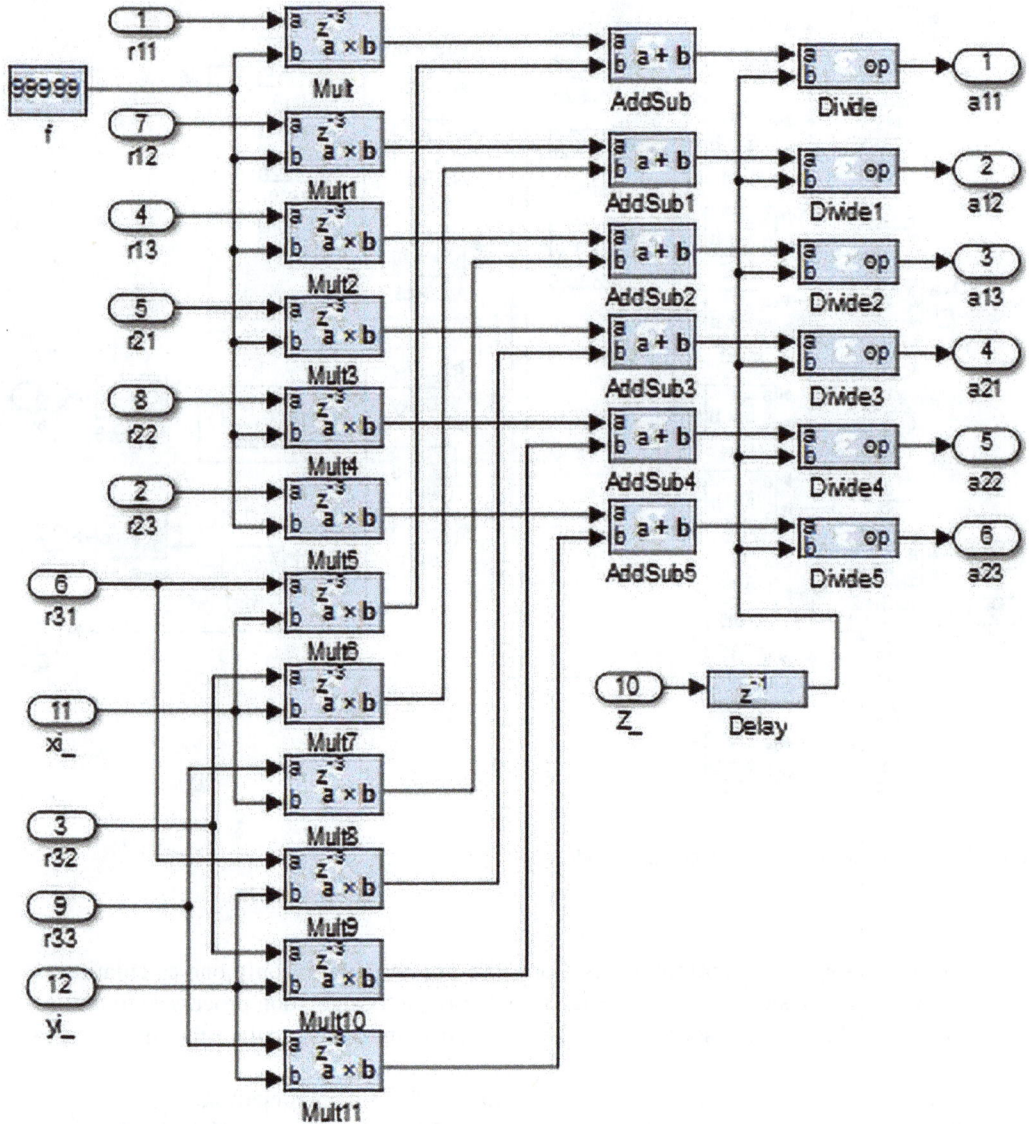

FIGURE 5.7 $a_{11}-a_{23}$ calculation module.

the column vectors formed by column matrix L need to be constructed. A^T is the transpose of A, so the row vectors of A correspond to the column vectors of A. The construction method of row vectors and column vectors is the same. Take the construction of the first-row vector A1 of matrix A^T as an example, which is shown in Figure 5.10 for the hardware implementation structure. The module uses the data selector (short for multiplexer, or Mux unit) to combine multiple incoming data signals into one signal output, and the Delay unit is used to control the sequence of incoming data entering the multiplexer. Counters (Counter units) is used to complete the cycle count in ascending order.

The matrix multiplier module can reference the parallel computing structure proposed by Tian et al. (2008). The computing structure reduces the computational complexity of the matrix multiplication, with good parallelism, which can effectively improve the calculation speed of the matrix multiplication. The matrix A^T is multiplied by the calculation of matrix A, that

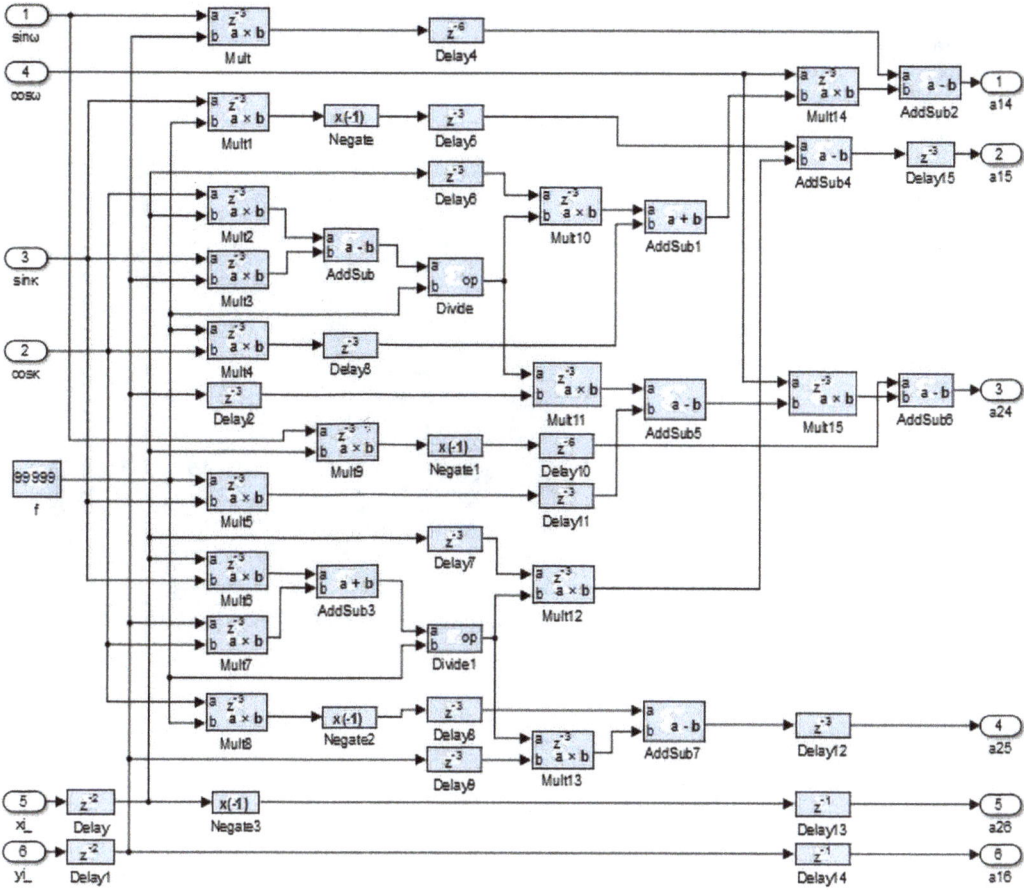

FIGURE 5.8 a_{14}–a_{26} calculation module.

is, the line vector of A^T is multiplied by each other. Since the matrix A^TA is a symmetrical matrix, only the upper triangular matrix is required. Set $B=(A^TA)$, the element of the matrix B is b_{ij} (i, j=1,2, 3,9), then the calculation module for multiplying matrix A^T and A shown in Figure 5.11 can be designed. The matrix multiplication module includes a control module, 6 row vector input ports, 21 multiplication and accumulation processing units (Processing Element – PE), and 21 output ports. The control module is combined by a counter, a comparator, and a constant unit, generating reset signal and enabling signals for controlling the calculation of the PE unit (Zhou et al. 2018).

The PE unit designed in this chapter can implement two vectors of multiplier, thereby completing the calculation of one element value of the target matrix, and the internal structure is shown in Figure 5.12. The advantage of using the PE unit is to save a large number of multipliers and adder resources. In this module, the control signal defaults to high levels, that is, the logical value of the signal is 1. The high level of the reset signal RST comes to at least one clock cycle than the high level of the enable signal EN, so that the register is set to the initial state (for example, setting the value stored in the register to 0). The process of PE implementation of the multiplying operation is: multiplying the input A and B data; the RST signal is set to the initial state; the next clock cycle, the initial value 0 is stored in the signal EN control register; 0. The value is added to the multiplier (a×b) after the Delay unit delays 1 clock cycle, and then outputs after the register; the first addition result is the input data of the next additional addition to the next addition. The reached product is added, and then output after the next clock cycle after the register. This operation is repeated after

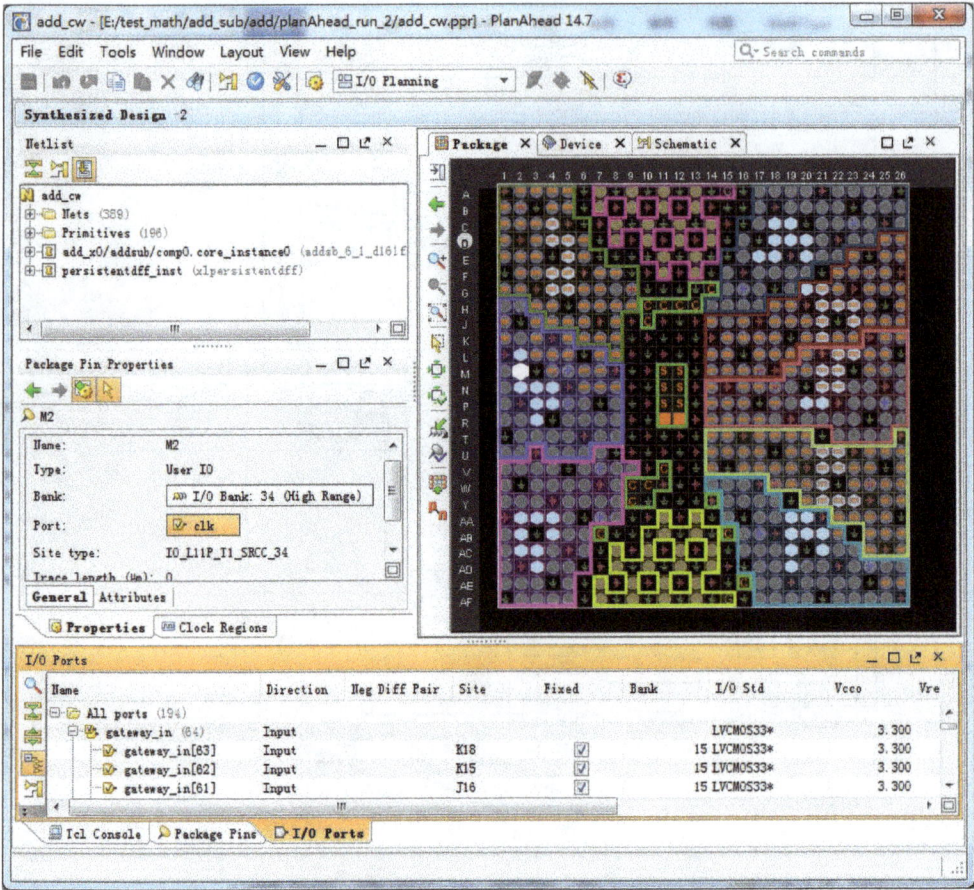

FIGURE 5.9　The I/O pin allocation interface using PlanAhead software.

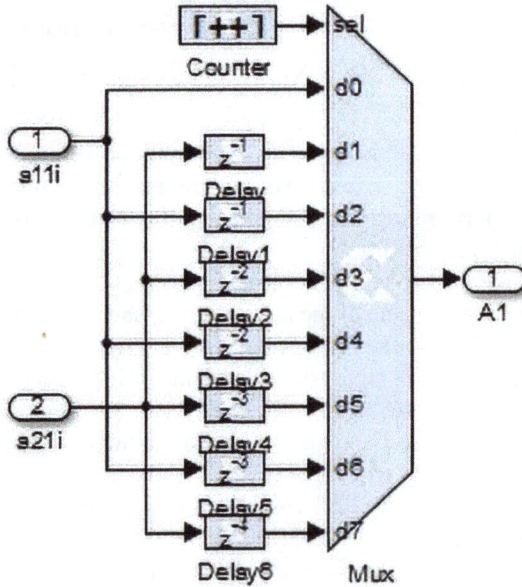

FIGURE 5.10　Building the implementation module of row vector A1.

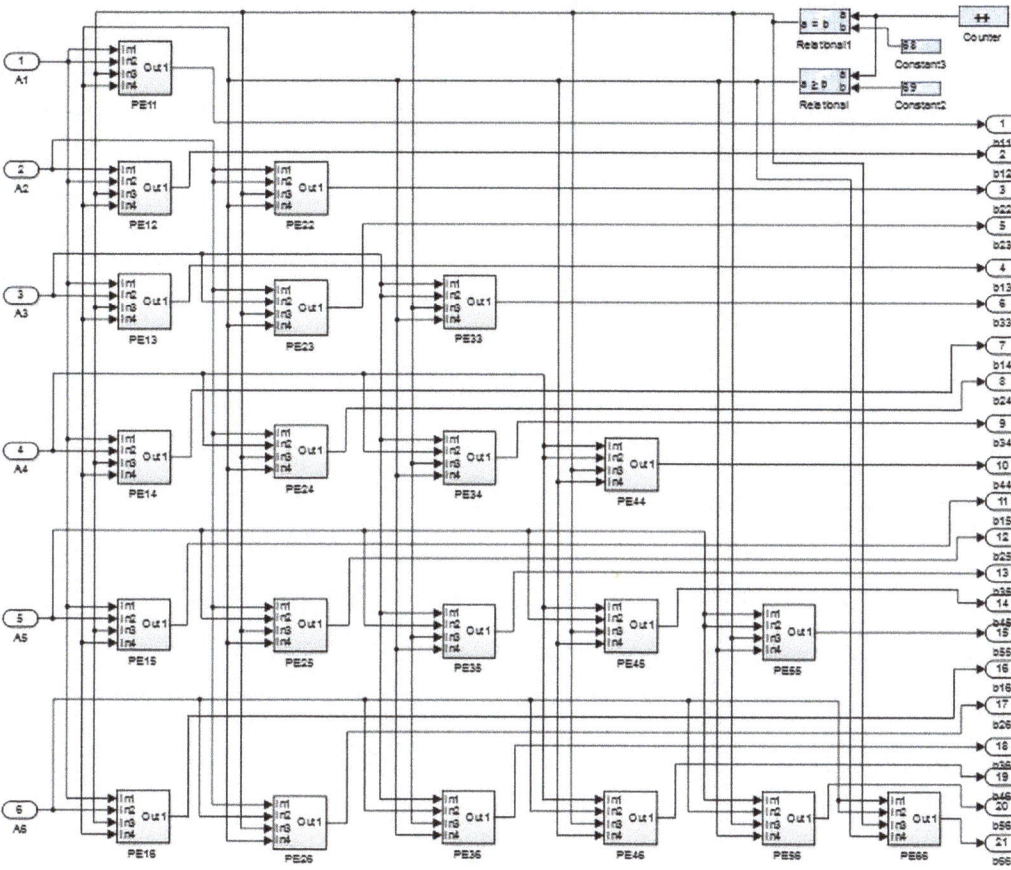

FIGURE 5.11 Parallel computing structure of the matrix A^T and A multiplication module.

the multiplication and addition calculation of the N times (N is the number of data containing data), and the corresponding element is completed(Xilinx 2014).

The implementation of the matrix A^T and L multiplication can be similar to the implementation structure of the $A^\mathrm{T}A$ matrix. However, six PE units are required. Set $C=A^\mathrm{T}L$, the implementation structure of the matrix A^T and L multiplication module is shown in Figure 5.13. The delay of the two clock cycles increasing for the column vector L is used to align the data stream, thereby obtaining the correct calculation result (Zhou et al. 2018).

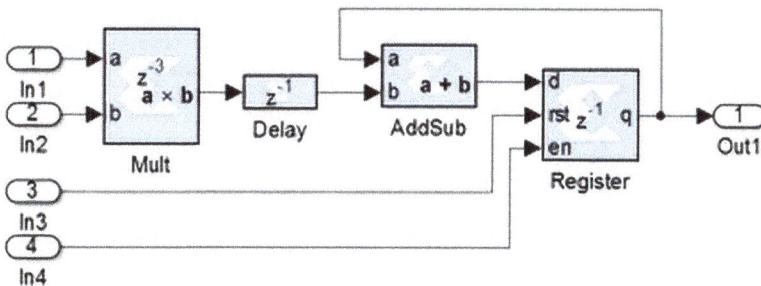

FIGURE 5.12 Internal structure of the multiplication and accumulation processing unit.

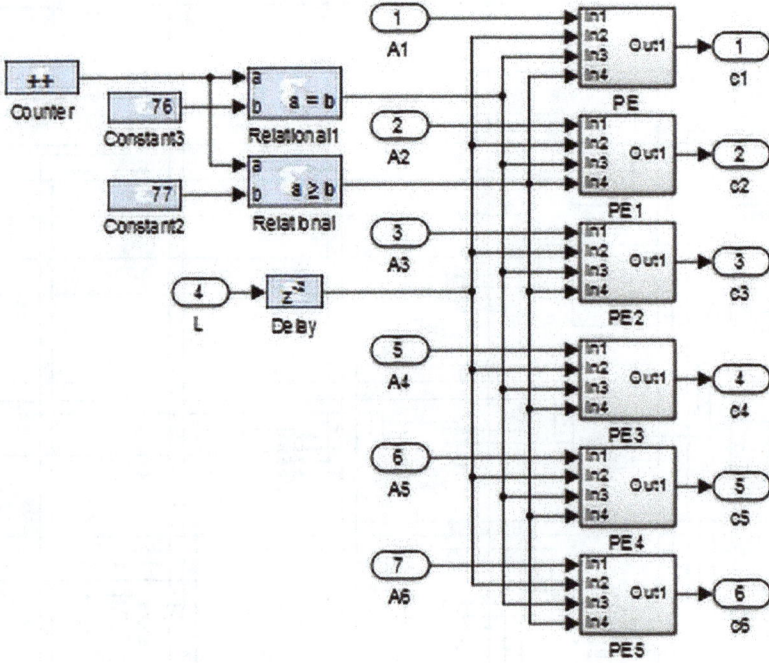

FIGURE 5.13   Parallel computing structure of matrix A^T and L multiplication module.

5.3.4   FPGA-BASED IMPLEMENTATION OF ON-BOARD CALIBRATION

5.3.4.1   The Hardware Implementation for Solution of Matrix Inverse

The hardware implementation of matrix inversion mainly adopts the decomposition method, that is, the original matrix is decomposed into a special matrix with certain characteristics and easy to invert, such as a triangular matrix or a unitary matrix, and then the decomposed matrix is inverted and multiplied. It can complete the inversion of the original matrix (Lin 2007). The commonly used method of decomposition method is inverse matrix with Cholesky decomposition, QR decomposition and LU decomposition. These types of matrix decomposition have different characteristics, and their scope of application is different. Cholesky decomposition is the simplest, but can only be used for triangle decomposition of Hermite positive matrix (or symmetric positive matrix), the application is limited. The QR decomposition is suitable for any matrix, but the calculation process is much complicated, and its hardware is difficult to achieve. The LU decomposition can be carried out by the trigonometric decomposition of matrices whose main and sub-expressions are not equal to zero. Most of the reversible matrices meet this condition, and its calculation is simpler than the QR decomposition, so the LU decomposition is in the hardware implementation of the matrix and is widely used (Lin 2007).

The matrix $A^T A$ is a symmetrical positive matrix, so the Cholesky decomposition method and the LU decomposition method can be preferentially employed. This chapter would describe the hardware-based method for matrix $A^T A$ using LU decomposition. Because $A^T A$ is a 6×6 matrix, its hardware implementation is still complicated if the general LU decomposition method is used. In this method, the original matrix is evenly divided into checkboard blocks, and then each submatrix is decomposed and inverted, respectively. Finally, the inverse matrix of the original matrix can be obtained by multiplying the submatrices. It has been demonstrated that this method can reach the inverse of big matrices with low computational complexity (Wang et al. 2007). The calculation is described below.

1. **LU decomposition method for reverse matrix**

For an $n{\times}n$ matrix A, decompose it into a product of a unit lower triangular matrix L and an upper triangular matrix U, that is A = LU, called the LU decomposition of matrix A (Shao 2010), i.e.,

$$\begin{pmatrix} a_{11} & a_{12} & \cdots & a_{1n} \\ a_{21} & a_{22} & \cdots & a_{2n} \\ \vdots & \vdots & \ddots & \vdots \\ a_{n1} & a_{n2} & \cdots & a_{nn} \end{pmatrix} = \begin{pmatrix} 1 & & & \\ l_{21} & 1 & & \\ \vdots & \ddots & \ddots & \\ l_{n1} & \cdots & l_{n,n-1} & 1 \end{pmatrix} \begin{pmatrix} u_{11} & u_{12} & \cdots & u_{1n} \\ & u_{22} & \cdots & u_{2n} \\ & & \ddots & \vdots \\ & & & u_{nn} \end{pmatrix} \tag{5.10}$$

we have

$$\begin{cases} a_{1j} = u_{1j} \ (j=1,2,\cdots n); \\ a_{i1} = l_{i1}u_{11}(i=2,3,\cdots n); \\ a_{kj} = \sum_{t=1}^{k-1} l_{kt}u_{tj} + u_{kj} \ (j=k,k+1,\cdots,n;k=2,3,\cdots,n); \\ a_{ik} = \sum_{t=1}^{k-1} l_{it}u_{tk} + l_{ik}u_{kk} \ (i=k+1,k+2,\cdots,n;k=2,\cdots,n). \end{cases} \tag{5.11}$$

where k represents the order of the sequential principles. The transformation is performed by the Formula (5.11), and the calculation formula of the LU decomposition of A can be obtained

$$\begin{cases} u_{1j} = a_{1j} \ (j=1,2,\cdots n); \\ l_{i1} = \dfrac{a_{i1}}{u_{11}}(i=2,3,\cdots n); \\ u_{kj} = a_{kj} - \sum_{t=1}^{k-1} l_{kt}u_{tj} \ (j=k,k+1,\cdots,n;k=2,3,\cdots,n); \\ l_{ik} = \dfrac{1}{u_{kk}}\left(a_{ik} - \sum_{t=1}^{k-1} l_{it}u_{tk} \right)(i=k+1,k+2,\cdots,n;k=2,\cdots,n). \end{cases} \tag{5.12}$$

Then the matrix U and the matrix L are inverted. Suppose, according to the elementary transformation method, the calculation formula of the inverse matrix V of U can be obtained, namely

$$v_{ij} = \begin{cases} 1/u_{ii}, & i=j, 1\le i \le n \\ -\sum_{k=i+1}^{j} v_{kj}u_{ik}/u_{ii}, & 1\le i \le n-1, 2 \le j \le n \\ 0, & j < i \end{cases} \tag{5.13}$$

For the three triangular matrix L, $Q=L^{-1}$, then:

$$Q = L^{-1} = \left((L^T)^T \right)^{-1} = \left((L^T)^{-1} \right)^T \tag{5.14}$$

we have

$$
\begin{bmatrix} q_{11} & & & \\ q_{21} & q_{22} & & \\ \vdots & \vdots & \ddots & \\ q_{n1} & q_{n2} & \cdots & q_{nn} \end{bmatrix} = \left[\begin{bmatrix} 1 & l_{21} & \cdots & l_{n1} \\ & 1 & \cdots & l_{n2} \\ & & \ddots & \vdots \\ & & & 1 \end{bmatrix}^{-1} \right]^{T}
\tag{5.15}
$$

The elements that are derived by deriving the alive matrix Q can be calculated from the formula

$$
q_{ji} = \begin{cases} 1, & i = j, 1 \le i \le n \\ -\sum_{k=i+1}^{j} q_{jk} \cdot l_{ki}, & 1 \le i \le n-1, 2 \le j \le n \\ 0, & j < i \end{cases}
\tag{5.16}
$$

Finally, multiply the matrix V and matrix Q to the inverse matrix of the original matrix A, as an example of a 3×3 matrix

$$
A^{-1} = \begin{bmatrix} v_{11} & v_{12} & v_{13} \\ & v_{22} & v_{23} \\ & & v_{33} \end{bmatrix} \begin{bmatrix} 1 & & \\ q_{21} & 1 & \\ q_{31} & q_{32} & 1 \end{bmatrix} = \begin{bmatrix} v_{11} + v_{12}q_{21} + v_{13}q_{31} & v_{12} + v_{13}q_{32} & v_{13} \\ v_{22}q_{21} + v_{23}q_{31} & v_{22} + v_{23}q_{32} & v_{23} \\ v_{33}q_{31} & v_{33}q_{32} & v_{33} \end{bmatrix}
\tag{5.17}
$$

2. **Billion LU decomposition seeking inversion matrix**
According to the principle of uniform block, the matrix $A^{T}A$ can be split into 4 3 × 3 sub-matrices. The steps consists of:
A. *Block LU decomposition of matrix $A^{T}A$*
Let B=$A^{T}A$, split the matrix B into 4 pieces, and then perform block LU decomposition into

$$
\begin{bmatrix} B_{11} & B_{12} \\ B_{21} & B_{22} \end{bmatrix} = \begin{bmatrix} L_{11} & \\ L_{21} & L_{22} \end{bmatrix} \begin{bmatrix} U_{11} & U_{12} \\ & U_{22} \end{bmatrix}
\tag{5.18}
$$

The specific cycle decomposition process is:
   a. L_{11} and U_{11} are obtained by LU decomposition of B_{11}, that is, $B_{11} = L_{11}U_{11}$;
   b. The inverse matrices L_{11}^{-1} and U_{11}^{-1} are calculated according to the triangular matrix inversion method;
   c. Matrix calculation for $L_{21} = B_{21}U_{11}^{-1}, U_{12} = L_{11}^{-1}B_{12}$;
   d. Matrix calculation for $\hat{B}_{22} = B_{22} - L_{21}U_{12}$, which is a symmetric matrix of 3×3 matrix;
   e. Perform LU decomposition of $\hat{B}_{22}$ to obtain L_{22} and U_{22}, namely $\hat{B}_{22} = L_{22}U_{22}$;
   f. The inverse matrix is calculated by the inverse method of triangular matrix L_{22}^{-1} and U_{22}^{-1};
B. *Block U matrix inversion*
Set $V = \begin{bmatrix} V_{11} & V_{12} \\ & V_{22} \end{bmatrix} = U^{-1} = \begin{bmatrix} U_{11} & U_{12} \\ & U_{22} \end{bmatrix}^{-1}$,

$$V_{ij} = \begin{cases} U_{ii}^{-1} & i = j, i = 1, \cdots, k \\ -V_{ii} \sum_{s=i+1}^{j} U_{is} \cdot V_{sj} & q = 1, \cdots, k-1; \ i = 1, \cdots, k-q; \ j = i+q \\ 0 & j < i \end{cases} \tag{5.19}$$

so

$$\begin{cases} V_{11} = U_{11}^{-1}, V_{22} = U_{22}^{-1} \\ V_{12} = -V_{11}U_{12}V_{22} \end{cases} \tag{5.20}$$

C. *Block U matrix inversion*

Set $M = \begin{bmatrix} M_{11} & \\ M_{21} & M_{22} \end{bmatrix} = L^{-1} = \begin{bmatrix} L_{11} & \\ L_{21} & L_{22} \end{bmatrix}^{-1},$

It can be known from $M = L^{-1} = ((L^T)^T)^{-1} = ((L^T)^{-1})^T$, we have

$$M_{ji} = \begin{cases} L_{ii}^{-1} & i = j, i = 1, \cdots, k \\ \sum_{s=i+1}^{j} M_{js} \cdot L_{si}(-M_{ii}) & q = 1, \cdots, k-1; \ i = 1, \cdots, k-q; \ j = i+q \\ 0 & j < i \end{cases} \tag{5.21}$$

so

$$\begin{cases} M_{11} = L_{11}^{-1}, M_{22} = L_{22}^{-1} \\ M_{21} = M_{22}L_{21}(-M_{11}) \end{cases} \tag{5.22}$$

D. *Sub-block matrix multiplication and accumulation*

After performing block LU decomposition and block matrix inversion of the matrix B decomposition, the inverse matrix V of the upper triangular block U matrix and the inverse matrix M of the lower triangular block L matrix are obtained, so there is $B^{-1} = V \cdot M$. Suppose $D = B^{-1}$, then

$$D = \begin{bmatrix} D_{11} & D_{12} \\ D_{21} & D_{22} \end{bmatrix} = \begin{bmatrix} V_{11} & V_{12} \\ & V_{22} \end{bmatrix}\begin{bmatrix} M_{11} & \\ M_{21} & M_{22} \end{bmatrix} = \begin{bmatrix} V_{11}M_{11} + V_{12}M_{21} & V_{12}M_{22} \\ V_{22}M_{21} & V_{22}M_{22} \end{bmatrix} \tag{5.23}$$

5.3.4.2 Hardware Implementation for Inverse Matrix Using Block LU Decomposition

According to the block LU decomposition algorithm for matrix A^TA, the matrix inverse module, as shown in Figure 5.14, is designed and constructed with the System Generator. In this module, each computation step is implemented separately, and then a complete computation module is obtained by connecting each sub-module. The module includes two LU factorization inverse of the 3×3 matrix and D_{11} module for adding the elements of the two matrices, and the remaining sub-module is the matrix multiplication module. The design and implementation of modules for solving inverse

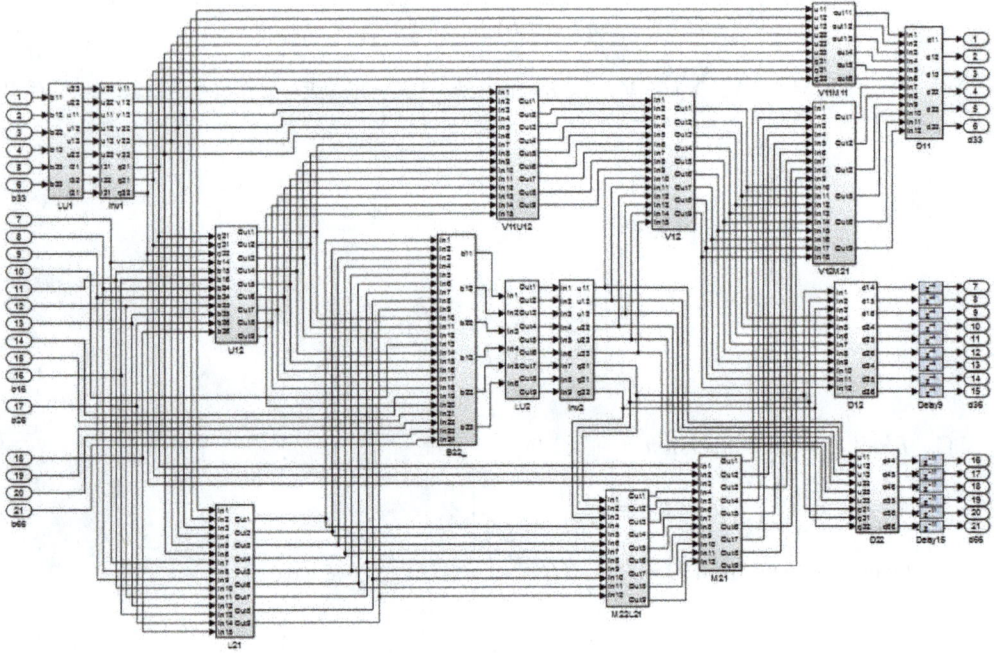

FIGURE 5.14 Implementation structure of block LU decomposition inverse module of matrix $A^T A$.

of 3×3 matrix LU factorization and multiplication of different types of 3×3 matrices are briefly described below.

1. **Construction of LU decomposition and inverse module of 3×3 matrix**

   For a 3×3 symmetric matrix B_{11}, compared with the matrix LU decomposition structure proposed by Lin (2007), it is better to directly construct its computing module in terms of computing speed and resource utilization. The LU decomposition calculation of the symmetric matrix B_{11} of the 3×3 can be obtained from Equation (5.12), i.e.,

   $$\begin{cases} u_{11} = b_{11},\ u_{12} = b_{12},\ u_{13} = b_{13} & l_{11} = l_{22} = l_{33} = 1 \\ u_{22} = b_{22} - l_{21}u_{12},\ u_{23} = b_{23} - l_{21}u_{13} & l_{21} = b_{12}\,/\,u_{11},\ l_{31} = b_{13}\,/\,u_{11} \\ u_{33} = b_{33} - (l_{31}u_{13} + l_{32}u_{23}) & l_{32} = (b_{23} - l_{31}u_{12})\,/\,u_{22} \end{cases} \tag{5.24}$$

   With Equation (5.24), the LU decomposition module, as shown in Figure 5.15, can be constructed.

   The inverse of the upper triangular matrix can be realized by using the pulsating array structure (Seige et al. 1998). Figure 5.15 shows the implemented structure of the inverse of a 4×4 upper triangular matrix, in which M_{ij} (i,j=1,2,3) represents the multiplication and addition element, and D_i represents the divider. A similar structure can be used in the FPGA implementation of the inverse of triangular matrix. According to Equations (5.13) and (5.16), the inverse formula of the 3×3 upper triangular matrix and the unit lower triangular matrix can be obtained by

   $$\begin{cases} v_{11} = 1\,/\,u_{11},\ v_{22} = 1\,/\,u_{22},\ v_{33} = 1\,/\,u_{33} & q_{11} = q_{22} = q_{33} = 1 \\ v_{12} = -v_{22}u_{12}\,/\,u_{11},\ v_{23} = -v_{33}u_{23}\,/\,u_{22} & q_{21} = -l_{21},\ q_{32} = -l_{32} \\ v_{13} = -(v_{23}u_{12} + v_{33}u_{13})\,/\,u_{11} & q_{31} = -(q_{32}l_{21} + q_{33}l_{31}) \end{cases} \tag{5.25}$$

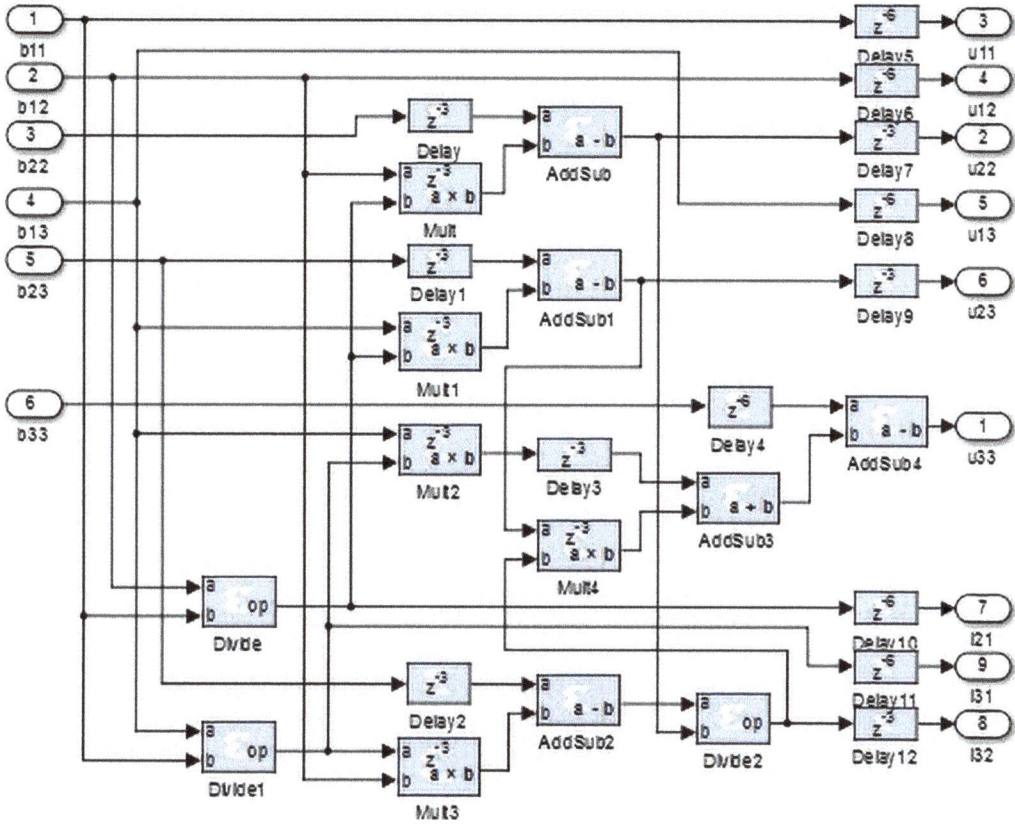

FIGURE 5.15   LU decomposition module of submatrix B_{11}.

According to Equation (5.25), the calculation module for solving the inverse matrix of the 3×3 upper triangular matrix and the unit lower triangular matrix, as shown in Figures 5.16 and Figure 5.17, respectively, is constructed by imitating the pulsating array structure. Multiplexers are used to build vector data. The calculation structure uses fewer multipliers and dividers than a directly implemented structure, and some other resources (data selectors, delay units, and counters) are used.

2. **Construction of different types of the 3×3 matrix multiplication modules**
   The parallel computing architecture proposed by Tian et al. (2008) can be used for the multiplication operation of a 3×3 matrix. However, the block LU decomposition inverse module in Figure 5.18 includes six different types of 3×3 matrix multiplications. Some matrix multiplication modules are not suitable for the parallel computing structure based on PE unit, but the direct computing structure can get better results in computing speed and resource consumption. The direct calculation structure realizes the basic mathematical operations through a calculation mode of one by one and connects them in serial or parallel structure. This means that this structure does not use PE unit and is suitable for simple calculation process. In the block LU decomposition inverse matrix module in Figure 5.18, the parallel computing structure based on PE unit is used for matrix multiplication, including the multiplication of general 3×3 matrices (B22_ and V12M21 modules), the multiplication of general 3×3 matrices and upper triangular matrices (L21 and V12 modules), and the multiplication of upper triangular matrices and general 3×3 matrices (V11U12 modules). The matrix multiplication operations using the direct calculation structure include the multiplication of the unit lower triangular matrix and the general 3×3 matrix (U12 and

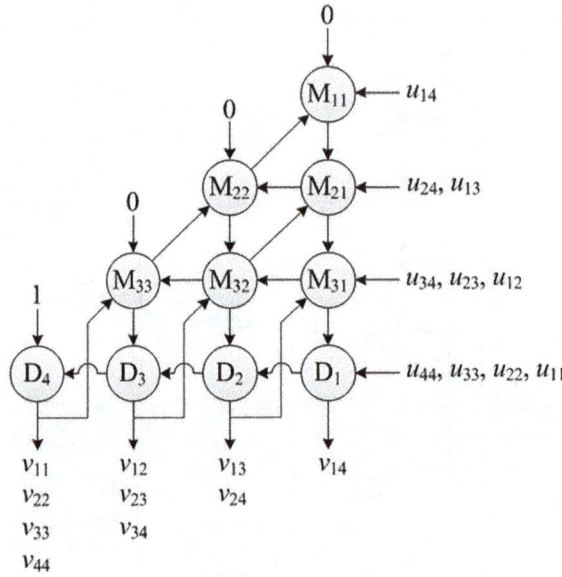

FIGURE 5.16 Pulsating array structure for inverse of the 4×4 upper triangular matrix.

M22L21 modules), the multiplication of the general 3×3 matrix and the unit lower triangular matrix (M21 and V12M22 modules), and the multiplication of the upper triangular
matrix and the unit lower triangular matrix (V11M11 and D22 modules). Modules B22_,
V12M21, V11M11, and D22 calculate the elements on and above the diagonal of the corresponding symmetric matrix. In the following, the construction of computing modules of
L21 and U12 is taken as an example to explain the characteristics of parallel computing
structure and direct computing structure based on PE unit, respectively.

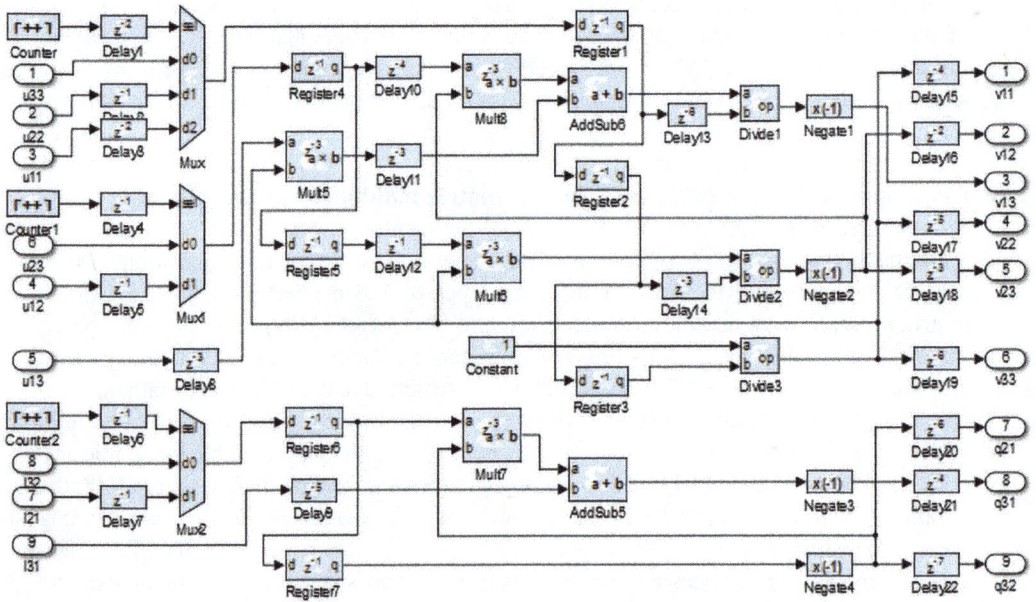

FIGURE 5.17 Internal structure of inverse module for the 3×3 upper triangular matrix and unit lower triangular matrix.

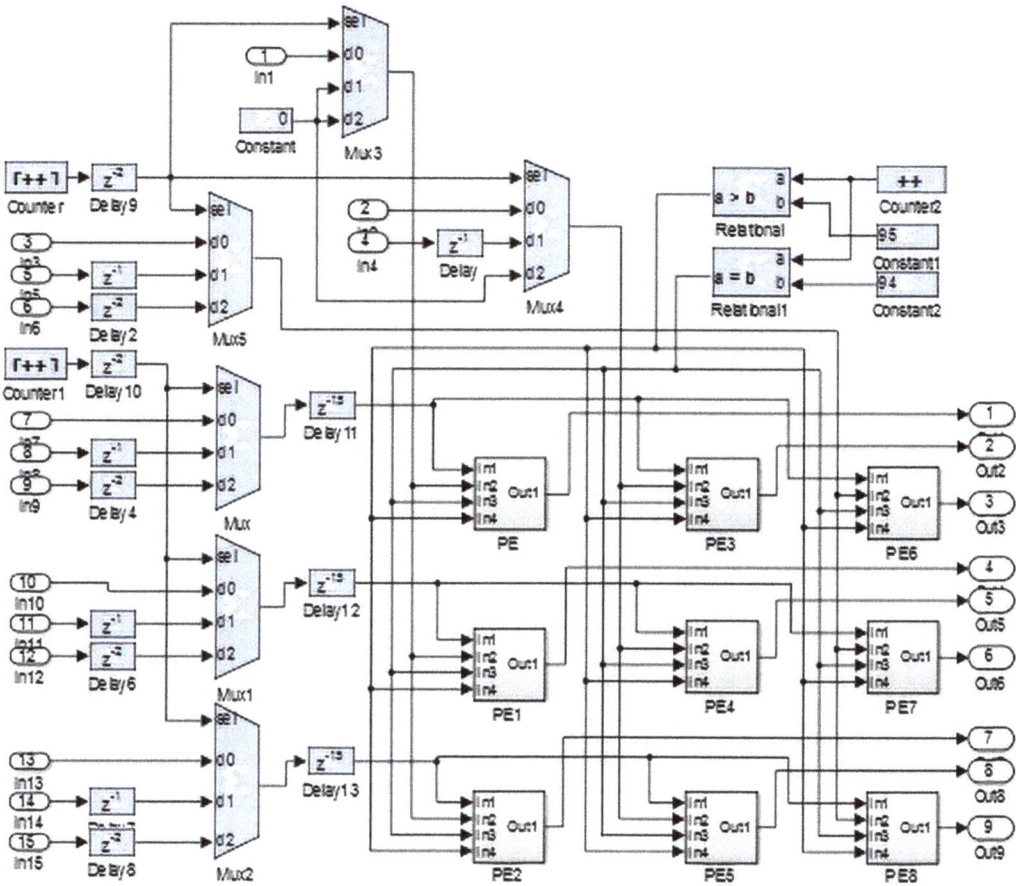

FIGURE 5.18 The 3×3 matrix multiplication module on the basis of PE element.

The parallel computing structure in a 3×3 matrix multiplication module represented by the L21 module is shown in Figure 5.18. The module mainly uses 9 PE units and 6 data selectors. The upper triangular matrix uses the constant 0 as the supplementary data of the vector data. Compared to the direct computing structure, the parallel computing module only takes a few more clock cycles to complete the computation, but the usage of both multipliers and adders is reduced by two-thirds (the usage of other resources is increased). Figure 5.19 shows the U12 calculation module built in accordance with the direct calculation structure, which realizes the multiplication operation of triangular matrix and ordinary 3×3 matrix under the third-order unit. In addition to the delay time introduced by the alignment data signal, the real calculation time of the module is only 3 clock cycles of the multiplier, and the use of hardware resources is reduced compared with the PE unit on the basis of computing structure.

5.4 SIMULATION AND VERIFICATION

5.4.1 EXPERIMENTAL DATA

An image taken by the frame aerial camera RC30 was used for the simulation experiment. The ground area corresponding to the image is the downtown Denver, Colorado, USA. The major objects on the image are high houses, buildings, and roads. The geodetic coordinates of the ground control points (GCPs) obtained from the image are derived from Triangulated Irregular Network (TIN)

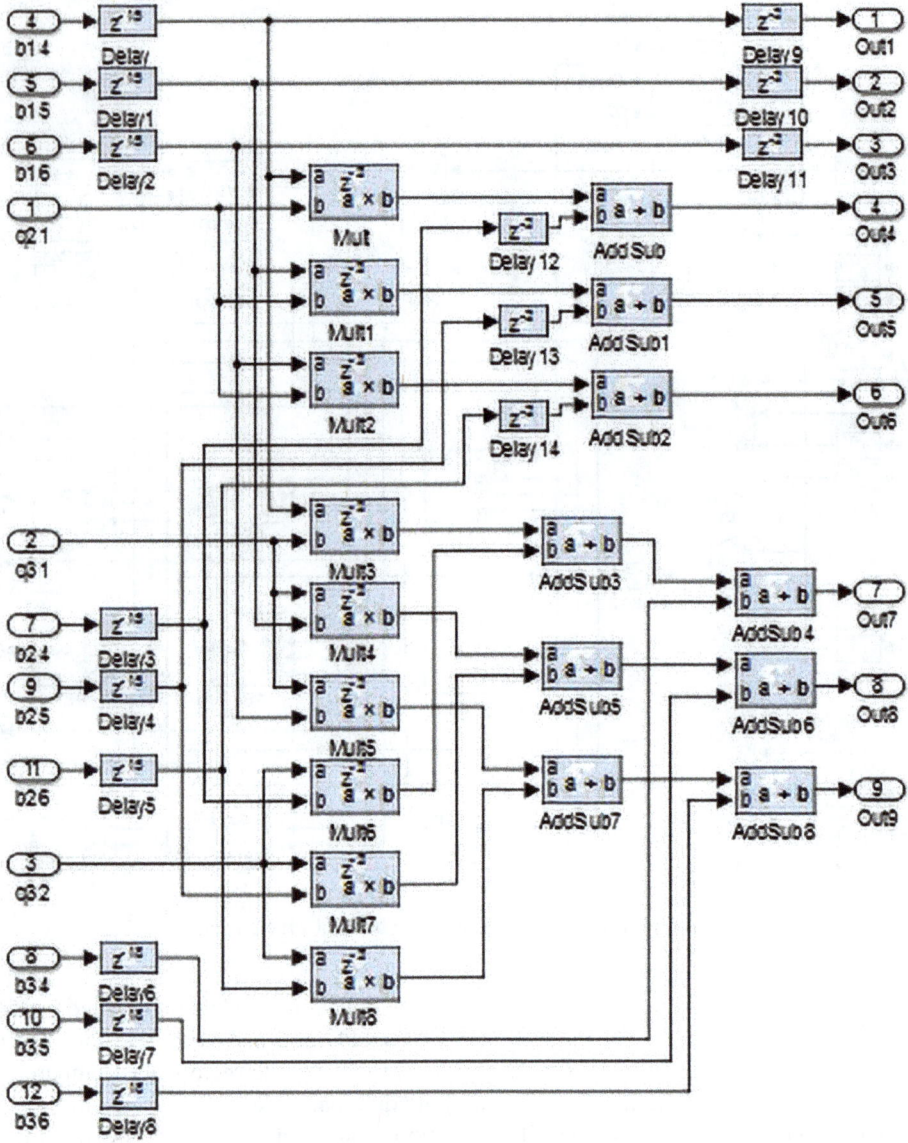

FIGURE 5.19 The module for calculating the multiplication of triangular matrices and 3×3 matrix.

data structure under the WGS84 coordinate system. Four GCPs are selected, all of which are corner of the building. Figure 5.20 shows the distribution of GCPs on the aerial photograph. The two lines connected their fiducial marks on the aerial photograph are constructed in the frame image coordinate system. Figure 5.21shows the distribution of GCPs in the TIN data.

The information of the four GCPs required for the experiment are listed in Table 5.1, including the image point coordinates under the frame coordinate system and the geodetic coordinates of the corresponding GCPs. The interior orientation parameters (IOPs) of the aerial photos are the focal length of the camera f = 153.022 mm and the principal point coordinates of photograph are x_0 = 0.002 m, y_0 = -0.004 m. The initial EOPs of the aerial photos are as follows: X_{S0} = 3,143,040.5559877 m, Y_{S0} = 1,696,520.9258295 m, Z_{S0} = 9072.2729450347 m, Phi = -0.0016069590546607 rad, Omega = -0.029855399214836 rad and Kappa = -1.5538531899589 rad.

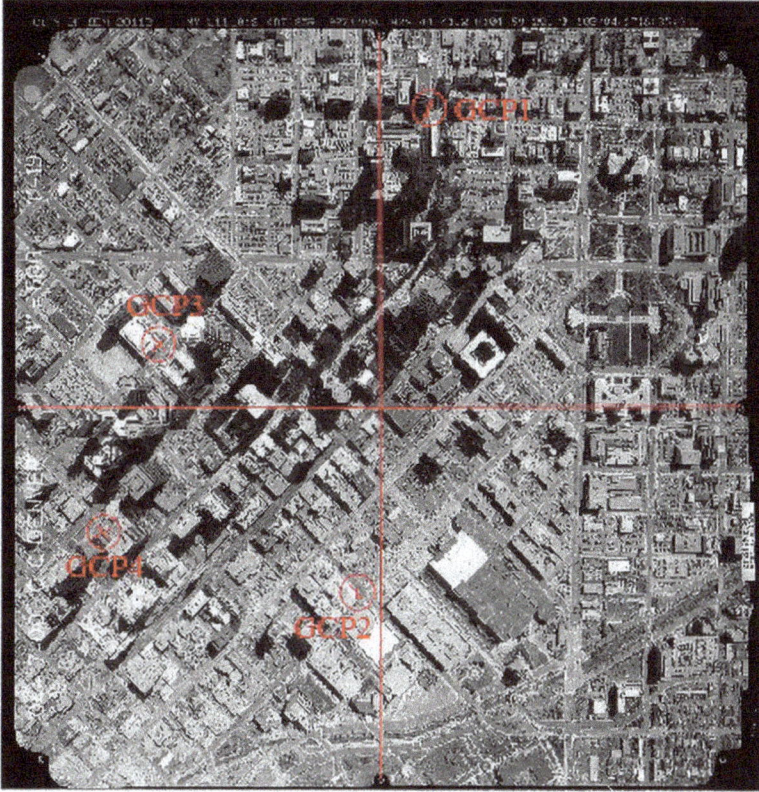

FIGURE 5.20 Distribution of 4 GCPs and the frame image coordinate system on the aerial graph photo.

FIGURE 5.21 Distribution of ground control points (GCPs) in digital terrain model (DTM).

TABLE 5.1

Ground Control Points (GCPs) Data

GCP Serial Number	Image Point Coordinate x (mm)	Image Point Coordinate y (mm)	Geodetic Coordinates X (m)	Geodetic Coordinates Y (m)	Geodetic Coordinates Z (m)
GCP1	14.863	89.707	3,145,274.591	1,696,075.849	5269.992
GCP2	-7.520	-55.916	3,141,628.343	1,696,569.504	5220.351
GCP3	-68.900	18.829	3,143,472.869	1,698,126.977	5218.668
GCP4	-86.273	-37.752	3,142,061.273	1,698,533.583	5212.150

5.4.2 Verification and Hardware Resource Utilization Analysis of Primary Calculation

By multiplying the inverse matrix of A^TA and column matrix A^TL, we can get the solution of the first iteration adjustment calculation, i.e., the correction of six EOPs. The multiplication module of matrix $(A^TA)^{-1}$ and A^TL can adopt the same implementation structure as the calculation module in Figure 5.18. Finally, six corrections of the EOPs are constructed into a vector through MUX module (a data selector, a counter, and multiple delay units are used internally) and output from one port. By connecting all sub-modules, a primary adjustment calculation system for the single image space resection is obtained (Figure 5.22).

Using the data in Section 5.3.6.1 to simulate all the sub-modules in Figure 5.22 step by step, and comparing the results with the reference values (the results calculated by the MATLAB program), it can be found that all design modules can complete the expected calculation functions. The entire computing system outputs the corrections of six EOPs in the 146th –151st clock cycle. The FPGA calculation results and software calculation results for primary adjustment are listed in Table 5.2. With comparison, it can be found that the simulation results of the adjustment calculation system are correct, and the results have high accuracy due to the use of double-precision floating-point numbers during operation.

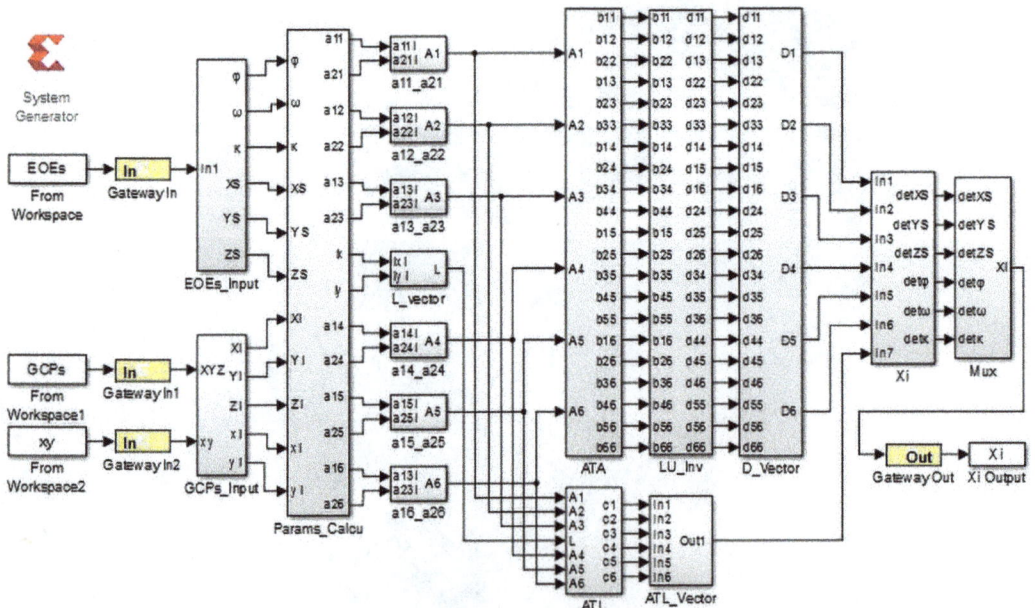

FIGURE 5.22 Primary adjustment calculation system for the space resection of single image.

TABLE 5.2

Comparison of FPGA-Based and MATLAB-Based Software Calculation Results

Parameter	FPGA-Based Calculation Results	MATLAB-Based Calculation Results	The Difference
ΔX_S/m	-99.165194612645730	-99.1651946127088	6.3068e−11
ΔY_S/m	-64.062338577704200	-64.0623385777028	1.3927e−12
ΔZ_S/m	-1.301166966155861	-1.30116696622042	6.4559e−11
$\Delta\varphi$/rad	0.0005978302725707929	0.000597830272574276	3.4831e−15
$\Delta\omega$/rad	0.003055123116735	0.0030551231167502	1.5200e−14
$\Delta\kappa$/rad	0.0006303149951943920	0.000630314995209887	1.5495e−14

Using the System Generator module in Figure 5.22 to automatically convert the primary adjustment calculation results into the files such as RTL code and IP core code, and automatically generate testbench code files for simulation. Because the RTL-level circuit obtained after conversion is too large, the computer memory available is insufficient when the ISim simulation tool is directly used under the ISE software platform (the available memory of the PC is 8G). Calling the third-party Modelsim software to perform the functional simulation of the design circuit, the simulation can be successfully completed, and the simulation waveform (hexadecimal representation) is obtained in Figure 5.23. Converting the hexadecimal number of the signal gateway_out_net in the waveform to a binary number, and then converting it to a decimal number according to the IEEE 754 standard calculation given above, the obtained values are consistent with the FPGA calculation result, which are listed in Table 5.2. This result demonstrated that the RTL level circuit is functionally correct.

Performance of the synthesis process on the RTL-level circuit to get the corresponding RTL-level netlist structure is shown in Figure 5.23. The netlist and the first iteration of adjustment calculation is shown in Figure 5.24. By observing the reports, a preliminary understanding of the hardware resource usage of the design circuit, which is listed in Tables 5.3 and 5.4, can be obtained, and it can be demonstrated that more hardware resources on the FPGA chip are used in the design of the circuit. However, we can only preliminarily understand the use of hardware resources of the design circuit from here, and cannot determine whether the hardware resources of the FPGA chip can meet the requirements.

Since the data of the input and output ports are with a double-precision floating point format (represented by 64-bit binary), 4 data input and output ports occupy 256 I/O pins, and one I/O pin is used as clock input port, in which the I/O pins are assigned to the input and output ports of the design circuit, and then the implementation process is executed. During the experiment, the translation step was successfully completed, but the software was forced to stop if the Slice unit and DSP48E1 unit used in the design circuit exceed the resource limit of the experimental FPGA chip. Looking at the hardware resource usage report of the mapping step, it can be found that the design circuit uses 39,738 slice units and 2035 DSP48E1 units, while the maximum number of Slice units on the experimental FPGA chip is 33,650, and the total number of the embedded DSP48E1 is

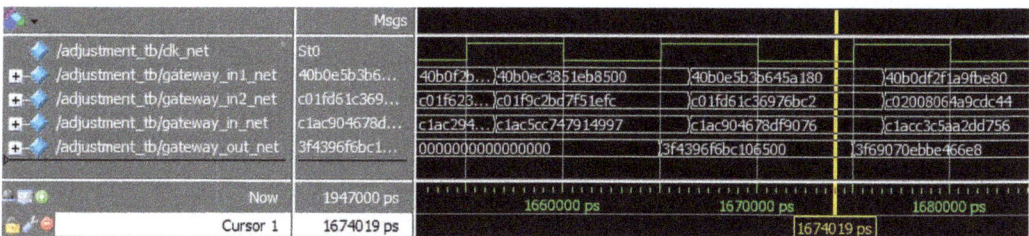

FIGURE 5.23 Functional simulation waveform of design circuit (local).

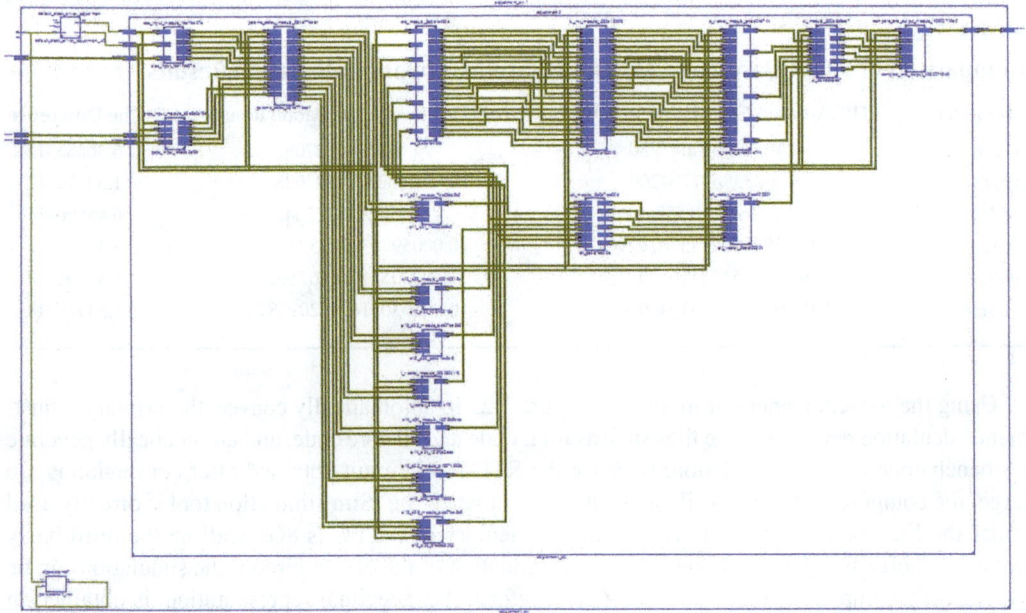

FIGURE 5.24 RTL-level netlist structure after integration.

TABLE 5.3
Hardware Primitives and Black Box Usage

Primitive and Black Box	Amount(Piece)	Primitive and Black Box	Amount(Piece)
BELS (Including GND、INV、LUT、MUX、VCC and other logical resources)	4982	Various counters	34
Flip flops/latches	34,914	Floating point to fixed point conversion core	3
Shift registers	19,990	Fixed point to floating point conversion core	6
Clock buffers	1	CORDIC core	3
IO buffers	256	Floating point divider	22
Block RAM	5	Floating point multiplier	185
Floating point adder and subtractor	173	comparator	3

TABLE 5.4
Logic Resource Usage on FPGA Chip (Estimated Value)

Logic Utilization	Used	Available	Utilization
Number of slice registers	34,914	269,200	12%
Number of LUTs	24,420	134,600	18%
Number of fully used LUT-FF pairs	20,439	38,895	52%
Number of bonded IOBs	257	400	64%
Number of BUFG/BUFGCTRLs	1	32	3%

740 (reference to Table 5.1). Due to layout and routing reasons, for some Slice cells, only part of its internal LUT and flip-flop (FF) resources are used, so more Slice cells are occupied. The number of DSP48E1 units used far exceeds the upper limit. It can be seen that there are too many hardware resources used in the design of the circuit, and an evaluation FPGA chip can no longer meet its requirements. The design circuit occupies too many hardware resources, which needs to be solved by modifying the design or using FPGA chips with rich hardware resources.

5.4.3 Implementation and Verification of Iterative Computing System

On the basis of the primary adjustment calculation above, the iterative adjustment calculation system can be constructed in Figure 5.25 in accordance with the FPGA implementation structure of the single-image space resection shown in Figure 5.8. The design system needs to transmit the correction obtained from the last computation to the initial value input module of the EOPs, and sum the corrections and the initial values to update the EOPs. The updated data can use the original module for the next iteration calculation. The system can be realized only through multiplexing of each computing module, which can greatly reduce the area of the circuit. However, this design system is difficult to obtain correct calculation results. To make each sub-module repeat the operation with the specific data within the specified clock cycle as required, it is necessary to modify the timing control inside all sub-modules related to timing (such as adding counters to obtain new control signals and adding more delay units to make the timing meet the requirements). Even in order to obtain the required timing (especially the timing of the data selector), the implementation structure of some sub-modules has to be redesigned. The computing system has more complicated and strict timing requirements, which increases the difficulty of implementation and increases the demand for hardware resources. If the multiplication and inversion of a large matrix are encountered, the two problems above may become more obvious.

Considering the complexity of iterative computation, one iteration calculation can be divided into two separate calculations, which can be realized through independent calculation modules, so as to

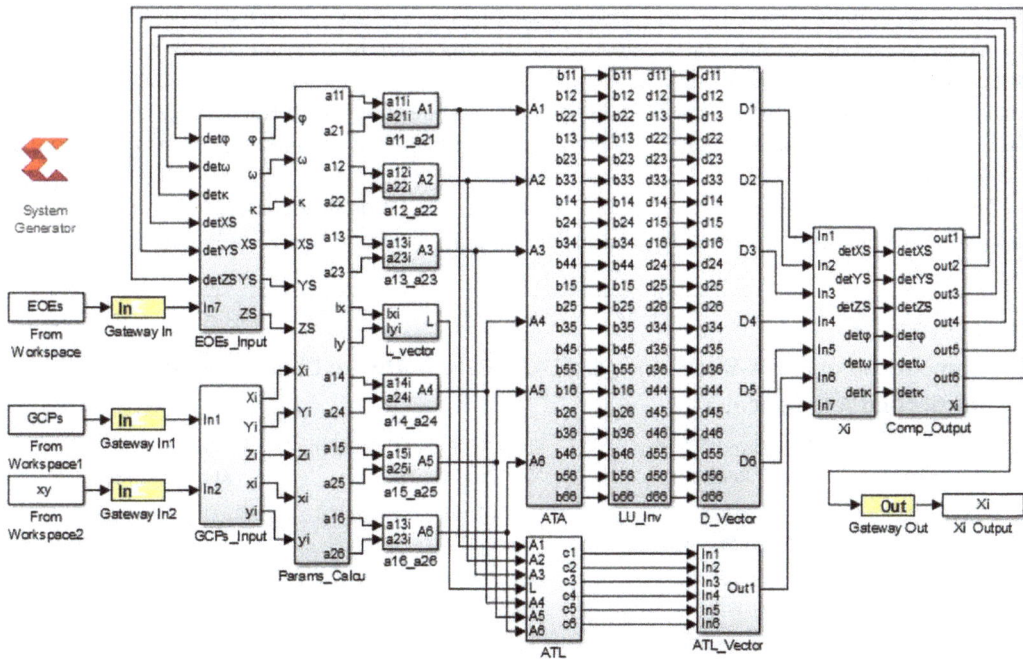

FIGURE 5.25   The first iterative calculation.

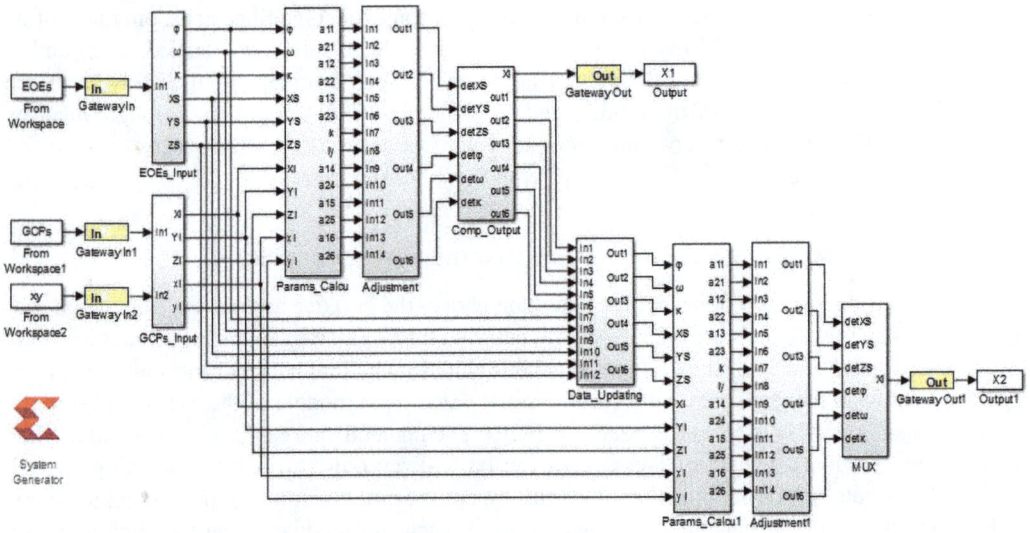

FIGURE 5.26   The second iterative calculation.

obtain the second iterative calculation system, which is shown in Figure 5.26. The second iterative calculation system uses about twice the hardware resources for a calculation system. The construction of the system is relatively simple. The internal structure of the parameter calculation module and the matrix calculation module are the same, and only the parameter settings of the constant module used for time sequence control in the matrix calculation module are different. The internal structure of the threshold comparison and the resulting output module is shown in Figure 5.27. The module compares the corrections of the three rotational angle elements with the preset threshold, and uses the logical value obtained from the comparison to control the output of the calculation results (if all are less than a given threshold, the logical value is 1, and the calculation result may not be taken as input data in the next iterative calculation, and if it is 0, the calculation result may be taken as input data in the next iterative calculation module). In accordance with the description in Section 5.3.3, the threshold of the corrections for three rotation angles is 0.1', which is approximately equal to 0.0000290888 rad. The data update module sums the initial value of the EOPs and the corrections to obtain the updated EOPs, which are then taken as input data in the next iterative calculation.

The comparison analysis between the FPGA-based calculation results and the MATLAB-based calculation results for the second iterative calculation are listed in Table 5.5. It can be found from

TABLE 5.5

Comparison of FPGA-Based and MATLAB-Based Software Calculation Results for the Second Adjustment

Parameter	FPGA-Based Calculation Results	MATLAB-Based Calculation Results	Absolute Difference
ΔX_S/m	0.0105216347425719	0.0104992485967856	2.2386e–05
ΔY_S/m	0.0180373155932267	0.0180254573216991	1.1858e–05
ΔZ_S/m	-0.0224123208315854	-0.0224172637980877	4.9430e–06
$\Delta\varphi$/rad	-2.08496488609610e-06	-2.07988797915225e-06	5.0769e–09
$\Delta\omega$/rad	-3.15369324220675e-06	-3.15051331093826e-06	3.1799e–09
$\Delta\kappa$/rad	-5.09185593581525e-08	-5.14193137756369e-08	5.0075e–10

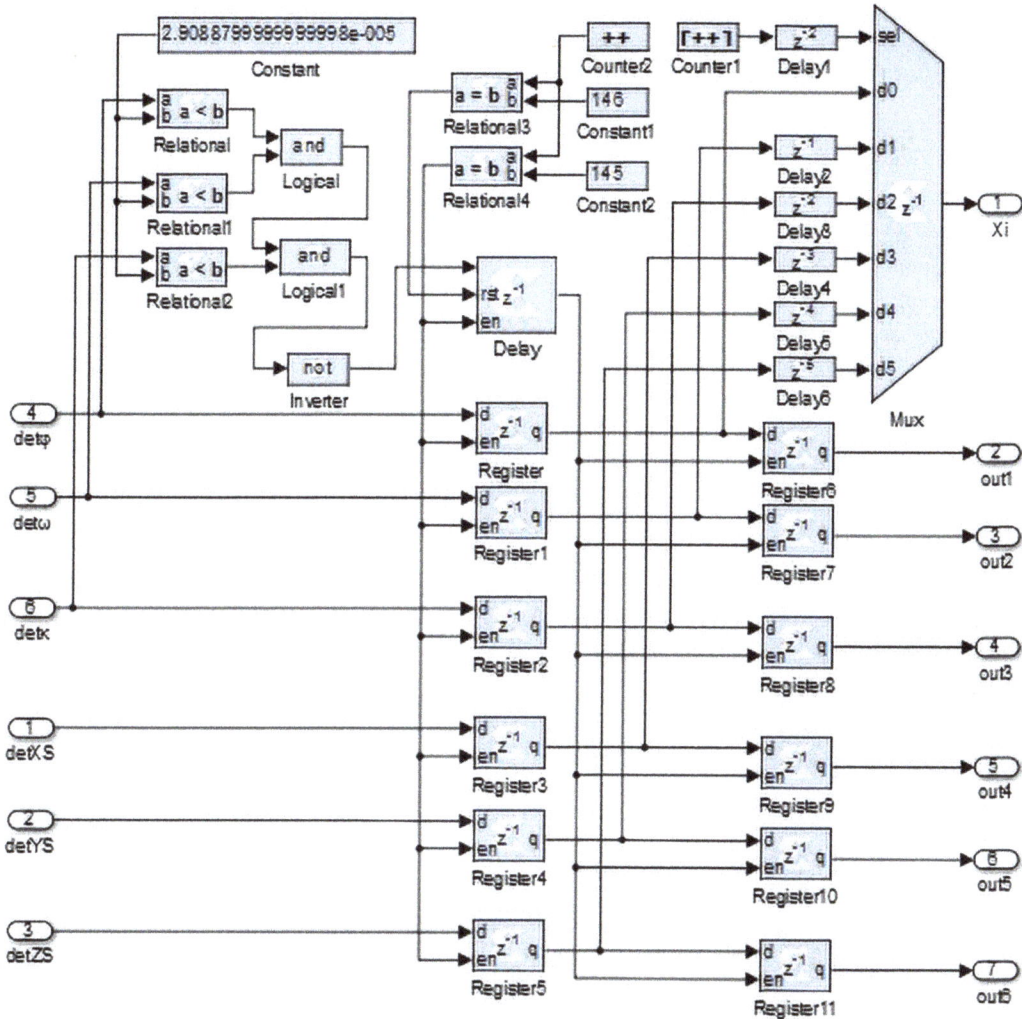

FIGURE 5.27 The internal implementation of the threshold comparison and the result output module.

Table 5.5 that the second iterative calculation system can correctly complete the calculation, but the accuracy is reduced (the corrections of the three line elements only has 2 to 4 significant digits consistent, and the corrections of three line elements are only 1–3 significant digits are the same). By checking the calculation results of each sub-module in the second iterative calculation system one by one, it is found that due to the error in the result of the first iterative calculation, after many operations of addition, subtraction, multiplication, and division in the second iterative calculation, this error is continuously amplified, which reduces the accuracy of the final calculation result. Due to the limited precision of data that are represented by double-precision floating-point numbers, rounding errors in calculation results caused by multiplication or division, both extremely large numbers or extremely small numbers may also cause a decrease in accuracy. However, even if the extended double-precision floating-point number with larger data bit width (floating-point number length is 80, mantissa length is 64, symbol bit is 1, exponential length is 15) is used, the accuracy of the results is not significantly improved. It can also be found that the accumulation of calculation errors is the main resources of the final accuracy decreases. In the result of the second iterative calculation, the corrections of the three rotation angle elements are all less than the given threshold, so it is unnecessary to perform iterative calculations. With changing the accuracy in the first two iterative calculation results, if the third iterative calculation is required, the accuracy of the results may be lower than those in the first two iterations.

5.5 CONCLUSIONS

Firstly, this chapter introduced FPGA-based design and implementation for on-board geometric calibration for a frame camera with fiducial marks, and explored the difference in the accuracy of mathematical operations under different data formats and the difference when using hardware resources of the FPGA chip. The FPGA implementation for least square adjustment calculation of a single image space resection for a frame image is detailedly described, and the geometric calibration calculation with solution of the EOPs of a frame image is described.

With the characteristics of the least square adjustment algorithm and the top-down design method, a system generator, parallel computing modules for data input, parameter calculation, matrix multiplication, matrix inversion, and other operations are designed and constructed. Especially, the construction of the block LU decomposition and inversion module for the symmetric positive definite matrix $A^{\mathrm{T}}A$, are implemented. The verification using real aerial photograph data is conducted. It has been demonstrated that the calculation with double-precision floating-point data format can obtain high-precision results. Finally, two FPGA-based hardware implementation structures for the iterative calculation system are described, and the iterative calculation system that separates the two calculations can obtain the correct calculation results.

REFERENCES

Brieß, K., Bärwald, W., Gill, E., Kayal, H., Lura, F., Montenbruck, O., Montenegro, S., Halle, W., Skrbek, W., Studemund, H., Terzibaschian, T., Venus, H., Technology demonstration by the BIRD mission. *Acta Astronautica*, 2005, 56: 51–54.

DLR, Mission objectives of BIRD. 2003, URL:c.

Halle, W., Brie, K., Schlicker, M., Skrbek, W., Venus, H., Autonomous onboard classification experiment for the satellite BIRD. *The International Archives of the Photogrammetry, Remote Sensing and Spatial Information Sciences*, 2002, 34(1): 63–68.

Jacobsen, K., Issues and method for in-flight and on-orbit calibration. *Workshop on Radiometric and Geometric Calibration*, 2003, Gulfport.

Jacobsen, K., Geometry of Satellite Images–Calibration and Mathematical Models. *Korean Society of Remote Sensing, ISPRS International Conference*, Korea, Jeju, 2005, pp. 182–185.

Jiang, L.J., Research on On-board Geometric Calibration of Linear Array CCD Satellite Images Based on FPGA, *Master Degree Thesis of Guilin: Guilin University of Technology*, April 2016.

Lin, H., The Realize of Matrix Operation based on FPGA. *Master Degree Thesis of Nanjing: Nanjing University of Science and Technology*, 2007.

Seige, P., Reinartz, P., Schroeder, M., The MOMS-2P mission on the MIR station. *The International Archives of Photogrammetry and Remote Sensing*, 1998, 32: 204–210.

Shao, Y., A Study of the Matrix Operation Harden Implementation on FPGA. *Master degree Thesis of PLA Information Engineering University*, 2010.

Tian, X., Zhou, F., Chen, Y.W., Liu, L., Chen, Y., Design of field programmable gate array based real-time double-precision floating-point matrix multiplier. *Journal of Zhejiang University (Engineering Science)*, 2008, 42(9): 1611–1615.

Wang, R., Hu, Y.H., Du, F.H., FPGA design and implementation of inverse of arbitrary dimensional matrix, *China Integrated Circuit*, April 2007, 95: 51–55.

Wang, R.X., Hu, S., Wang, X.Y., and Yang, J.F., The construction and application of mapping satellite-1 engineering. *Journal of Remote Sensing*, 2012, 16 (supplement): 2–5.

Xilinx. 7 Series FPGAs Overview v1.15, February 2014. URL: *http://china.xilinx.com/support/documentation/data_sheets/ds180_7Series_Overview.pdf*.

Zhang, J.Q., Pan, L., Wang, S.G., *Photogrammetry*, Wuhan: Wuhan University Press, 2009.

Zhou, G., Baysal, O., Kaye, J., Habib, S., Wang, C., Concept design of future intelligent earth observing satellites. *International Journal of Remote Sensing*, 2004, 25(14): 2667–2685.

Zhou, G., Li, C., Yue, T., Jiang, L., Liu, N., Sun, Y., Li, M. 2015a. An overview of in-orbit radiometric calibration of typical satellite sensors, *The 2015 Int. Workshop on Image and Data Fusion (IWIDF)*, 21–23 July, 2015, Kona, Hawaii, USA. DOI: 10.5194/ISPRS Archives-XL-7-W4-235-2015.

Zhou, G., Li, C., Yue, T., Liu, N., Jiang, L., Sun Y., Li, M. 2015b. FPGA-based data processing module design of onboard radiometric calibration in visible and near infrared bands, *Proceedings of SPIE on 2015 International Conference on Intelligent Earth Observing and Applications*, 0277-786X, Vol. 9808, 23–24 October 2015, Guilin, China.

Zhou, G., Li, C., Yue, T., Jiang, L., Liu, N., Sun, Y., Li, M. 2015c, FPGA-based data processing module design of on-board radiometric calibration in visible/near infrared band, *The 2015 International Workshop on Image and Data Fusion (IWIDF2015)*, 21–23 July 2015, Kona Hawaii, USA.

Zhou, G., Liu, N., Li, C., Jiang L., Sun, Y., Li, M., Zhang R. 2015d. FPGA-based remotely sensed imagery denoising, *2015 IEEE International Geoscience and Remote Sensing Symposium (IGARSS)*, 26–31 July 2015, Milan, Italy

Zhou, G., Jiang, L., Huang, J., Zhang, R., Liu, D., Zhou, X., Baysal, O., FPGA-based on-board geometric calibration for linear CCD imager. *Sensors*, 2018, 18: 1794. doi:10.3390/s18061794.

6 On-Board Geometric Calibration for Linear Array CCD Sensor

6.1 INTRODUCTION

Chapter 5 focuses on the on-board geometric calibration for frame camera with eight fiducial marks. This is a very classic and typical camera calibration method, which corresponds to collinearity equation. However, different imaging cameras or sensors have different modes, which corresponds to different imaging models, i.e., calibration camera. This chapter presents field programmable gate array (FPGA)-based method for on-board implementation of geometric calibration for a linear array charge-coupled device (CCD) scanner. This is because a camera geometric calibration is one of the most important works for high quality of acquiring high-resolution satellite imagery (Jacobsen 2006; Robertson et al. 2009; Wang et al. 2014) and is also a prerequisite for direct georeferencing of remotely sensed images (Honkavaara 2004; Radhadevi and Solanki 2008; Zhou 2020)). To this end, an architecture of FPGA-based implementation of on-board geometric calibration is depicted in Figure 6.1. The architecture consists of five major modules, which are Template Image Selection, Image Matching (Huang and Zhou (2017)), Initial value EOEs (Exterior Orientation Elements), Bundle Adjustment and Timing Control (Zhou et al. 2018).

All the functional modules in Figure 6.1 are managed by the Timing Control module. The Template Images Selection module is used for selecting target template images from a template image library that is created from many template images, each of which has a unique ID number. The principle of this module is carried out through matching between the geodetic coordinates centralized at an imaged area and the coordinates of a georeferenced template image. The Image Matching module accurately determines the image coordinates (from spaceborne) and geodetic coordinates (from template images) by matching their sub-image windows. The Initial EOEs (external orientation elements) module computes the initial values of EOEs of the spaceborne sensor. The Bundle Adjustment module accurately computes six external orientation elements. Because of the limitation of hardware resources during on-board processing, it is usual to apply external memory to store multiple template images and GCPs (ground control points) data. Random Access Memory (RAM) is used to store the data stream of a template image and/or an imaged scene temporarily. The row buffers of image data are generated using multiple RAMs for further image matching.

6.2 RELEVANT EFFORTS

Traditional geometric calibration for linear array CCD sensor has been investigated over several decades and a number of papers have been published in the computer vision community (Healey and Kondepudy 1994), image processing (Gupta and Hartley 1997; Strecha et al. 2008), robotic vision (Lenz and Tsai 1988; Heikkila 2000; Zhang 2000; Zhou and Li 2000; Li and Zhou 2002; Zhou et al. 2004; Chen and Tu 2011). However, a spaceborne-based geometric calibration on-board implementation has not yet been reported globally so far, but spaceborne-based in-lab and inflight (also called *on-orbit*) geometric calibrations for various satellites have been reported in the past two decades (Valorge 2003; Poli et al. 2010; Bouillon and Gigord 2004), such as SPOT1-5 (Westin 1992), IKONOS (Jacke and James 2005), ALOS (Gruen et al. 2007), Orbview-3 (David 2000),

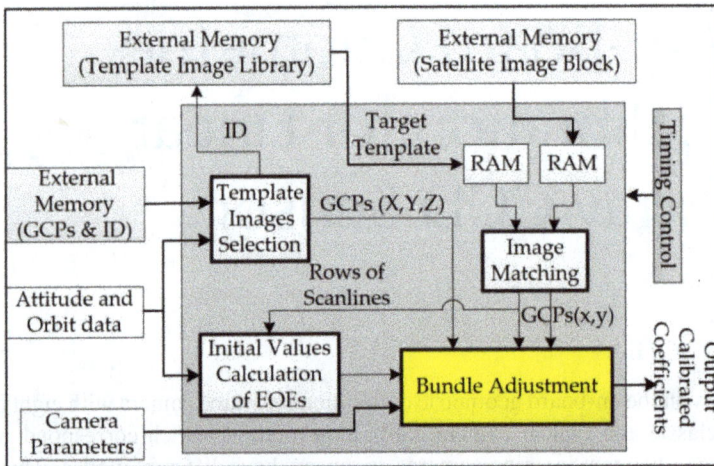

FIGURE 6.1   An FPGA architecture of on-board geometric calibration.

IRS-1C (Srivastava and Alurkar 1997), GeoEye-1 (Crespi et al. 2010), MOMS-2P (Kornus et al. 1998; Kornus et al. 1999), CBERS-02B (Flavio et al. 2008), TH-1 (Li et al. 2012) and ZY-3 (Lei 2011; Li 2012; Li and Wang 2012).

Jacobsen (Jacobsen 2003; Jacobsen 2005) implemented geometric calibration of the IRS-1C satellite using self-calibration bundle block adjustment with additional parameters. Crespi et al. (2010) used the block adjustment method to investigate the inflight calibration of the GeoEye-1 satellite, with which the position accuracy reached 3.0 m without GCPs. Lei (2011) presented a self-calibration bundle block adjustment on the basis of known interior orientation parameters with line array CCD. Li (2012) and Li et al. (2012) proposed a geometric calibration model and step-by -step methods on the basis of an imaging model of the ZY-3 satellite. Their results indicated that the positioning accuracy of ZY-3 can meet the requirement of the accuracy of 1:50,000 mapping. Yang and Wang (2013) also presented an on-orbit geometrical calibration model for ZY-1 02C panchromatic camera and the experimental results demonstrated that the positioning accuracy of the calibrations with and/or without GCPs is better than 0.3 pixels. Zhang et al. (2014) proposed an on-orbit geometric calibration and validation of ZY-3 linear array sensors. Zhou and Li (2000) and Li et al. (2002) constructed an on-orbit geometric calibration model through performing reasonable optimizing and parameter selection of the rigorous geometric imaging model. The experiment results discovered that the geometric accuracy without GCPs is significantly improved after on-orbit geometric calibration. Cao et al. (2014) proposed a look-angle calibration method for on-orbit geometric calibration of ZY-3 satellite imaging sensors. The analysis found that the object positioning accuracy of nadir images is better than ±2.7 m after calibrating satellite sensor without laboratory calibration parameters and with only five GCPs and the object positioning accuracy of nadir images is better than ±11 m when using CCD look angles for extrapolative object positioning. Cao et al. (2015) proposed a simple and feasible orientation method for calibrating the CCD-detector look angles in the attitude determination reference for the three-line cameras of ZY-3. The results found that the positioning accuracy of planimetric and height is 3.4 and 1.6 m, respectively, after a bundle adjustment with four GCPs. Wang et al. (2016) proposed an on-orbit geometric calibration method for TH-1 three-line-array camera based on a rigorous geometric model. The experiment results showed that the planimetric and height accuracies of three stereo pairs are approximately 6.93 and 3.96 m after calibration, respectively.

Although on-board geometric calibration has little investigation so far, articles relevant to FPGA-based on-board data processing for spaceborne has been reported. For instance, the German small satellite BIRD used an on-board data processing system consisting of DSP, FPGA and network co-processor to implement radiation correction, rectification of partial systematic geometric and image classification based on a neural network algorithm (Winfried 2001). Surrey Satellite Technology Limited in United Kingdom used FPGA as an on-board data processing chip in a small satellite (Vladimirova 2004). The French planned to use FPGA to realize on-board data processing for Pleiades image (Arnaud et al. 2006). Stuttgart University designed an on-board computing system based on FPGA for the small satellite Flying Laptop to realize real-time processing of attitude control, housekeeping, data compression and image processing (Kuwahara et al. 2009). Jet Propulsion Laboratory used an FPGA chip from Xilinx Inc. to implement on-board hyperspectral image classification based on Support Vector Machine (SVM) algorithm (Pingree 2010). González et al. (2012) conducted the investigation on an FPGA implementation of N-FINDR algorithm for hyperspectral image analysis. Williams et al. (2002) also investigated an FPGA-based implementation of a real-time cloud detection system for on-board remote sensing platform. Hihara et al. (2015) analyzed an on-board image processing system for hyperspectral sensor. The work mentioned above has given a promising for FPGA-based on-board data processing in satellite. This chapter presents an FPGA-based implementation of on-board geometric calibration.

6.3 GEOMETRIC CALIBRATION MODEL FOR LINEAR ARRAY CCD SENSOR

6.3.1 Brief Overview of Geometric Calibration Model

For the geometric calibration algorithm of a linear array pushbroom imaging mode, the different linear imaging systems may have their own imaging modes. This chapter takes MOMS-2P as an example (Li et al. 1998).

The geometric calibration algorithm of the MOMS imaging system was performed at the laboratories of the German Aerospace company DASA, where the MOMS was developed and manufactured. A rigorous model of describing geometry calibration is composed of five parameters (also see Table 6.1):

- Principal point coordinates of each sensor (x_0, y_0),
- Rotation parameter k of CCD array in the image plane,
- Deviation of the focal length df and distortion parameter Kc of the sensor curvature. The sensor curvature is modeled by a second order polynomial equation. The parameter Kc here indicates the along track deviation at the edges of the CCD array at 3000-pixel distance from the array center, caused by the sensor curvature.

1. **Interior orientation**

   Interior orientation transforms screen coordinates (i and j in Figure 6.2) into image coordinates (x and y in Figure 6.2) and corrects for lens distortion (symmetric and tangential) and CCD array curvature distortion. Thus, the following tasks should be performed.
   - Define the screen coordinate system and image coordinate systems,

TABLE 6.1

Lab-Calibrated Camera Parameters for Moms-2p (Li et al. 1998; Ebner et al. 1996)

	HR5A	HR5B	ST6	ST7
F(mm)	660.256	660.224	237.241	237.246
x_0(pixel)	0.1	0.2	-7.2	-0.5
y_0(pixel)	-0.4	0.1	8.0	19.2
Kc(pixel)	-0.3	-0.4	-1.1	1.7
K(mdeg)	-2.9	5.4	-1.5	-1.4

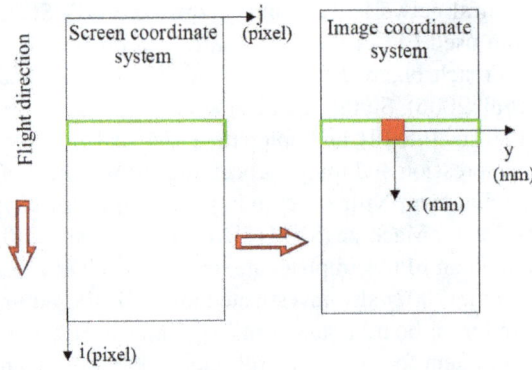

FIGURE 6.2   Screen and image coordinate system.

- Apply the principal offset that should be estimated from a laboratory or in-flight based calibration, and
- Correct various distortions.

2. **Transformation from image to reference coordinate system**

   Each of the fore-, nadir- and aft-looking array has its own image coordinate system (x_c, y_c and z_c, right-handed). An image reference coordinate system (x_R, y_R and z_R, right-hanged) is defined to unify image coordinates from all three arrays (Figure 6.3). The transformation from an image coordinate system to the reference coordinate system involves a translation (dx, dy and dz) and three rotations ($\omega(t)$, $\varphi(t)$ and $\kappa(t)$). We define a counterclockwise rotation angle as positive.

   The transformation equation is

   $$
   \begin{pmatrix} x_R \\ y_R \\ z_R \end{pmatrix} = R_G^R \begin{pmatrix} x_c \\ y_c \\ z_c \end{pmatrix} + \begin{pmatrix} d_x \\ d_y \\ d_z \end{pmatrix} = R_G^R \begin{pmatrix} x \\ y \\ -f \end{pmatrix} + \begin{pmatrix} d_x \\ d_y \\ d_z \end{pmatrix}
   \tag{6.1}
   $$

3. **Geometric model**

   For any image point within a CCD array, its image reference coordinates are (x_R, y_R and z_R). The coordinates of the exposure center of the array in the ground coordinate system at the

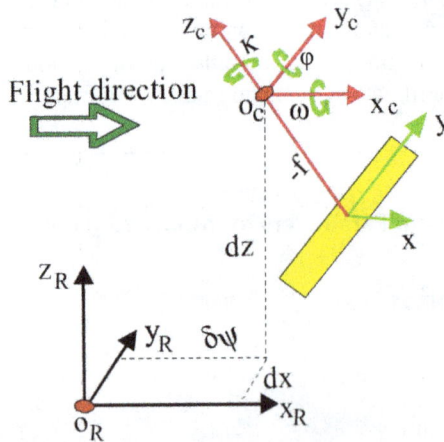

FIGURE 6.3   From image coordinate to image reference coordinate system (Zhou et al. 2002).

imaging epoch t are $(X_C(t), Y_C(t), Z_C(t))$. The corresponding ground point coordinates are (X_G, Y_G, Z_G). The collinearity condition states that all these three points must lie on the same straight line.

$$
\begin{cases}
x_R = z_R \dfrac{r_{11}\left(X_G - X_C(t)\right) + r_{12}\left(Y_G - Y_C(t)\right) + r_{13}\left(Z_G - Z_C(t)\right)}{r_{31}\left(X_G - X_C(t)\right) + r_{32}\left(Y_G - Y_C(t)\right) + r_{33}\left(Z_G - Z_C(t)\right)} \\[4mm]
y_R = z_R \dfrac{r_{21}\left(X_G - X_C(t)\right) + r_{22}\left(Y_G - Y_C(t)\right) + r_{23}\left(Z_G - Z_C(t)\right)}{r_{31}\left(X_G - X_C(t)\right) + r_{32}\left(Y_G - Y_C(t)\right) + r_{33}\left(Z_G - Z_C(t)\right)}
\end{cases}
\tag{6.2}
$$

where r_{ij} $(i, j=1, 2, 3)$ are the elements of rotation matrix $R_G{}^R$, which also can be expressed as follows.

$$
R_G^R = \begin{pmatrix}
\cos\varphi(t)\cos\kappa(t) & \cos\omega(t)\sin\kappa(t) + \sin\omega(t)\sin\varphi(t)\cos\kappa(t) & \sin\omega(t)\sin\kappa(t) - \cos\omega(t)\sin\varphi(t)\cos\kappa(t) \\
-\cos\varphi(t)\sin\kappa(t) & \cos\omega(t)\cos\kappa(t) - \sin\omega(t)\sin\varphi(t)\sin\kappa(t) & \sin\omega(t)\cos\kappa(t) + \cos\omega(t)\sin\varphi(t)\sin\kappa(t) \\
\sin\varphi(t) & -\sin\omega(t)\cos\varphi(t) & \cos\omega(t)\cos\varphi(t)
\end{pmatrix}
\tag{6.3}
$$

where the rotation angles $\omega(t)$, $\varphi(t)$ and $\kappa(t)$ are defined for each CCD array at the epoch t.

A separate image coordinate system is defined for each CCD array which is related to an image reference coordinate system by a 3D transformation. It has five constants for each CCD array. They are a focal length f, a principal point offset (x_0, y_0), lens distortion and a curvature parameter Kc. The screen coordinates (i, j) are transformed to image coordinates in the following steps:

- *Step 1:* Transformation from screen coordinates to image coordinates by $x'=0$; $y'=(j-\text{column}/2)\times10e^{-3}$(mm);
- *Step 2:* CCD curvature correction by $Cx=Kc\times y'\times y'$;
- *Step 3:* Lens distortion correction is unavailable;
- *Step 4:* Final image coordinates computed by $x= cx-x_0$, $y=y'-y_0$.

To reduce the amount of calculation and further save the FPGA resources, a first-order polynomial equation is adopted to compute the six EOEs based on a short distance of two scanning lines, i.e.,

$$
\begin{cases}
X_C(t) = X_C^0 + a_0 & \varphi_C(t) = \varphi_C^0 + d_0 + d_1 t \\
Y_C(t) = Y_C^0 + b_0 & \omega_C(t) = \omega_C^0 + e_0 + e_1 t \\
Z_C(t) = Z_C^0 + c_0 & \kappa_C(t) = \kappa_C^0 + f_0 + f_1 t
\end{cases}
\tag{6.4}
$$

where $\left(X_C^0(t), Y_C^0(t), Z_C^0(t), \varphi_C^0(t), \omega_C^0(t), \kappa_C^0(t)\right)$ are initial values of EOEs at epoch t. The initial values of EOEs are provided by DLR, Germany.

In one epoch t, Equation. (6.2) is linearized by Taylor Series. Substitute Equation (6.4) into Equation (6.2) and then linearize it by Taylor Series; it yields

$$
\begin{cases}
v_x = a_{11}\Delta a_0 + a_{12}\Delta b_0 + a_{13}\Delta c_0 + a_{14}\Delta d_0 + a_{15}\Delta e_0 + a_{16}\Delta f_0 + a_{17}\Delta d_1 + a_{18}\Delta e_1 + a_{19}\Delta f_1 - l_x \\
v_y = a_{21}\Delta a_0 + a_{22}\Delta b_0 + a_{23}\Delta c_0 + a_{24}\Delta d_0 + a_{25}\Delta e_0 + a_{26}\Delta f_0 + a_{27}\Delta d_1 + a_{28}\Delta e_1 + a_{29}\Delta f_1 - l_y
\end{cases}
\tag{6.5}
$$

The vector form of Equation (6.5) is described as

$$
V_t = A_t X_t - l_t
\tag{6.6}
$$

where $V_t = \begin{bmatrix} v_x & v_y \end{bmatrix}^T$, $X_t = \begin{bmatrix} \Delta a_0 & \Delta b_0 & \Delta c_0 & \Delta d_0 & \Delta e_0 & \Delta f_0 & \Delta d_1 & \Delta e_1 & \Delta f_1 \end{bmatrix}^T$, which are the unknowns; $l_t = [l_x \quad l_y]^T$; A_t is coefficient vector, which is expressed by

$$A_t = \begin{bmatrix} a_{11} & a_{12} & a_{13} & a_{14} & a_{15} & a_{16} & a_{17} & a_{18} & a_{19} \\ a_{21} & a_{22} & a_{23} & a_{24} & a_{25} & a_{26} & a_{27} & a_{28} & a_{29} \end{bmatrix} \tag{6.7}$$

The detailed derivation of A_t can be referenced to (Li et al. 2002).

When the number of GCPs are greater than 5, the observation equation is constructed as follows.

$$V = AX - L \tag{6.8}$$

where $V = \begin{bmatrix} V_1 & V_2 & \cdots & V_n \end{bmatrix}^T$, $X = \begin{bmatrix} X_1 & X_2 & \cdots & X_n \end{bmatrix}^T$, $A = \begin{bmatrix} A_1 & A_2 & \cdots & A_n \end{bmatrix}^T$, $L = \begin{bmatrix} l_1 & l_2 & \cdots & l_n \end{bmatrix}^T$.
The Equation (6.8) is solved by least squares algorithm, and the solutions are

$$X = \left(A^T A \right)^{-1} \left(A^T L \right) \tag{6.9}$$

Solution for Equation (6.9) is carried out by an iterative process and the iteration will be ended when it meets the requirement of a given threshold.

6.3.2 FPGA-Based Implementation of Geometric Calibration

6.3.2.1 The Hardware Architecture

An FPGA-based hardware architecture for on-board geometric calibration is proposed in Figure 6.4. The FPGA architecture consists of four functional modules, Input Data module, Coefficient Calculation module, Adjustment Computation module and Comparison module. The details of the four modules are described as follows:

1. Initial data and updating data are stored in a RAM of the Input Data module. When receiving an enable signal, the data are sent to the Coefficient Calculation module at the same clock cycle.

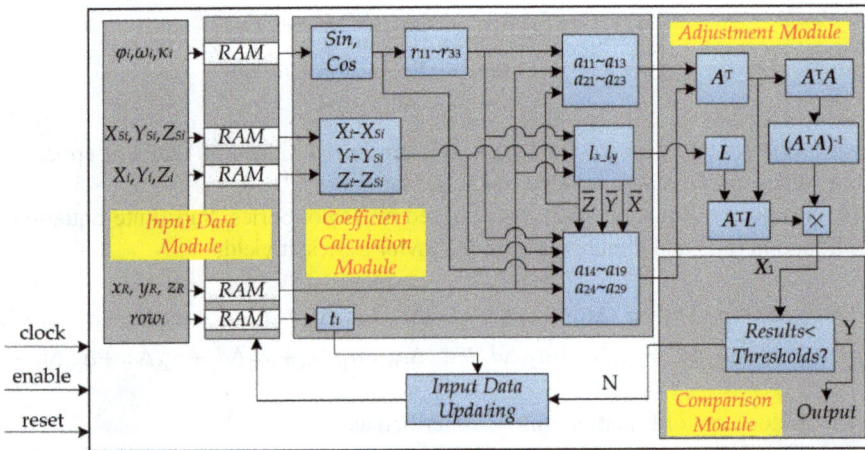

FIGURE 6.4 Hardware architecture of on-board geometric calibration.

2. The elements of matrixes A and L (expressed by Equation (6.8)) are calculated by the Coefficient Calculation module and then sent to the Adjustment Computation module at the same clock cycle.
3. The solution of matrix X in Equation (6.9) is calculated by matrixes A and L in the Adjustment Computation module.
4. If the increments of solutions X meet the requirement of a given threshold, the iteration computation is stopped and the solution a_0, b_0, c_0, d_0, e_0, f_0, d_1, e_1 and f_1 are outputted as the final results. Otherwise, the X is updated and replaced for iteration computation until the X meets the requirement of the given threshold.

6.3.2.2 FPGA Computation for Matrixes R_G^R, l_t and A_t

To compute the rotation matrix R_G^R in Equation (6.3), a parallel computational method is presented in Figure 6.5(a), where r_{11} through r_{33} are computed by the sine and cosine functions of three rotational

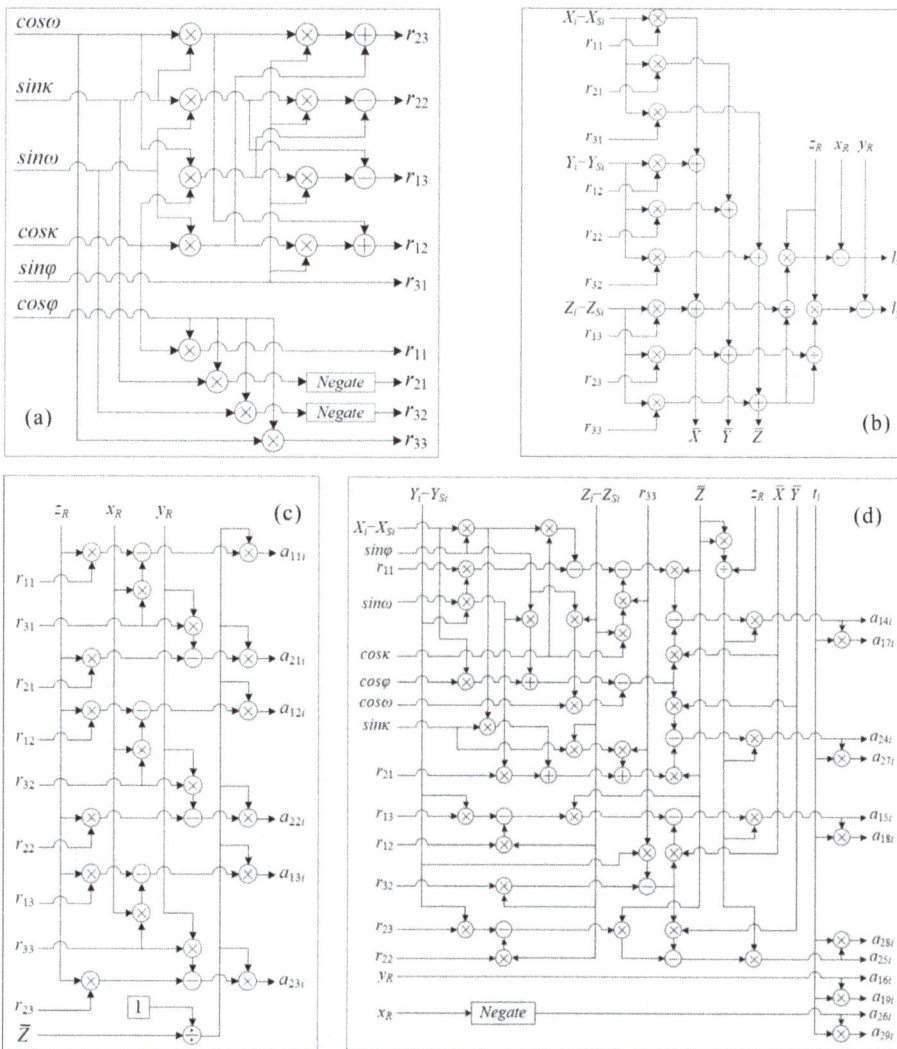

FIGURE 6.5 FPGA-based computation for coefficients; (a) is for rotation matrix R^R_G computation (i.e., r_{11} to r_{33}) in which the negate unit means the negation of value; (b) is for matrix l computation (i.e., lx and ly); (c) is for computing a_{11} through a_{13} and a_{21} through a_{23}; and (d) is for parallel computing of a_{14} through a_{19} and a_{24} through a_{29}.

angles, φ, ω and κ. The implementations of sine and cosine functions are carried out by a CORDIC IP core. To ensure all of the intermediate results are outputted at the same clock, the delay units are adopted for some intermediate results, which need less clock period for computation. This module includes 12 multipliers, 2 adders and 2 subtractors.

To compute l_t in Equation (6.6), the initial data stored in the Input Data module and rotation matrix calculated by Figure 6.5(b) are used to compute the l_x and l_y of Equation (6.6). In the computational process, the delay units are also adopted to ensure the results into the next module in the same clock cycle.

Six elements of A_t (i.e., a_{11i}, a_{12i}, a_{13i}, a_{21i}, a_{22i}, a_{23i}) are implemented using x_R, y_R, z_R and R_G^R in a parallel computing mode (see Figure 6.5(c)), where 18 multipliers, 1 divider and 6 subtractors are employed. The rest of the elements in matrix A_t (i.e., a_{14i}, a_{15i}, a_{16i}, a_{17i}, a_{18i}, a_{19i}, a_{24i}, a_{25i}, a_{26i}, a_{27i}, a_{28i}, a_{29i}) in Figure 6.5(d) are computed in a parallel mode which includes 39 multipliers, 10 subtractors, 3 adders and 1 divider.

6.3.2.3 FPGA-Based Computation for AᵀA and AᵀL

To compute the multiplication of two matrixes (i.e., $A^T A$), an optimum method through modifying the traditional equation is presented and expressed by

$$B_{9\times9} = A_{9\times2n}^T A_{2n\times9} = \begin{bmatrix} A_1 A_1 & A_1 A_2 & \cdots & A_1 A_9 \\ A_2 A_1 & A_2 A_2 & \cdots & A_2 A_9 \\ \vdots & \vdots & \ddots & \vdots \\ A_9 A_1 & A_9 A_2 & \cdots & A_9 A_9 \end{bmatrix} \tag{6.10}$$

Considering the limited resource of the FPGA and the symmetry of $A^T A$, an FPGA-based parallel computing for the upper triangular matrix is proposed and depicted in Figure 6.6(a), which includes many processing elements (PEs) units to reduce the complexity of computation (Tian et al. 2008). A PE is also defined as a multiply adder unit. All of the PE units are with the same structure, i.e., "$a_1 b_1 + a_2 b_2$", which is enlarged in Figure 6.6(b). Similarly, the computing method for $A^T L$ in Equation (6.8) is the same as that of the $A^T A$.

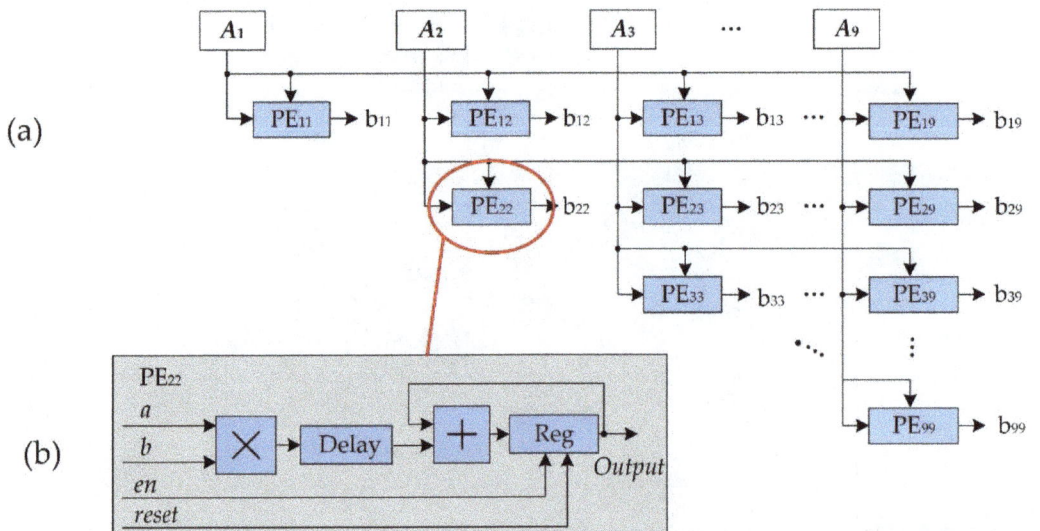

FIGURE 6.6 Parallel computing method for *ATA*.

6.3.2.4   FPGA-Based Computation for B^{-1}

To use FPGA to compute the B^{-1}, Cholesky decomposition method (Yang et al. 2012) is modified by

$$B = LDL^T \tag{6.11}$$

where L is a lower triangular matrix, D is a diagonal matrix and L^T is L's transpose. The details can be referenced to Chapter 5. The solutions of L and D are expressed by

$$
\begin{cases}
d_{11} = b_{11}, \ d_{ii} = b_{ii} - \displaystyle\sum_{k=1}^{i-1} l_{ik} d_{kk} l_{ik} & (2 \le i \le n) \\[4mm]
l_{i1} = \dfrac{b_{i1}}{d_{11}} \ (2 \le i \le n), \quad l_{ij} = \left(b_{ij} - \displaystyle\sum_{k=1}^{j-1} l_{ik} d_{kk} l_{jk} \right) \Big/ d_{jj} & (1 \le j < i \le n)
\end{cases}
\tag{6.12}
$$

In accordance with the characteristics of the LDL^T, d_{11} and l_{i1} are first calculated, then d_{ii} and l_{ij} are calculated on the basis of d_{11} and l_{i1}. In other words, the latter results are calculated on the basis of the computed former results. Since Equations (6.11) and (6.12) avoid the computation of square root and alleviate the dependency of the data, the modified equations in Equation (6.11) and Equation (6.12) are able to speed up the computation (Yang et al. 2009).

The hardware architecture of the LDL^T algorithm is presented in Figure 6.7. As observed in Figure 6.7, PE_i ($i = 1, 2, 3$ and 4) is the multiply adder units that calculates d_{ii} and l_{ij}, and 1 driver and 8 PE units are used. Thus, the computation of B^{-1} can be divided into two steps: (1) Decompose B into a LDL^T; (2) Compute the inversion of LDL^T by

$$B^{-1} = \left(LDL^T \right)^{-1} = \left(L^T \right)^{-1} (LD)^{-1} = \left(L^{-1} \right)^T D^{-1} L^{-1} \tag{6.13}$$

The FPGA-based B^{-1} computation for Equation (6.13) is depicted in Figure 6.8, where five functional modules are proposed. The computations are explained in detail as follows.

MUX module at the left hand of Figure 6.8 is to construct the column elements of B.

1. The LDL^T decomposition module is to calculate the elements of L and D.
2. The third module, consisting of two MUX modules, L^{-1} and D^{-1}, is to compute the inversion of L and D, in which *(i)* the second MUX is to construct the vector of L; *(ii)* L^{-1} is for computing the inversion of L using a systolic array architecture that is used for fast inversion of unit lower triangular matrix (El-Amawy 1989); *(iii)* the third MUX is to construct the vector of $(L^{-1})^T$; and *(iv)* D^{-1} is to construct the row vectors composed of the elements of D and calculate the reciprocal of the elements of D by a divider; D^{-1} is the outputted result.
3. $(L^{-1})^T D^{-1}$ module means that $(L^{-1})^T$ matrix multiplies with D^{-1}.
4. $(L^{-1})^T D^{-1} L^{-1}$ module denotes that $(L^{-1})^T D^{-1}$ matrix multiplies with L^{-1}.
5. B^{-1} is the outputted result through the last MUX.

6.4   VERIFICATIONS AND PERFORMANCE ANALYSIS

6.4.1   TEST AREA AND DATA

The test area used for geometric calibration of MOMS-2P in DLR, Germany is presented in Figure 6.9. The test area is comprised of Scenes 27 through Scenes 30 from southeast of Germany to about 160 km beyond the Austrian border. Its area is about 178×50 km^2. The ground pixel size of imagery is 5.9 and 17.7 m at 390 km orbit height.

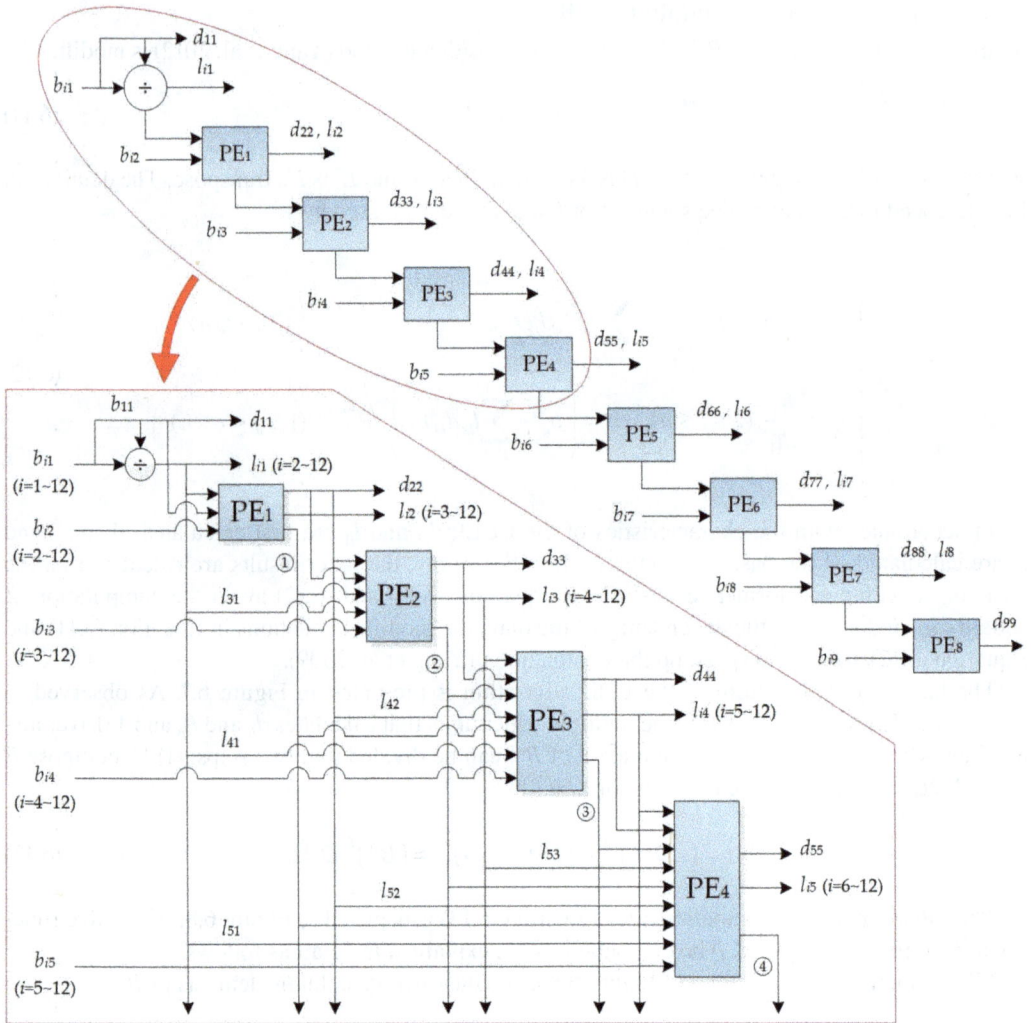

FIGURE 6.7 FPGA-based computation of B^{-1} through LDL^T decomposition, where the dimension of B at 9×9 (i.e., n = 9) is taken as an example.

Figure 6.10 depicts the geographic location of test data (reference to (Li et al. 1998) for the details). In test area (Scenes 27 through 30), the ground coordinates of 10 GCPs and 24 check points were obtained by topographic maps at a scale of 1: 50,000 with an accuracy of 1.5 m in X, Y and Z and they were in the 4th zone of Gauss-Krueger coordinate system. The navigation data of Orientation Lines (OLs), the ground coordinates of GCPs and the corresponding image coordinates were all provided by Institute of Photogrammetry at University of Stuttgart (Kornus and Lehner 1998).

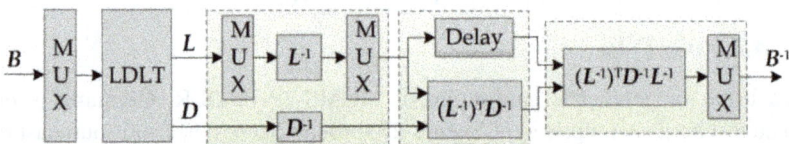

FIGURE 6.8 Hardware architecture for the FPGA-based B^{-1} computation.

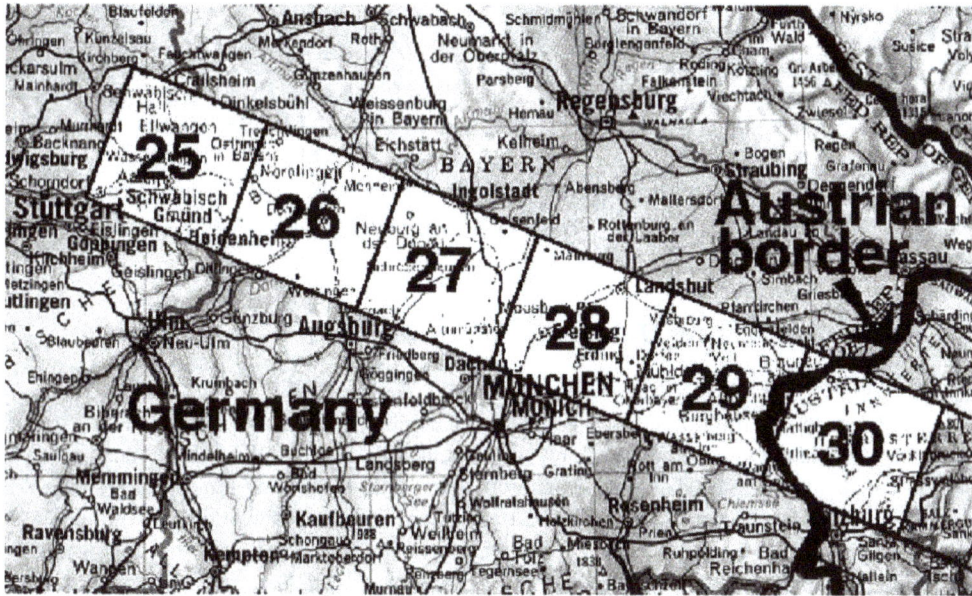

FIGURE 6.9 Geographical location of test area. (Kornus and Lehner 1998.)

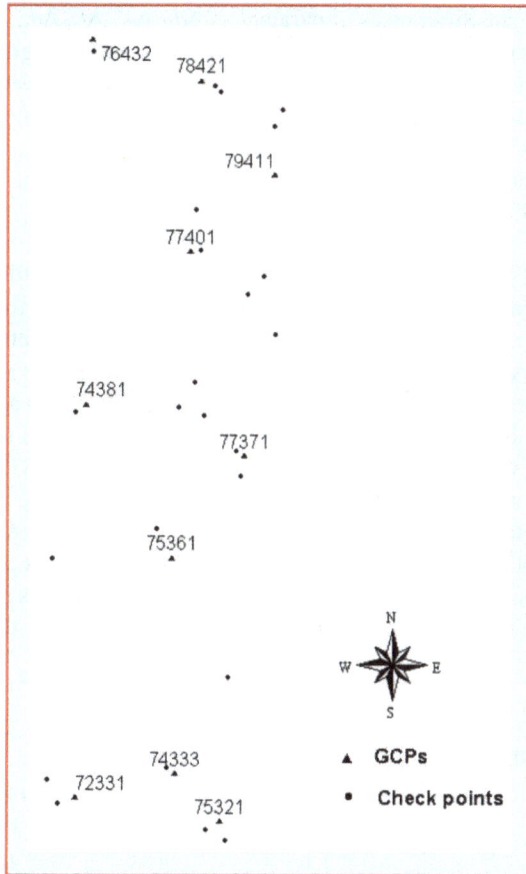

FIGURE 6.10 Distribution of 10 GCPs and check points in test field. (Li et al. 1998.)

6.4.2 HARDWARE ENVIRONMENT

A Virtex-7 FPGA VC707 board produced by Xilinx Corporation is connected to a Lenovo ThinkStation computer equipped with Intel Xeon E5645 CPU (2.4 GHz), an 8G RAM and a 64-bits Windows 7 operation system. The key chip, XC7VX485T FFG1761, has 485,760 logic cells, 507,200 CLB Flip-Flops, 2,800 DSP DSP48E1, 37,080 kb total block RAM and 700 Single-Ended I/O. A Vivado software (2014.2 version) and a System Generator software (2014.2 version) developed by Xilinx Corporation are adopted as the development tools and the hardware description language is Verilog HDL.

6.4.3 DETERMINATION OF THE DATA WIDTH OF FLOATING-POINT

6.4.3.1 Relationship between the Data Width and the Accuracy

Because of the resource limitation of the selected FPGA chip, this chapter wants to study the relationship between data width of floating-point vs. the accuracy of geometric calibration. The experiment was conducted using 6 GCPs under different data width of floating-point (i.e., 44-bit, 48-bit, 50-bit, 54-bit and 64-bit). According to the IEEE standard 754, a data width of floating-point consists of the sign part, exponent part and fractional part. The five types of floating-point data are described in Table 6.2.

The results computed using both FPGA and PC computer with different data width are presented in Table 6.3 As seen in Table 6.3, when the data width increases from 44-bit to 64-bit, the maximum differences of parameters, Δb_0 and Δc_0 decrease from 2.8684 to 5.8651E-04 for Δb_0, and from 0.9676 to 0.0021 for Δc_0. The differences of parameters Δd_0, Δe_0, Δf_0, Δd_1, Δe_1 and Δf_1 are approximately 1E-7 in 44-bit decrease to 1E-10 in 64-bit, which is small enough so that we can neglect them in computation. When the data width reaches 64-bit, the FPGA-based computational results of 9 parameters are almost exactly the same as that of the PC-based computation (data width with 64-bit). With the analysis above, it can be concluded that FPGA-based computational results for geometric calibration parameters will not lose the accuracy relative to the results of the PC-based computations (Li et al. 1998; Li et al. 2002).

6.4.3.2 Relationship between the Data Width and the Consumption of FPGA Resources

As concluded above, when applying a larger data width, the accuracy of the calibrated parameters can reach higher. However, it is very necessary to know the relationship between the data width and the consumption of FPGA resources. Here, the five different data widths of floating-point and the consumptions of FPGA resources, including FF, LUT, Memory LUT, I/O and DSP48 are analyzed. The results are displayed in Figure 6.11. As seen from Figure 6.11, when the data width increases from 44-bit to 54-bit, the utilization of FF increases from 14.98% to 16.37%, LUT from 61.31% to 75.33%, Memory LUT from 9.32% to 10.86%, I/O from 33.57% to 40.71% and DSP48 from 55.71% to 92.86%. It can also be found that the DSP48 unit consumption increases from 6 to 10 when data width changes from 50-bit to 54-bit. This means that the DSP48 resources consumes very fact when data width increases. When the data width reaches 64-bit, the utilizations of FPGA resources fails to computation operation, which means the usage of DSP48 reaches over 100%. It can therefore be

---

TABLE 6.2

Format of Floating-Point

	44-bit	48-bit	50-bit	54-bit	64-bit
Sign part	1	1	1	1	1
Exponent part	8	8	9	9	11
Fractional part	35	39	40	44	52

TABLE 6.3

Differences of the Results Computed by FPGA-Based and PC-Based with Various Data Width

|PC-FPGA|	44-bit	48-bit	50-bit	54-bit	64-bit
Δa_0	0.1026	0.0408	0.0050	4.3282E-04	1.8679E-05
Δb_0	2.8684	0.1184	0.0418	0.0019	5.8651E-04
Δc_0	0.9676	0.0301	0.0043	0.0018	0.0021
Δd_0	2.018E-07	1.995E-08	2.281E-09	9.935E-10	8.365E-10
Δe_0	3.359E-07	1.031E-08	1.154E-09	5.487E-11	7.704E-11
Δf_0	7.660E-07	2.798E-08	3.685E-09	4.254E-10	1.630E-10
Δd_1	1.316E-08	6.464E-10	1.685E-10	1.819E-12	5.190E-12
Δe_1	9.478E-11	7.295E-11	7.087E-12	1.687E-12	3.684E-13
Δf_1	2.776E-08	1.912E-09	1.833E-10	9.606E-12	8.882E-14

concluded that the data width with 54-bit, not 64-bit, is recommended for the FPGA given in this chapter to ensure the redundancy of FPGA resources.

6.4.3.3 Relationship between the Data Width and the Computational Time

Considering that computational time is one of the most important indexes in FPGA-based on-board implementation, this section discusses the relationship between the data width and the computational time using the five types of floating-point data. For a purpose of comparison, the results computed with the same algorithm and with PC-based MATLAB software (R2014a version) are taken a reference. The compared results are presented in Table 6.4.

As observed from Table 6.4, the computational time at 44-bit, 48-bit, 50-bit and 54-bit can reach a 217-clock period (approximately 0.01736 ms under the clock frequency of 12.5 MHz) when using FPGA, and the speedup of FPGA-based implementation is 6,347 times (approximately 6000 times) when compared with PC-based implementation. The computational time at data width with 64-bit fails to calculate since the utilization of the DSP48 unit is greater than 100% (see Figure 6.11). As a result, the data width with 64-bit is not recommended. With the analysis above, it can be concluded that the FPGA-based computational time is capable of increasing approximately 6000 times than PC-based one for all of the data widths.

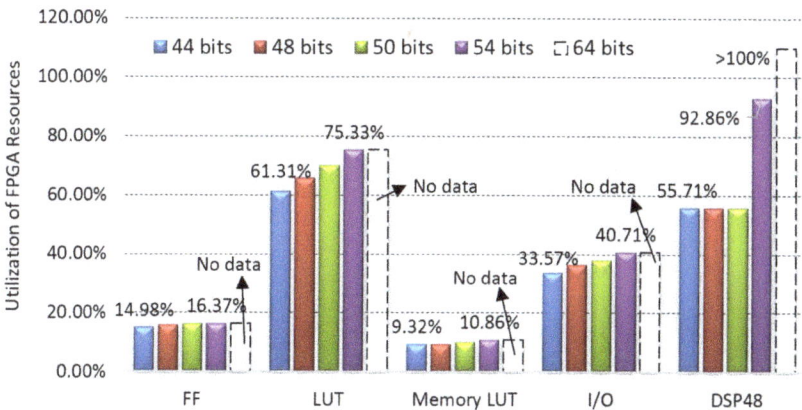

FIGURE 6.11 Comparison analysis of the utilizations of several types of FPGA resources vs. different data width.

TABLE 6.4

Comparison of FPGA-Based and PC-Based Computational Time

FPGA-Based Implementation		PC-Based Implementation (64-bit)	Speedup
44-bit	0.01736 (ms)	110.187 (ms)	6,347 (times)
48-bit			
50-bit	(217 clock, 12.5 MHz)		
54-bit			
64-bit	Fail to cal.		Fail to cal.

With the results from Table 6.3, Table 6.4 and Figure 6.11, it can be concluded that more than 90% of the DSP48 unit is consumed when the data width is selected as either 54-bit or 64-bit. With considering the redundancy of FPGA resources, a data width of floating point with 50-bit or 48-bit is recommended for FPGA-based on-board geometric calibration. Thereby, the data width with 50-bit is recommended for the FPGA selected in this chapter.

6.4.4 ANALYSIS OF THE OPTIMUM NUMBER OF GCPs FOR ON-BOARD CALIBRATION

6.4.4.1 Relationship between the Number of GCPs and the Accuracy

In general, the more number of GCPs, the higher accuracy of geometric calibration. However, in a FPGA-based implementation of on-board calibration, many GCPs will add the burden of computational time and the consumption of FPGA resources. Thereby, it is necessary to optimize the number of GCPs for on-board geometric calibration. With the data width of floating-point as 50-bit, the different numbers of GCPs (i.e., 5, 6, 8 and 10) are selected for FPGA-based computation. For the purpose of comparison, PC-based MATLAB software (R2014a version) is applied for the same GCPs. The differences of calibration parameters between the FPGA-based and PC-based computations are shown in Table 6.5. As seen in Table 6.5, with increasing number of GCPs, the differences of calibration parameters become smaller and smaller. The maximum difference reaches 0.0148 m for a_0, when the number of GCPs is 5; even decreases to 0.0075 m when the number of GCPs reaches 10. The differences of the other parameters are similar (e.g., see Table 6.5 for b_0, c_0, d_0, e_0, f_0, d_1, e_1 and f_1).

TABLE 6.5

Differences of Calibration between FPGA-Based and PC-Based Calculations

|PC-FPGA|	5 GCPs	6 GCPs	8 GCPs	10 GCPs
Δa_0	0.01481804	0.00502983	0.00648983	0.00749144
Δb_0	1.12103E-05	0.04178977	0.00766325	0.00091729
Δc_0	0.00077522	0.00427705	0.00244571	0.00394770
Δd_0	3.3568E-09	2.28135E-09	2.19579E-10	2.72209E-10
Δe_0	2.99778E-10	1.15354E-09	1.19552E-09	5.38698E-10
Δf_0	1.5894E-09	3.68499E-09	7.32229E-10	2.71596E-09
Δd_1	5.21121E-11	1.68484E-10	1.68234E-11	3.37348E-11
Δe_1	1.39511E-11	7.08677E-12	1.05921E-11	1.29294E-11
Δf_1	9.92273E-12	1.8332E-10	9.54881E-11	1.29121E-10

TABLE 6.6
The Utilizations of FPGA Resources under Different Numbers of GCPs

GCPs	BRAM (%)	DSP48 (%)	BUFG (%)	FF (%)	LUT (%)	Memory LUT (%)
5	1.31	55.71	3.12	16.14	69.67	9.46
6	1.31	55.71	3.12	16.37	70.04	10.09
8	1.31	55.71	3.12	16.96	70.63	11.48
10	1.31	55.71	3.12	17.58	71.78	12.92

6.4.4.2 Relationship between the Number of GCPs and the Consumption of FPGA Resources

In order to select an optimum number of GCPs, a relationship between the number of GCPs and the consumption of the FPGA resources is analyzed in this section. With the selected data wide of 50 bit, 5 GCPs, 6 GCPs, 8 GCPs and 10 GCPs are selected for experiments to test the consumption of FPGA resources. The results are displayed in Table 6.6. As observed from Table 6.6, when the number of GCPs varies from 5 to 10, the consumptions of the BRAM, DSP48 and BUFG remain the same, but the consumptions of FF, LUT and Memory LUT increase a little bit. As a result, it can be concluded that the increasing number of GCPs will not significantly consume the FPGA resources.

6.4.4.3 Relationship between the Number of GCPs and the Computational Time

From Vivado software's instruction, the clock periods span from 75.94 ns, 76.76 ns, 75.34 ns to 75.82 ns. To ensure enough clock delay when executing a floating-point calculation, the clock period is set at 80 ns (the corresponding clock frequency is 12.5 MHz) in this group of experiments. The computational time under the different number of GCPs is listed in Table 6.7. As observed in Table 6.7, the computational time changes from 0.0170 ms to 0.0186 ms when the number of GCPs increases from 5 to 10, while the computational time from PC-based computation increases from 109.8631 ms to 110.8381 ms. In other words, the maximum speedup of the computational time reaches approximately 6,447 times when the number of GCPs is 5; the minimum speedup is approximately 5,946 times when the number of GCPs increases to 10. Thereby, approximately 6000 times speedup can be reached when applying FPGA-based implementation relative to the PC-based implementation.

6.4.5 ACCURACY COMPARISON BETWEEN FPGA-BASED AND INFLIGHT-BASED COMPUTATION

To compare the accuracy between FPGA-based and inflight-based implementations of calibration, the calibration model, associated with the numbers of GCPs, and other conditions, are the same. The computed results are presented in Table 6.8. As shown in Table 6.8, accuracy of 11 m for x-coordinate, 8 m for y-coordinate and 11 m for z-coordinate can be reached. The maximal

TABLE 6.7
The FPGA-Based Computational Time under Different Numbers of GCPs

GCPs	By FPGA (ms)	By PC (ms)	Speedup
5	0.0170 (213 clock, 12.5 MHz)	109.8631	6,447
6	0.0174 (217 clock, 12.5 MHz)	110.1870	6,347
8	0.0180 (225 clock, 12.5 MHz)	110.5228	6,140
10	0.0186 (233 clock, 12.5 MHz)	110.8381	5,946

TABLE 6.8

Accuracy of Ground Coordinates Calculated by Inflight-Based and FPGA-Based Computation

	GCP	Check Points	σ_0 (μm)	σ_x (m)	σ_y (m)	σ_z (m)	RMS_x (m)	RMS_y (m)	RMS_z (m)
Inflight-based (González et al. 2012)	10	24	6.990	7.140	5.324	6.010	11.144	8.276	10.758
FPGA-based	10	24	6.990	7.287	5.493	6.158	11.308	8.466	10.872
Δ	0	0	0	0.147	0.169	0.148	0.164	0.190	0.114

difference of RMS between inflight-based and FPGA-based computation is 0.19 m for y-coordinate (i.e., RMS_Y) and the minimal difference of RMS is 0.114 m for z-coordinate (i.e., RMS_Z).

6.5   CONCLUSIONS

This chapter first develops an FPGA-based implementation for on-board geometric calibration method for a linear array CCD sensor. A Xilinx Virtex-7 FPGA VC707 board is selected as hardware and the experimental data from DLR, Germany, is employed to validate the method. The main contributions of this chapter are as follows:

1. An FPGA-based on-board geometric calibration computation is designed and implemented.
2. FPGA-based parallel computing for coefficient matrixes building (e.g., matrix A and matrix L), matrix multiplication (e.g., A^TA and A^TL), matrix decomposition (e.g., $B=LDL^T$), and matrix inversion (e.g., B^{-1}) are developed, which are demonstrated to save large amounts of the FPGA resources.
3. Different experiments are designed to validate the characteristics of the proposed method. From the experimental results, a few conclusions can be drawn up as follows:
   a. With increasing data width, FPGA resources consume increasingly. For example, the utilization of DSP48 unit suddenly increase from 55.71% to 92.86% (near 100%), this fact demonstrates that the data width of 64-bit is impropriate to be set in the selected FPGA chip duo to the limitation of the selected FPGA chip.
   b. The computing time executed by FPGA reaches as approximately 6,447 times fast as that executed by PC computer when the number of GCPs is 5; and approximately 5,946 times fast when the number of GCPs increases to 10. It can therefore be concluded that the computing time executed by FPGA is up to approximately 6000 times faster than that executed by PC-based MATLAB software.
   c. More than 90% of the DSP48 unit is consumed when the data width of 54-bit or 64-bit is selected. Considering the limitation of FPGA resources, a data width of 50-bit floating point is recommended for FPGA-based on-board geometric calibration.
   d. When the data width of floating-point is fixed at 50-bit and the number of GCPs varies from 5 to 10, the utilizations of the FPGA's BRAM, DSP48, and BUFG remain unchanged, the utilizations of FF, LUT and Memory LUT slightly increase and the computational time increase a little bit. It can therefore be concluded that increasing the number of GCPs will not significantly increase the consumption of the FPGA resources and the computational time.
4. Because two bundle adjustment models are the same, except the data width and the orders of polynomial equation, the difference between inflight-based and FPGA-based computations is very close.

REFERENCES

Arnaud, M., Boissin, B., Perret, L., Boussarie, E., Gleyzes, A., The Pleiades optical high-resolution program, *Proceeding of the 57th IAC/IAF/IAA, Valencia: International Astronautical Congress*: IAC-06-B1.1.04, 2006.

Bouillon, A., Gigord, P., SPOT 5 HRS location performance tuning and monitoring principles, the international archives of photogrammetry, *Remote Sensing and Spatial Information Sciences*, 2004, 35(B1): 379–384.

Cao, J., Yuan, X., Gong, J., In-orbit geometric calibration and validation of zy-3 three-line cameras based on CCD-detector look angles, *The Photogrammetric Record*, 2015, 30(150): 211–226.

Cao, J., Yuan, X., Gong, J., Duan, M., The look-angle calibration method for on-orbit geometric calibration of zy-3 satellite imaging sensors, *Acta Geodaetica Et Cartographica Sinica*, 2014, 43(10): 1039–1045.

Chen, G., Tu, L., A., Stereo camera calibration based on robotic vision, *Cognitive Informatics & Cognitive Computing, 2011 10th IEEE International Conference on*, 18–20 Aug. 2011.

Crespi, M., Colosimo, G., Vendictis, L.D., et.al., GeoEye-1: Analysis of radiometric and geometric capability, *Lecture Notes of the Institute for Computer Sciences, Social Informatics and Telecommunications Engineering*, 2010, 43(7): 354–369.

David, M., Preparations for the on-orbit geometric calibration of the orbview-3 and 4 satellites, *International Archives of Photogrammetry and Remote Sensing*, Amsterdam, 2000, XXXIII, Part B1.

Ebner, H., Ohlhof, T., Putz, E., Orientation of MOMS-02/D2 and MOMS-2P imagery, *International Archives of Photogrammetry and Remote Sensing*, 1996, 31: 158–164.

El-Amawy, A., A systolic architecture for fast dense matrix inversion, *IEEE Transactions on Computers*, 1989, 38(3): 449–455.

Flavio, J. P., Junior, J.Z., Lamparelli, R.A., In-flight absolution calibration of the cbers-2 CCD sensor data, *Anais Da Academia Brasileira De Ciências*, 2008, 80(2): 373–380.

González, C., Mozos, D., Resano, J., Plaza, A., FPGA implementation of the n-findr algorithm for remotely sensed hyperspectral image analysis, *IEEE Transactions on Geoscience and Remote Sensing*, 2012, 50(2): 374–388.

Gruen, A., Kocaman, S., Wolff, K., Calibration and validation of early ALOS/PRISM images, *The Journal of the Japan Society of Photogrammetry and Remote Sensing*, 2007, 46(1): 24–38.

Gupta, R., Hartley, R.I., Linear pushbroom cameras, *IEEE Transactions on Pattern Analysis and Machine Intelligence*, 1997, 19(9): 963–975.

Healey, G.E., Kondepudy, R., Radiometric CCD camera calibration and noise estimate, *IEEE Transactions on Pattern Analysis and Machine Intelligence*, 1994, 16(3): 267–276.

Heikkila, J., Geometric camera calibration using circular control points, *IEEE Transactions on Pattern Analysis and Machine Intelligence*, 2000, 22(10): 1066–1077.

Hihara, H., Moritani, K., Inoue, M., Hoshi, Y., Iwasaki, A., Takada, J., et al. Onboard image processing system for hyperspectral sensor, *Sensors*, 2015, 15(10): 24926–24944.

Honkavaara, E., In-flight camera calibration for direct georeferencing. *International Archives of Photogrammetry, Remote Sensing and Spatial Information Sciences*, 2004, 35(B1): 166–172.

Huang, J., Zhou, G., On-board detection and matching of feature points, *Remote Sensing*, 2017, 9(6): 601.

Grodecki, J., and James L., IKONOS geometric calibrations, *Presented at ASPRS 2005*, Baltimore, Maryland, 2005, pp. 7–11.

Jacobsen, K., Calibration of optical satellite sensors, *In International Calibration and Orientation Workshop EuroCOW*, International Archives of Photogrammetry, Remote Sensing and Spatial Information Sciences (ISPRS), Vol. XXXVII, Part B1, 2006.

Jacobsen, K., Issues and method for in-flight and on-orbit calibration, *Workshop on Radiometric and Geometric Calibration*, CRC Press, Gulfport, Mississippi USA 2003.

Jacobsen, K., Geometry of satellite images-calibration and mathematical models, *Korean Society of Remote Sensing, ISPRS International Conference*, Korea, Jeju, pp. 182–185, 2005.

Kornus, W., Lehner, M., Photogrammetric point determination and dem generation using moms-2p/priroda three-line imagery, *International Archives of Photogrammetry and Remote Sensing*, Stuttgart, Germany, 1998, vol. 32, Part 4, pp. 321–328.

Kornus, W., Lehner, M., Schroeder, M., Geometric inflight calibration of the stereoscopic CCD-linescanner MOMS-2P, *ISPRS Commission I Symposium,* Bangalore, IntArchPhRS, 1998, vol XXXII-1, pp. 148–155.

Kornus, W., Lehner, M., Schroeder, M., Geometric inflight-calibration by block adjustment using moms-2p-imagery of three intersecting stereo-strips, *ISPRS Workshop on Sensors and Mapping from Space 1999*, Hanover, 1999, pp. 27–30.

Kuwahara, T., Bohringer, F., Falke, A., Eickhoff, J., Huber, F., Roser, H.P., FPGA-based operational concept and payload data processing for the flying laptop satellite, *Acta Astronautica*, 2009, 65(s11–12): 1616–1627.

Lei, R., Study on theory and algorithm of the in-flight geometric calibration of spaceborne linear array sensor, Ph.D. Dissertation, PLA Information Engineering University, Zhengzhou, China, 2011.

Lenz, R.K., Tsai, R.Y., Techniques for calibration of the scale factor and image center for high accuracy 3D machine vision metrology, *IEEE Transactions on Pattern Analysis and Machine Intelligence*, 2009, 10(5): 713–720.

Li, R., Zhou, G., Schmidt, N.J., Schmidt, N.J., Fowler, C., Tuell, G., Photogrammetric processing of high-resolution airborne and satellite linear array stereo images for mapping applications, *International Journal of Remote Sensing*, 2002, 23(20): 4451–4473.

Li, D. China's first civilian tree-line-array stereo mapping satellite: Zy-3, *Acta Geodaetica Et Cartographica Sinica*, 2012, 41(3): 317–322.

Li, D., Wang, M., On-orbit geometric calibration and accuracy assessment of ZY-3, *Spacecraft Recovery & Remote Sensing*, 2012, 33(3): 1–6.

Li, J., Wang, R., Zhu, L., Haile, H., In-flight geometric calibration for mapping satellite-1 surveying and mapping camera, *Journal of Remote Sensing*, 2012, 16(Supplement): 35–39.

Li, R., Zhou, G., Photogrammetric processing of high-resolution airborne and satellite linear array stereo images for mapping applications, *International Journal of Remote Sensing*, 2002, 23(20): 4451–4473. doi: 10.1080/01431160110107662

Li, R., Zhou, G., Felus, Y., Coastline mapping and change detection using one-meter resolution satellite imagery, *Project Report submitted to Sea Grant/NOAA*, 1998.

Pingree, P.J., Advancing NASA's Onboard processing capabilities with reconfigurable FPGA technologies, *Parallel & Distributed Processing, Workshops and PhD Forum (IPDPSW), 2010 IEEE International Symposium on*, Atlanta, GA, 2010.

Poli, D., Angiuli, E., Remondino, F., Radiometric and geometric analysis of WorldView-2 stereo scenes, *ISPRS, Commission I*, WG I/4, 2010.

Radhadevi, P., Solanki, S., In-flight geometric calibration of different cameras of IRS-P6 using a physical sensor model, *The Photogrammetric Record*, 2008, 23(121): 69–89.

Robertson, B., Beckett, K., Rampersad, C., Putih, R., Quantitative geometric calibration & validation of the rapid eye constellation, *IEEE International Geoscience & Remote Sensing Symposium, IGARSS 2009, July 12–17, 2009*, University of Cape Town, Cape Town, South Africa, Proceedings, 2009, pp. 192–195.

Srivastava, P., Alurkar, M.S., Inflight calibration of IRS-1C imaging geometry for data products, *ISPRS Journal of Photogrammetry & Remote Sensing*, 1997, 52(5): 215–221.

Strecha, C., Hansen, V.W., Gool, L.V., Fua, P., Thoennessen, U., On benchmarking camera calibration and multi-view stereo for high resolution imagery, *Computer Vision and Pattern Recognition*, CVPR IEEE Conference on 2008, 23–28 June.

Tian, X., Zhou, F., Chen, Y.W., Liu, L., Chen, Y., Design of field programmable gate array based real-time double-precision floating matrix multiplier, *Journal of Zhejiang University (Engineering Science)*, 2008, 42(9): 1611–1615.

Valorge, C., et al.40 years of experience with spot in-flight calibration, *International Workshop on Radiometric and Geometric Calibration*, CRC Press, Gulfport, Mississippi, USA, 2003, pp. 2–5.

Vladimirova, T., ChipSat – A system-on-a-chip for small satellite data processing and control architectural study and FPGA implementation, *Microelectronics Presentation, ESTEC*, ESA, Noordwijk, The Netherlands, 2004, pp. 4–5.

Wang, C., Cheng, T., Wang, H., Zha, X., Qin, X., On-orbit geometric calibration for mapping satellite-1, *Hydrographic Surveying and Charting*, 2016, 36(4): 31–34.

Wang, M., Yang, B., Hu, F., Zang, X., On-orbit geometric calibration model and its applications for high-resolution optical satellite imagery, *Remote Sensing*, 2014, 6(5): 4391–4408.

Westin, T., Inflight calibration of SPOT CCD detector geometry, *Photogrammetric Engineering and Remote Sensing*, 1992, 58(9): 1313–1319.

Williams, J.A., Dawood, A.S., Visser, S.J., FPGA-based cloud detection for real-time onboard remote sensing, *In Field-Programmable Technology, IEEE International Conference on* 2002, pp. 110–116, Dec.

Winfried, H., Thematic data processing on board the satellite BIRD, *Proceeding of SPIE*–2001, 4540, pp. 412–419.

Yang, D., Peterson, G., Li, H., Compressed sensing and Cholesky decomposition on FPGAs and GPUs, *Parallel Computing*, 2012, 38(8): 421–437.

Yang, B., Wang, M., On-orbit geometric calibration method of ZY1-02C panchromatic camera, *Journal of Remote Sensing*, 2013, 17(5): 1175–1190.

Yang, D., Peterson, G.D., Li, H., Sun, J., An FPGA Implementation for solving least square problem, *The 17th IEEE Symposium on Field Programmable Custom Computing Machines (FCCM)*, Napa, California, April, 2009.

Zhang, Z., A flexible new technique for camera calibration, *IEEE Transactions on Pattern Analysis and Machine Intelligence*, 2000, 22(11): 1330–1334.

Zhang, G., Jiang, Y., Li, D., Huang, W., Pan, H., Tang, X., Zhu, X., In-orbit geometric calibration and validation of ZY-3 linear array sensors, *Photogrammetric Record*, 2014, 29(145): 68–88.

Zhou, G., *Urban High-Resolution Remote Sensing: Algorithms and Methods*, Taylor & Francis/CRC Press, 2020, ISBN: 9780367857509, 468 pages.

Zhou, G., Li, R., Accuracy evaluation of ground points from high-resolution satellite imagery IKONOS, *Photogrammetry Engineering & Remote Sensing*, 2000, 66(9): 1103–1112.

Zhou G., Jezek, K., Wright, W., Rand, J., Granger, J., Ortho-rectification of 1960s satellite photographs covering Greenland. *IEEE Transactions Geoscience Remote Sensing*, 2002, 40, 1247–1259.

Zhou, G., Baysal, O., Kaye, J., Concept design of future intelligent earth observing satellites, *International Journal of Remote Sensing*, 2004, 25(14): 2667–2685.

Zhou, G., Jiang, Linjun, Huang, Jingjin, Zhang, Rongting, Liu, D., Zhou, X., Baysal, O. FPGA-based on-board geometric calibration for linear CCD array sensors. *Sensors*, 2018, 18: 1794. doi:10.3390/s18061794.

7 On-Board Georeferencing Using Optimized Second-Order Polynomial Equation

7.1 INTRODUCTION

Georeferencing is to establish the relationship between image coordinates and ground coordinates through various functions, such as collinearity equation models, polynomial function models, rational function models and direct linear transformation models (Zhou et al. 2005; Zhou 2010; Ziboon and Mohammed 2013; Kartal et al. 2017; Wang et al. 2014; Chen et al. 2010). However, these traditional algorithms were developed for serial instruction systems based on personal computers (PC). As a result, these methods are very difficult to meet the response time demands of time-critical disasters ((Zhou 2009; Gill et al. 2010; González et al. 2016; Zhou et al. 2017; Lee et al. 2011; Dawood et al. 2002; Zhang and Kerle 2008; Qi et al. 2018; Zhou 2020; Zhou et al. 2018; Huang 2019).

To figure out the processing speed of the remotely sensed images, many researchers proposed effective alternative processing methods. High-performance computing (HPC) was widely used in remotely sensed images processing system. Cluster computing has already offered access to greatly increased computational power at a low cost in a few hyperspectral imaging applications (Plaza et al. 2006; 2011). Although the algorithms of remotely sensed images processing generally map quite nicely to multi-processor systems composed of clusters or networks of CPUs, these systems are usually expensive and difficult to adapt to the on-board remotely sensed images processing scenarios. For these reasons above, the specialized integrated hardware devices of low weight and low power consumption are essential to reduce mission payload and obtain analysis results in a real-time mode. Field-programmable gate arrays (FPGAs) and graphics processing units (GPUs) exhibit a good potential to allow for on-board real-time analysis of remotely sensed images. Fang et al. (2014) presented near real-time approach for CPU/GPU based preprocessing of ZY-3 satellite images. Van der Jeught et al. (2012) proposed a new method to generate undistorted images by implementing the required distortion correction algorithm on a commercial GPU. Thomas et al. (2011) described the georeferencing of an airborne hyperspectral imaging system based on pushbroom scanning. Reguera-Salgado et al. (2012) presented a method for real-time geo-correction of images from airborne pushbroom sensors using the hardware acceleration and parallel computing characteristics of modern GPU. López-Fandiño et al. (2015) proposed a parallel function with GPU to accelerate the extreme learning machine (ELM) algorithm which is performed on remotely sensed images for land cover applications and achieved competitive accuracy results. Lu et al. (2008) mapped the fusion method of remotely sensed images to GPU. The result shown that the image fusion speed based on GPU was much quicker than that based on CPU when the image size was getting bigger. Now, the use of GPUs to accelerate remotely sensed images processing has resulted in many related research achievements. However, most parallel processing methods based on GPU multitasking are not alone capable of surmounting the shortcomings of serial instrument methods (Zhu et al. 2012; Ma et al. 2016). Additionally, FPGAs exhibit lower power dissipation figures than GPUs (Lopez et al. 2013). FPGAs have been consolidated as the standard choice for on-board remotely sensed images processing due to their smaller size, weight and power consumption when compared to other HPC systems (Zhou et al. 2004; Lopez et al. 2013; Huang et al. 2018a; Pakartipangi et al.2015; Yu et al. 2009; Huang et al. 2018b). Zhou et al. (2004) presented the concept of the "on-board processing system". Huang et al. (2018a) proposed an FPGA architecture that consisted of corner detection, corner matching, outlier rejection and sub-pixel precision localization. Pakartipangi et al. (2015) analyzed the camera array on-board

DOI: 10.1201/9781003319634-7

data handling using an FPGA for nano-satellite application. Huang et al. (2018b) proposed a new FPGA architecture that considered the reuse of sub-image data. Yu et al. (2009) presented a new on-board image compression system architecture for future disaster monitoring by LEO satellites. Zhou et al. (2017) proposed an on-board ortho-rectification for remote sensing images based on an FPGA. Qi et al. (2018) presented an FPGA and DSP co-processing system for an optical RS image preprocessing algorithm. Long et al. (2016) focused on an automatic matching technique for the specific task of georeferencing RS images and presented a technical frame to match large remotely sensed images efficiently using the prior geometric information of the images. Williams et al. (2002) discussed the design and implementation of a real-time cloud detection system for on-board remote sensing platform. González et al. (2012) presented an N-FINDR algorithm implementation using FPGA for hyperspectral image analysis.

This chapter presents an on-board georeferencing method using FPGA for remote sensing image. By decomposing the georeferencing algorithm, the proposed FPGA-based method consists of three modules: a data memory, coordinate transformation (including the transformation from geodetic coordinates to the raw image coordinates and the raw image coordinates to the scanning coordinates) and bilinear interpolation.

7.2 OPTIMIZATION FOR ON-BOARD GEOREFERENCING

The georeferencing method can be classified into direct and indirect methods (Swann et al. 1988; Jensen and Lulla 2007; Savoy et al. 2016). The direct method applies the raw image coordinates to compute the georeferenced coordinates, and the indirect method applies the georeferenced image coordinates to compute the raw image coordinates. In this chapter, the indirect method is adopted with a second-order polynomial equation.

7.2.1 A Brief Review of Georeferencing Algorithm

7.2.1.1 Traditional Second-Order Polynomial Equations

The second-order polynomial equations are expressed by (Bannari et al. 1995; Toutin 2004; Chen et al. 2006; Raffa et al. 2016):

$$x = a_0 + a_1 X + a_2 Y + a_3 X^2 + a_4 XY + a_5 Y^2 \tag{7.1}$$

$$y = b_0 + b_1 X + b_2 Y + b_3 X^2 + b_4 XY + b_5 Y^2 \tag{7.2}$$

where (x, y) are the coordinates of the raw image, (X, Y) are the corresponding ground coordinates (longitude, latitude), and a_i and b_i (i=0, 1, ..., 5) are the unknown coefficients of the second-order polynomial equation. After choosing a suitable mathematical distortion model, the unknown coefficients a_i and b_i can be obtained from Equations (7.1), (7.2) with the ground control points (GCPs). GCPs are an important parameter in the geometric calibration, which affect the accuracy of subsequent correction (Zhou and Li 2000). To ensure accuracy, many GCPs, such as ten GCPs, are selected in the study area. Let the coordinate pairs of ten GCPs are represented as: (x_1, y_1, X_1, Y_1), (x_2, y_2, X_2, Y_2), (x_3, y_3, X_3, Y_3),..., $(x_{10}, y_{10}, X_{10}, Y_{10})$. The matrix form of Equation (7.1) can be expressed as:

$$\begin{bmatrix} 1 & X_1 & Y_1 & X_1^2 & X_1Y_1 & Y_1^2 \\ 1 & X_2 & Y_2 & X_2^2 & X_2Y_2 & Y_2^2 \\ 1 & X_3 & Y_3 & X_3^2 & X_3Y_3 & Y_3^2 \\ 1 & X_4 & Y_4 & X_4^2 & X_4Y_4 & Y_4^2 \\ \vdots & \vdots & \vdots & \vdots & \vdots & \vdots \\ 1 & X_{10} & Y_{10} & X_{10}^2 & X_{10}Y_{10} & Y_{10}^2 \end{bmatrix} * \begin{bmatrix} a_0 \\ a_1 \\ a_2 \\ a_3 \\ a_4 \\ a_5 \end{bmatrix} = \begin{bmatrix} x_1 \\ x_2 \\ x_3 \\ x_4 \\ \vdots \\ x_{10} \end{bmatrix} \tag{7.3}$$

According to the least square method (Zhou et al. 2017), Equation (7.3) can be simplified as

$$(A^T PA)\Delta_a = A^T PL_x, \tag{7.4}$$

where A is the matrix of the ground coordinates of GCPs in the georeferenced image, Δ_a is the coefficients matrix of the second-order polynomial Equation (7.1), L_x is the coordinates matrix of GCPs in the raw image, and P is the weight matrix.

$$A = \begin{bmatrix} 1 & X_1 & Y_1 & X_1^2 & X_1Y_1 & Y_2^2 \\ 1 & X_2 & Y_2 & X_2^2 & X_2Y_2 & Y_2^2 \\ 1 & X_3 & Y_3 & X_3^2 & X_3Y_3 & Y_3^2 \\ 1 & X_4 & Y_4 & X_4^2 & X_4Y_4 & Y_4^2 \\ \vdots & \vdots & \vdots & \vdots & \vdots & \vdots \\ 1 & X_{10} & Y_{10} & X_{10}^2 & X_{10}Y_{10} & Y_{10}^2 \end{bmatrix}, \tag{7.5}$$

$$\Delta_a = \begin{bmatrix} a_0 \\ a_1 \\ a_2 \\ a_3 \\ a_4 \\ a_5 \end{bmatrix}, \tag{7.6}$$

$$L_x = \begin{bmatrix} x_1 \\ x_2 \\ x_3 \\ x_4 \\ \vdots \\ x_{10} \end{bmatrix}, \tag{7.7}$$

$$P = \begin{bmatrix} P_1 & & & & & \\ & P_2 & & & & \\ & & P_3 & & & \\ & & & P_4 & & \\ & & & & \ddots & \\ & & & & & P_{10} \end{bmatrix}, \tag{7.8}$$

Typically, the weight of each GCP is the same. So, $P = I$. Equation (7.4) can be expressed as:

$$\Delta_a = (A^T A)^{-1}(A^T L_x). \tag{7.9}$$

In the same way, the coefficients of b can be solved by the formula $\Delta_b = (A^T A)^{-1}(A^T L_y)$.

7.2.1.2 Coordinate Transformation

After establishing the second-order polynomial equation and solving the coefficients of Equations (7.1) (7.2), the raw image can be georeferenced at pixel-to-pixel. The steps are as follows (Zhou et al. 2016):

1. **Determination of the size of the georeferenced image**

   To properly obtain the georeferenced image, the storage area of georeferenced image must be computed in advance. As shown in Figure 7.1 (Qi et al. 2018), *abcd* is the raw image, with corners *a*, *b*, *c* and *d* in the *o-xy* image coordinate system in the Figure 7.1(a). Figure 7.1(b) shows the correct range of output image. *a'b'c'd'* is the georeferenced image with corners *a'*, *b'*, *c'* and *d'* in the *O-XY* ground coordinate system, and *ABCD* is the storage area for georeferenced image. Obviously, if the storage range is not correctly defined, the georeferenced image is improper with a large blank area, as shown in Figure 7.1(c). Therefore, the right boundary should include the entire georeferenced image with minimum exterior blank rectangle as possible.

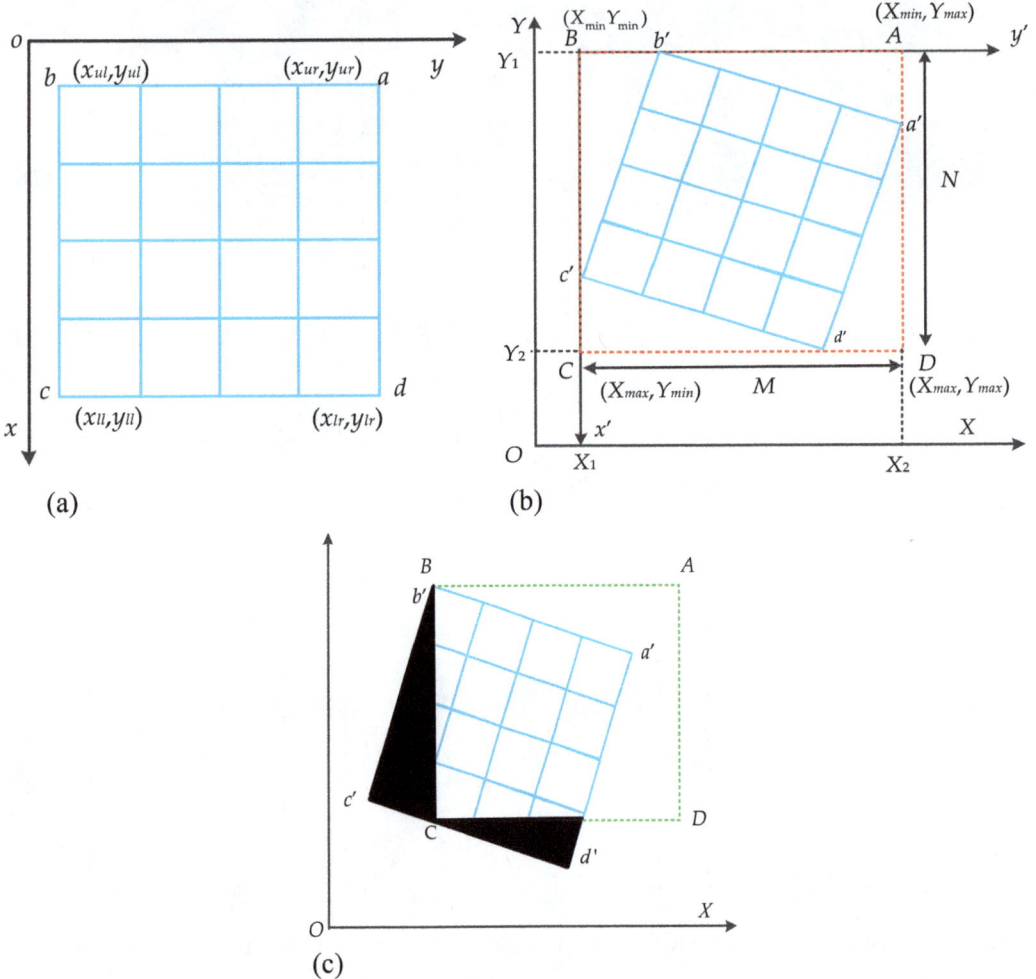

FIGURE 7.1 The georeferenced image size. (a) Input image. (b) Output image. (c) Incorrect output image. *ur*, upper right; *ul*, upper left; *ll*, lower left; *lr*, lower right. *abcd* is the input image. *ABCD* is the storage area for the georeferenced image. *a'b'c'd'* is the georeferenced image.

2. **Calculation of the coordinates of the four corners**

(x_{ul}, y_{ul}), (x_{ur}, y_{ur}), (x_{lr}, y_{lr}) and (x_{ll}, y_{ll}) are the coordinates of the four corners in the raw image (input image), (X_{ul}, Y_{ul}), (X_{ur}, Y_{ur}), (X_{lr}, Y_{lr}) and (X_{ll}, Y_{ll}) are the corresponding ground coordinates of the georeferenced image. The maximum and minimum ground coordinates of the storage area for the georeferenced image (X_{min}, X_{max}, Y_{min} and Y_{max}) are calculated from the (X_{ul}, Y_{ul}), (X_{ur}, Y_{ur}), (X_{lr}, Y_{lr}) and (X_{ll}, Y_{ll}).

$$
\begin{aligned}
X_{min} &= \min(X_{ur}, X_{ul}, X_{lr}, X_{ll}) \\
X_{max} &= \max(X_{ur}, X_{ul}, X_{lr}, X_{ll}) \\
Y_{min} &= \min(Y_{ur}, Y_{ul}, Y_{lr}, Y_{ll}) \\
Y_{max} &= \max(Y_{ur}, Y_{ul}, Y_{lr}, Y_{ll}).
\end{aligned}
\tag{7.10}
$$

3. **Calculation of the row (M) and column (N) of the georeferenced image.**

M and N can be solved by Equation (7.11).

$$
\begin{aligned}
M &= (X_{max} - X_{min}) / X_{GSD} + 1 \\
N &= (Y_{max} - Y_{min}) / Y_{GSD} + 1,
\end{aligned}
\tag{7.11}
$$

where X_{GSD} and Y_{GSD} are the ground-sampling distances (GSDs) of the output image. Hence, the location of pixel can be described by the row and column in the B-$x'y'$ coordinate system (x'=1, 2, 3,..., M; y'=1, 2, 3,..., N).

4. **Calculation of the coordinate transformation.**

The georeferenced model only expresses the relationship between the ground coordinates (X, Y) and the (x, y) coordinates of the raw image. To further express the relationship between the raw image coordinates and scanning coordinates of the georeferenced image, it is necessary to convert the ground coordinate into the row and the column of the georeferenced image, i.e.,

$$
\begin{aligned}
X_g &= X_{min} + X_{GSD}(x' - 1) \\
Y_g &= Y_{max} - Y_{GSD}(y' - 1)
\end{aligned}
\tag{7.12}
$$

where x' and y' are the row and column of the georeferenced image, respectively, and (X_g, Y_g) are the corresponding ground coordinate.

7.2.1.3 Resampling Using Bilinear Interpolation

As estimated, the defined grid center from the georeferenced image will not usually project to the center location of a pixel in the raw image after coordinate transformation. To solve this problem, nearest-neighbor interpolation, bilinear interpolation and cubic interpolation (Shlien 1979) are commonly used. To balance the accuracy and complexity of the preprocessing function, bilinear interpolation is chosen in this chapter. The algorithm obtains the pixel value by taking a weighted sum of the pixel values of the four nearest neighbors surrounding the calculated location (Gribbon and Bailey 2004) (Figure 7.2) is

$$
I(x, y) = I(i, j)(1-u)(1-v) + I(i+1, j)u(1-v) + I(i, j+1)(1-u)v + I(i+1, j+1)uv
\tag{7.13}
$$

where (x, y) are the coordinates of the georeferenced image, $I(x, y)$ is the gray value of a pixel with (x, y) coordinates in the georeferenced image; i and j are the integer part of the corresponding coordinate in the raw image, respectively; u and v are corresponding coordinate in the raw image, respectively. $I(i, j)$ is the gray value at (i, j) position in the raw image.

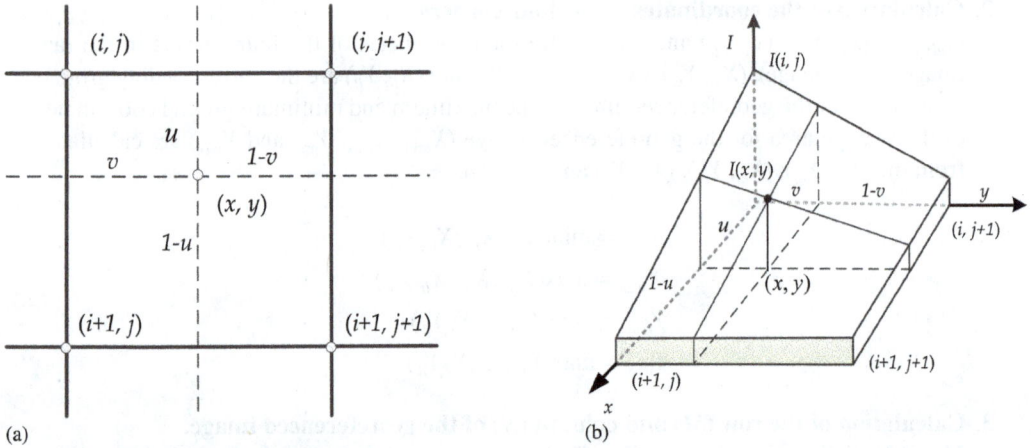

FIGURE 7.2 The bilinear interpolation method: (a) distribution of the four nearest neighbor pixels; (b) 3D spatial distribution of the bilinear interpolation.

7.2.2 Optimized Georeferencing Algorithm

7.2.2.1 Optimized Second-Order Polynomial Model

To implement georeferencing of an image using FPGA chip at a parallel mode, the second-order polynomial equations are optimized with

$$
\begin{aligned}
x &= a_0 + a_1 X + a_2 Y + a_3 X^2 + a_4 XY + a_5 Y^2 = a_0 + (a_1 + a_3 X)X + (a_2 + a_4 X + a_5 Y)Y \\
&= a_0 + p_1(X) + p_2(X,Y),
\end{aligned}
\tag{7.14}
$$

$$
\begin{aligned}
y &= b_0 + b_1 X + b_2 Y + b_3 X^2 + b_4 XY + b_5 Y^2 = b_0 + (b_1 + b_3 X)X + (b_2 + b_4 X + b_5 Y)Y \\
&= b_0 + q_1(X) + q_2(X,Y),
\end{aligned}
\tag{7.15}
$$

where $p_1(X)=(a_1+a_3X)\,X$, $p_2(X,\ Y) = (a_2+a_4X+\ a_5Y)\ Y$, $q_1(X)=(b_1+b_3X)\ X$, and $q_2(X,\ Y) = (b_2+b_4X+\ b_5Y)\ Y$.

In Equation (7.1), eight multipliers and five adders are required to complete a transformation. However, in Equation (7.14), five multipliers and five adders are required to complete a transformation. Comparatively, six multipliers are reduced in all. The multipliers and adders required to complete the coordinate transformation are compared in Table 7.1.

TABLE 7.1

The Number of Multiplier and Adder Comparison with Equations (7.1), (7.2), (7.14) and (7.15)

Coordinate	Traditional 2nd Polynomial		Optimized 2nd Polynomial		Reduce	
	Multiplier	Adder	Multiplier	Adder	Multiplier	Adder
X	8	5	5	5	6	0
Y	8	5	5	5		

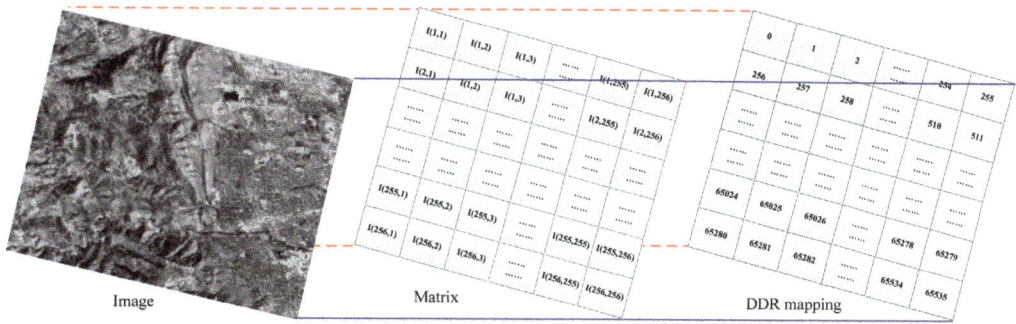

FIGURE 7.3   Raw image mapping with DDR.

7.2.2.2   Optimized Bilinear Interpolation

As observed to Equation (7.13), it requires eight multiplications, but careful factorization to exploit separability can reduce the multiplication by (Bailey 2011),

$$
\begin{cases}
I_1 = I(i,j) + u(I(i+1,j) - I(i,j)) \\
I_2 = I(i,j+1) + v(I(i+1,j+1) - I(i,j+1)) \\
I(x,y) = I_1 - v(I_2 - I_1).
\end{cases}
\tag{7.16}
$$

To compute the Equation (7.16), the gray values of four nearest-neighbors must be read from the memory. The principle of the mapping storage method is to optimize and balance the line and block data access rate (Qi et al. 2018). In order to improve memory access efficiency, a multi-array memory storage method is designed in Figure 7.3. The RS image size is $m \times n$. As such $m \times n$ memory units are designed in DDR. The relationship between the address of DDR and (i, j) coordinates can be described by Formula (7.17).

$$
addr_{ij} = (i-1) \times m + (j-1) \times n,
\tag{7.17}
$$

where i and j are the row and column of the raw image, respectively. $addr_{ij}$ is the corresponding address of the memory.

7.3   FPGA-BASED IMPLEMENTATION OF THE GEOREFERENCING

An FPGA hardware architecture for georeferencing the remotely sensed imagery is designed in Figure 7.4, which consists of five modules: (1) GCP data module; (2) input image module (IIM); (3) coefficient calculation module (CCM); (4) coordinate transformation module (CTM); (5) bilinear interpolation module (BIM). The details of five modules are described as follows.

- The GCPs data is stored in RAM of the GCP data module. The parameters are sent to the CCM, when the enable signal is being received.
- The gray values of the input image are saved to the ROM through the IIM, when the enable signal is being received.
- The coefficients of a_0, a_1, a_2, a_3, a_4, a_5, b_0, b_1, b_2, b_3, b_4 and b_5 are calculated in the CCM when the GCP data are arrived. The matrices $A^T A$, $A^T L_x$, $A^T L_y$ and $(A^T A)^{-1}$ are parallel computed.

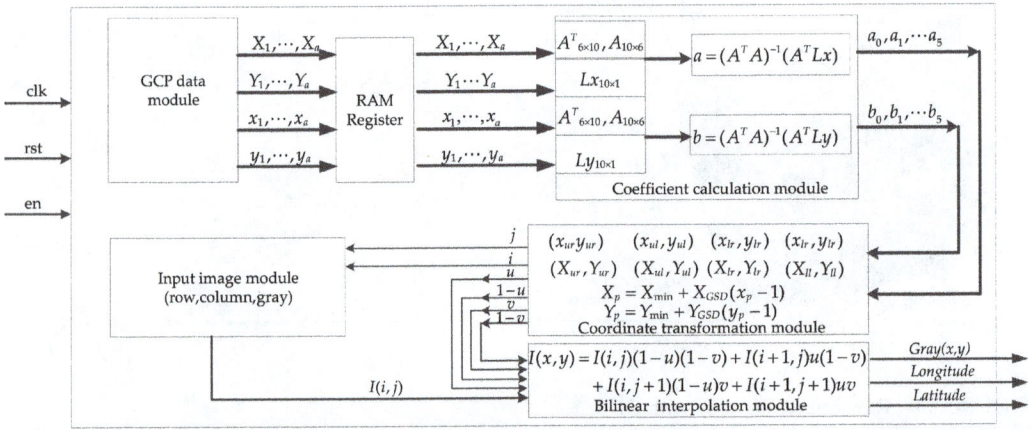

FIGURE 7.4 An FPGA architecture for the georeferenced remotely sensed images with second-order polynomial equation and bilinear interpolation scheme. RAM, random access memory; clk, clock; rst, reset; en, enable.

- The coordinate transformation is carried out in the CTM when the coefficients of second-order polynomial are arrived. The values of i, j, u, $1\text{-}u$, v and $1\text{-}v$ are simultaneously sent to the BIM.
- According to Equations (7.13) and (7.16), the interpolation values are calculated. At last, the gray values, latitude and longitude of the georeferenced image are outputted at the same clock cycle.

7.3.1　FPGA-Based Solution of the Second-Order Polynomial Equation

To implement the second-order polynomial equation, the processing of solving the coefficient is decomposed into three modules (Figure 7.5): (1) Form the matrices A, L_x and L_y; (2) Calculate $A^T A$, $(A^T A)^{-1}$, $A^T L_x$ and $A^T L_y$; and (3) perform $(A^T A)^{-1}(A^T L_x)$ and $(A^T A)^{-1}(A^T L_y)$.

7.3.1.1　Calculation of Matrix $A^T A$

According to Equation (7.9) and the principle of matrix multiplication, the coefficients of a_{1j} (j=1, 2, ..., 6) are calculated by

$$
A^T A =
\begin{bmatrix}
A_{11} & A_{21} & A_{31} & A_{41} & A_{51} & A_{61} & A_{71} & A_{81} & A_{91} & A_{a1} \\
A_{12} & A_{22} & A_{32} & A_{42} & A_{52} & A_{62} & A_{72} & A_{82} & A_{92} & A_{a2} \\
A_{13} & A_{23} & A_{33} & A_{43} & A_{53} & A_{63} & A_{73} & A_{83} & A_{93} & A_{a3} \\
A_{14} & A_{24} & A_{34} & A_{44} & A_{54} & A_{64} & A_{74} & A_{84} & A_{94} & A_{a4} \\
A_{15} & A_{25} & A_{35} & A_{45} & A_{55} & A_{65} & A_{75} & A_{85} & A_{95} & A_{a5} \\
A_{16} & A_{26} & A_{36} & A_{46} & A_{56} & A_{66} & A_{76} & A_{86} & A_{96} & A_{a6}
\end{bmatrix}
\begin{bmatrix}
A_{11} & A_{12} & A_{13} & A_{14} & A_{15} & A_{16} \\
A_{21} & A_{22} & A_{23} & A_{24} & A_{25} & A_{26} \\
A_{31} & A_{32} & A_{33} & A_{34} & A_{35} & A_{36} \\
A_{41} & A_{42} & A_{43} & A_{44} & A_{45} & A_{46} \\
A_{51} & A_{52} & A_{53} & A_{54} & A_{55} & A_{56} \\
A_{61} & A_{62} & A_{63} & A_{64} & A_{65} & A_{66} \\
A_{71} & A_{72} & A_{73} & A_{74} & A_{75} & A_{76} \\
A_{81} & A_{82} & A_{83} & A_{84} & A_{85} & A_{86} \\
A_{91} & A_{92} & A_{93} & A_{94} & A_{95} & A_{96} \\
A_{a1} & A_{a2} & A_{a3} & A_{a4} & A_{a5} & A_{a6}
\end{bmatrix},
$$

$$(7.18)$$

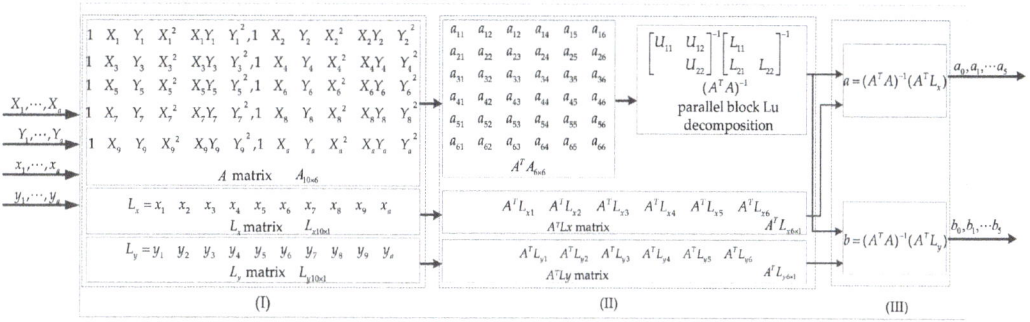

FIGURE 7.5 The proposed parallel implementation of the solving coefficient.

$$
\begin{bmatrix}
A_{11} \\
A_{21} \\
A_{31} \\
A_{41} \\
A_{51} \\
A_{61} \\
A_{71} \\
A_{81} \\
A_{91} \\
A_{a1}
\end{bmatrix}^{T}
*
\begin{bmatrix}
A_{11} & A_{12} & A_{13} & A_{14} & A_{15} & A_{16} \\
A_{21} & A_{22} & A_{23} & A_{24} & A_{25} & A_{26} \\
A_{31} & A_{32} & A_{33} & A_{34} & A_{35} & A_{36} \\
A_{41} & A_{42} & A_{43} & A_{44} & A_{45} & A_{46} \\
A_{51} & A_{52} & A_{53} & A_{54} & A_{55} & A_{56} \\
A_{61} & A_{62} & A_{63} & A_{64} & A_{65} & A_{66} \\
A_{71} & A_{72} & A_{73} & A_{74} & A_{75} & A_{76} \\
A_{81} & A_{82} & A_{83} & A_{84} & A_{85} & A_{86} \\
A_{91} & A_{92} & A_{93} & A_{94} & A_{95} & A_{96} \\
A_{a1} & A_{a2} & A_{a3} & A_{a4} & A_{a5} & A_{a6}
\end{bmatrix}
=
\begin{bmatrix}
a_{11} \\
a_{12} \\
a_{13} \\
a_{14} \\
a_{15} \\
a_{16}
\end{bmatrix}^{T}
. \qquad (7.19)
$$

Generally, ten multipliers and nine adders are needed to compute a_{11}. Therefore, it takes about 324 additions and 360 multiplications to calculate all coefficients. However, with the limited resources on an FPGA chip, a combination of serial and parallel computation is proposed in this chapter. To improve the processing speed, multiplier-adder (MD) modules are used. Meanwhile, the other coefficients of a_{1j}, a_{2j}, a_{3j}, a_{4j}, a_{5j} and a_{6j}, (j= 1, 2, 3, …, 6) are parallel computed at the same clock cycle. Figure 7.6 shows the architecture of $A^{T}A$ on the basis of an FPGA chip.

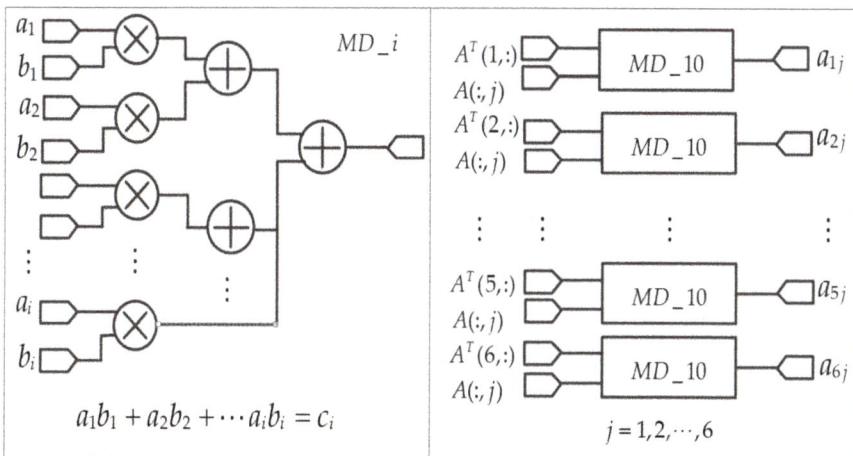

FIGURE 7.6 The architectures of $A^{T}A$ using FPGA chip.

7.3.1.2 LU Decomposition of the Matrix $A^T A$

This section proposes a parallel structure that uses the floating-point block LU decomposition to solve the inverse of $A^T A$ (Govindu et al. 2004) (a detail can be referenced to Chapter 5), i.e.,

$$A^T A = \begin{bmatrix} a_{11} & a_{12} & a_{13} & | & a_{14} & a_{15} & a_{16} \\ a_{21} & a_{22} & a_{23} & | & a_{24} & a_{25} & a_{26} \\ a_{31} & a_{32} & a_{33} & | & a_{34} & a_{35} & a_{36} \\ - & - & - & - & - & - & - \\ a_{41} & a_{42} & a_{43} & | & a_{44} & a_{45} & a_{46} \\ a_{51} & a_{52} & a_{53} & | & a_{54} & a_{55} & a_{56} \\ a_{61} & a_{62} & a_{63} & | & a_{64} & a_{65} & a_{66} \end{bmatrix} = \begin{bmatrix} A_{11} & A_{12} \\ A_{21} & A_{22} \end{bmatrix} = \begin{bmatrix} L_{11} & \\ L_{21} & L_{22} \end{bmatrix} \begin{bmatrix} U_{11} & U_{12} \\ & U_{22} \end{bmatrix}, \quad (7.20)$$

where $A_{11}, A_{12}, A_{21}, A_{22}, L_{21}$ and U_{12} are 3×3 matrices; L_{11} and L_{22} are 3×3 lower triangular matrices, U_{11} and U_{22} are 3×3 upper triangular matrices. From the Equation (7.20), additional equations can be yielded:

$$A_{11} = L_{11} U_{11}, \quad (7.21)$$

$$A_{12} = L_{11} U_{12}, \quad (7.22)$$

$$A_{21} = L_{21} U_{11}, \quad (7.23)$$

$$A_{22} = L_{21} U_{12} + L_{22} U_{22}. \quad (7.24)$$

The steps of LU decomposition are as follows:

Step 1: A_{11} is performed by block LU decomposition; L_{11} and U_{11} are obtained.
Step 2: From Equation (7.22), U_{12} can be solved by Equation (7.25):

$$U_{12} = L_{11}^{-1} A_{12}. \quad (7.25)$$

Step 3: From Equation (7.23), L_{21} can be calculated by the product of A_{21} and $(U_{11})^{-1}$ (see Equation (7.26)):

$$L_{21} = A_{21} U_{11}^{-1}. \quad (7.26)$$

Step 4: $A_{22}-A_{21}U_{12}$ matrix is performed by LU decomposition; L_{22} and U_{22} matrices are obtained.

1. FPGA-based parallel computation for the LU decomposition of matrix A_{11}
   To implement the LU decomposition of matrix A_{11} on the FPGA, Equation (7.21) can be modified as Equation (7.27). The steps of the LU decomposition of matrix A_{11} are as follows:

$$A_{11} = \begin{bmatrix} a_{11} & a_{12} & a_{13} \\ a_{21} & a_{22} & a_{23} \\ a_{31} & a_{32} & a_{33} \end{bmatrix} = \begin{bmatrix} 1 & & \\ l_{21} & 1 & \\ l_{31} & l_{32} & 1 \end{bmatrix} \begin{bmatrix} u_{11} & u_{12} & u_{13} \\ & u_{22} & u_{23} \\ & & u_{33} \end{bmatrix} = L_{11} U_{11}, \quad (7.27)$$

Step 1: $u_{11} = a_{11}, u_{12} = a_{12}, u_{12} = a_{13}, l_{11} = l_{22} = l_{33} = 1, l_{21} = a_{21}/u_{11} = a_{21}/a_{11}, l_{31} = a_{31}/a_{11}.$

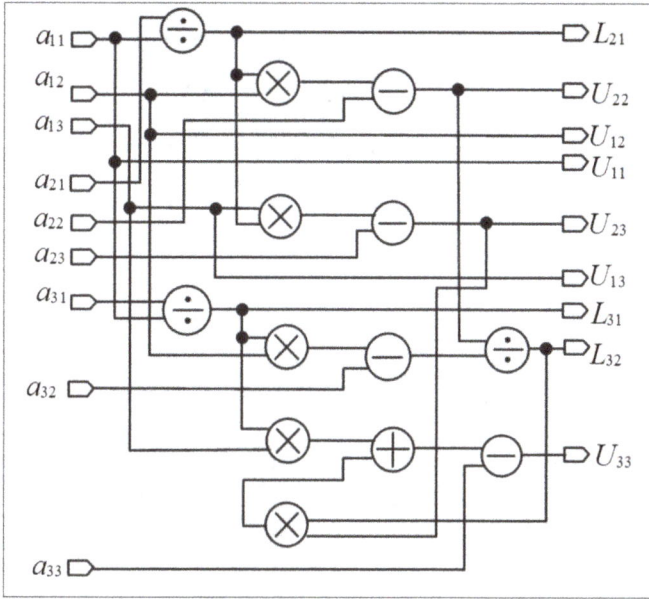

FIGURE 7.7   Parallel computational block LU decomposition of A_{11}.

> *Step 2:* $u_{22} = a_{22} - l_{21}u_{12}$, $u_{23} = a_{23} - l_{21}u_{13}$.
> *Step 3:* $l_{31} = (a_{32} - l_{31}u_{12})/u_{22}$.
> *Step 4:* $u_{33} = a_{33} - l_{31}u_{13} - l_{32}u_{23} = a_{33} - (l_{31}u_{13} + l_{32}u_{23})$.

According to Equation (7.27), a parallel computation architecture for matrices L_{11} and U_{11} is presented in Figure 7.7. Five multipliers, three divisors, one adder and four subtractors are used. Additionally, a few control signals are used to ensure that the results are outputted at the same clock cycle.

2. **FPGA-based parallel computation for matrices $(L_{11})^{-1}$ and $(U_{11})^{-1}$**

   The inverse of a lower triangular matrix is another lower triangular matrix, and the inverse of an upper triangular matrix is also another upper triangular matrix. Thereby, the inversion of matrices L_{11} and U_{11} can be rewritten as, respectively.

$$
\begin{bmatrix} 1 & & \\ IL_{21} & 1 & \\ IL_{31} & IL_{32} & 1 \end{bmatrix}
\begin{bmatrix} 1 & & \\ L_{21} & 1 & \\ L_{31} & L_{32} & 1 \end{bmatrix}
= \begin{bmatrix} 1 & & \\ & 1 & \\ & & 1 \end{bmatrix},
\tag{7.28}
$$

where $IL_{11} = -L_{21}$, $IL_{32} = -L_{32}$, and $IL_{31} = -IL_{32}L_{21} - L_{31}$.

$$
\begin{bmatrix} IU_{11} & IU_{12} & IU_{13} \\ & IU_{22} & IU_{23} \\ & & IU_{33} \end{bmatrix}
\begin{bmatrix} U_{11} & U_{12} & U_{13} \\ & U_{22} & U_{23} \\ & & U_{33} \end{bmatrix}
= \begin{bmatrix} 1 & & \\ & 1 & \\ & & 1 \end{bmatrix},
\tag{7.29}
$$

where $IU_{11} = 1/U_{11}$, $IU_{22} = 1/U_{22}$, $IU_{33} = 1/U_{33}$, $IU_{12} = -(IU_{11}U_{12})/U_{22} = -2U_{12}/(U_{11} U_{22})$, $IU_{23} = -(IU_{22}U_{23})/U_{33} = -U_{23}/(U_{22}U_{33})$ and $IU_{13} = -(IU_{11}U_{13} + IU_{12}U_{23})/U_{33} = \{(U_{12}U_{23}/(U_{22}U_{33}) - U_{13}/U_{33}\}/U_{11}$.

The parallel computation structures for calculating $(L_{11})^{-1}$ and $(U_{11})^{-1}$ are shown in Figure 7.8. Six multipliers, four divisors, one subtractor ("—" in the circle) and three reciprocals ("/" in the circle) are used. The reverse operation ("-" in the circle) for floating-point

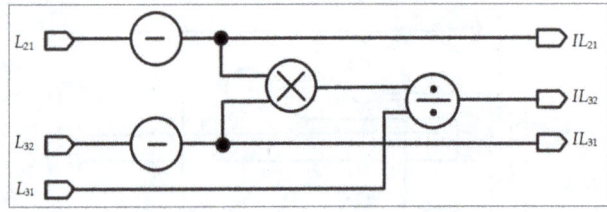

(a)

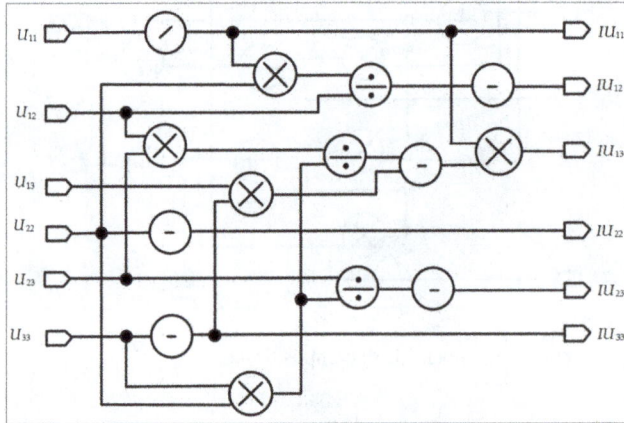

(b)

FIGURE 7.8 The architecture of FPFA-based computation for $(L_{11})^{-1}$ and $(U_{11})^{-1}$ matrices: (a) for solving $(L_{11})^{-1}$; (b) for solving $(U_{11})^{-1}$.

is simply to reverse the symbol bit and require very few resources. Additionally, some control signals are used to ensure that the results are outputted at the same clock cycle.

3. **FPGA-based implementation of matrices U_{12} and L_{21}**

   To implement U_{11} and L_{21} based on the FPGA, Equations (7.25), (7.26) can be rewritten into Equations (7.30), (7.31). In Equation (7.30), the elements of the matrix contain three formats, i.e., a_{14}, $IL_{21}a_{14} + a_{24}$ and $IL_{31}a_{14} + IL_{32}a_{24} + a_{34}$. In Equation (7.31), the elements of the matrix contain three formats, i.e., $a_{41}IU_{11}$, $a_{41}IU_{12} + a_{42}IU_{22}$ and $a_{41}IU_{13} + a_{42}IU_{23} + a_{43}IU_{33}$. The proposed parallel architecture for calculation of L_{21} and U_{12} is depicted in Figure 7.9. Twelve multipliers, fifteen MD modules and three adders are used.

$$
U_{12} = \left(L_{11}\right)^{-1} A_{12} = \begin{bmatrix} 1 & & \\ IL_{21} & 1 & \\ IL_{31} & IL_{32} & 1 \end{bmatrix} \begin{bmatrix} a_{14} & a_{15} & a_{16} \\ a_{24} & a_{25} & a_{26} \\ a_{34} & a_{35} & a_{36} \end{bmatrix}
$$

$$
= \begin{bmatrix} a_{14} & a_{15} & a_{16} \\ IL_{21}a_{14} + a_{24} & IL_{21}a_{15} + a_{25} & IL_{21}a_{16} + a_{26} \\ IL_{31}a_{14} + IL_{32}a_{24} + a_{34} & IL_{31}a_{15} + IL_{32}a_{25} + a_{35} & IL_{31}a_{16} + IL_{32}a_{26} + a_{36} \end{bmatrix} \quad (7.30)
$$

$$
= \begin{bmatrix} U12\_11 & U12\_12 & U12\_13 \\ U12\_21 & U12\_22 & U12\_23 \\ U12\_31 & U12\_32 & U12\_33 \end{bmatrix}.
$$

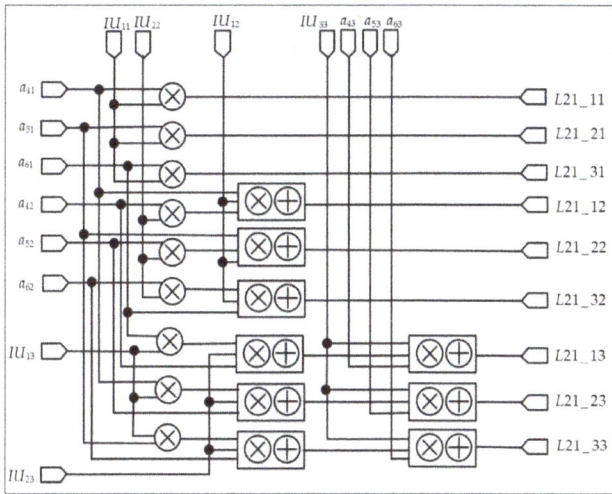

(a)

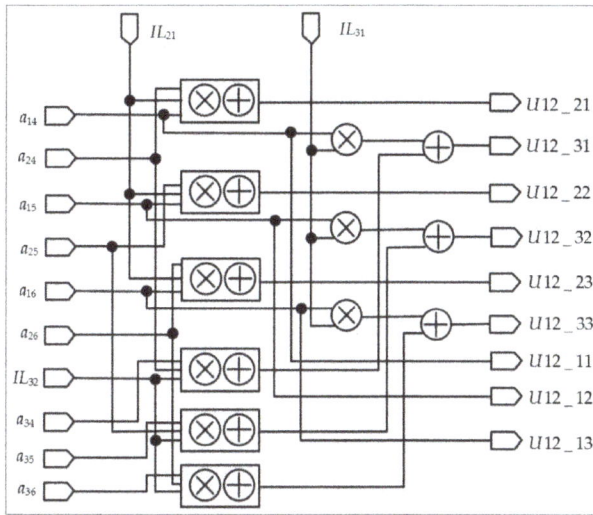

(b)

FIGURE 7.9   Parallel computation for L_{21} and U_{12}. (a) The architecture of L_{21}. (b) The architecture of U_{12}.

$$L_{21} = A_{21}U_{11}^{-1} = \begin{bmatrix} a_{41} & a_{42} & a_{43} \\ a_{51} & a_{52} & a_{53} \\ a_{61} & a_{62} & a_{63} \end{bmatrix} * \begin{bmatrix} IU_{11} & IU_{12} & IU_{13} \\ & IU_{22} & IU_{23} \\ & & IU_{33} \end{bmatrix}$$

$$= \begin{bmatrix} a_{41}IU_{11} & a_{41}IU_{12} + a_{42}IU_{22} & a_{41}IU_{13} + a_{42}IU_{23} + a_{43}IU_{33} \\ a_{51}IU_{11} & a_{51}IU_{12} + a_{52}IU_{22} & a_{51}IU_{13} + a_{52}IU_{23} + a_{53}IU_{33} \\ a_{61}IU_{11} & a_{61}IU_{12} + a_{62}IU_{22} & a_{61}IU_{13} + a_{62}IU_{23} + a_{63}IU_{33} \end{bmatrix} \quad (7.31)$$

$$= \begin{bmatrix} L21\_11 & L21\_12 & L21\_13 \\ L21\_21 & L21\_22 & L21\_23 \\ L21\_31 & L21\_32 & L21\_33 \end{bmatrix},$$

4. FPGA-based implementation of L_{22} and U_{22}

To implement L_{22} and U_{22} computations, Equation (7.24) can be described by

$$NewA22 = A_{22} - L_{21}U_{12}$$

$$= \begin{bmatrix} a_{44} & a_{45} & a_{46} \\ a_{54} & a_{55} & a_{56} \\ a_{63} & a_{65} & a_{66} \end{bmatrix} - \begin{bmatrix} L21\_11 & L21\_12 & L21\_13 \\ L21\_21 & L21\_22 & L21\_23 \\ L21\_31 & L21\_32 & L21\_33 \end{bmatrix} \begin{bmatrix} U12\_11 & U12\_12 & U12\_13 \\ U12\_21 & U12\_22 & U12\_23 \\ U12\_31 & U12\_32 & U12\_33 \end{bmatrix} \quad (7.32)$$

$$= \begin{bmatrix} 1 & & \\ L22\_21 & 1 & \\ L22\_31 & L22\_32 & 1 \end{bmatrix} \begin{bmatrix} U22\_11 & U22\_12 & U22\_13 \\ & U22\_22 & U22\_23 \\ & & U22\_33 \end{bmatrix}$$

where $NewA22$ is a 3×3 matrix, with a similar format of A_{11}. Therefore, the LU decomposition of $NewA22$ is not deduced in details. The parallel implementation is designed and implemented by (see Figure 7.10)

It takes great deal of time to inverse the matrix if using a traditionally method with FPG. For this reason, the Block LU is decomposed, the number of multiplications is about $n^3/3+(n\%d-0.5)\,n^2$ and division is $(d+1)\,n/2$ (where d is the dimension of each block, n is the size of matrix). In the standard LU decomposition, the number of multiplication operations is about $n(2n-1)\,(n-1)/6$, and division is $n(n-1)/2$ (Chen et al. 2009). The multipliers and dividers based block LU decomposition are approximately reduced 1.02 and 1.25 times than that the traditional LU decomposition.

7.3.1.3 FPGA-Based Implantation of the Matrix $(A^TA)^{-1}$

The description above implies that the block LU decomposition of matrix A^TA has been completed, when L_{11}, L_{21}, L_{22}, U_{11}, U_{12} and U_{22} matrices are solved. The processing of inversion matrix A^TA is based on the Equation (7.33) (Zhou and Li 2000; Zhou et al. 2016).

$$(A^TA)^{-1} = B = \left(\begin{bmatrix} L_{11} & \\ L_{21} & L_{22} \end{bmatrix} \begin{bmatrix} U_{11} & U_{12} \\ & U_{22} \end{bmatrix} \right)^{-1} = \begin{bmatrix} U_{11} & U_{12} \\ & U_{22} \end{bmatrix}^{-1} \begin{bmatrix} L_{11} & \\ L_{21} & L_{22} \end{bmatrix}^{-1}$$

$$= \begin{bmatrix} U_{11}^{-1}L_{11}^{-1} + M_{12}^{-1}N_{21}^{-1} & M_{12}^{-1}L_{22}^{-1} \\ U_{22}^{-1}N_{21}^{-1} & U_{22}^{-1}L_{22}^{-1} \end{bmatrix}, \quad (7.33)$$

where $N_{21}^{-1} = - L_{22}^{-1}L_{21}L_{11}^{-1}$ and $M_{12}^{-1} = - U_{11}^{-1}U_{12}U_{22}^{-1}$.

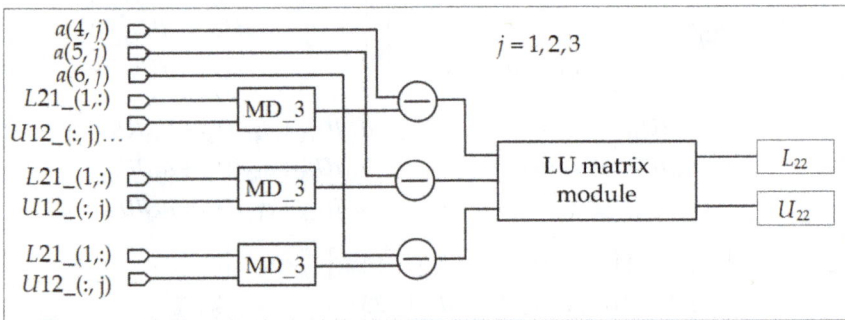

FIGURE 7.10 The architecture of L_{22} and U_{22}.

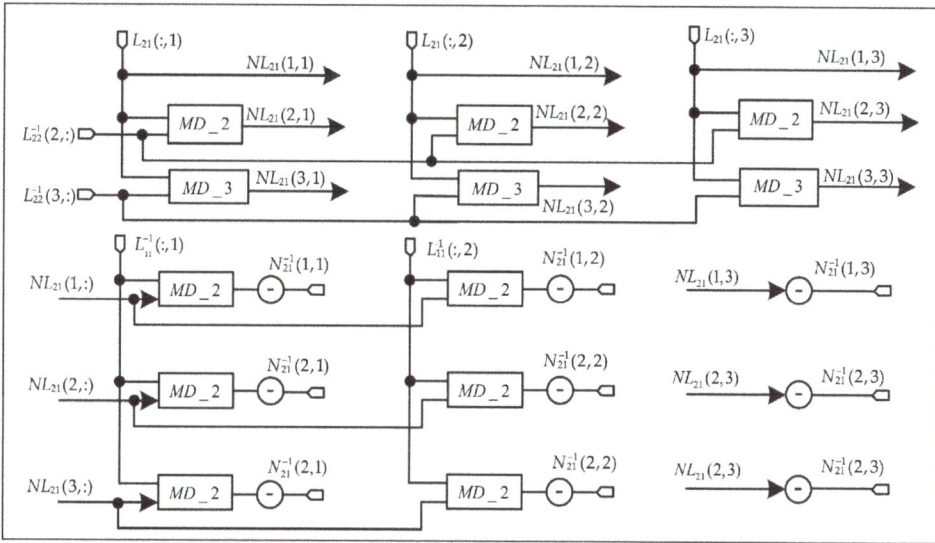

(a)

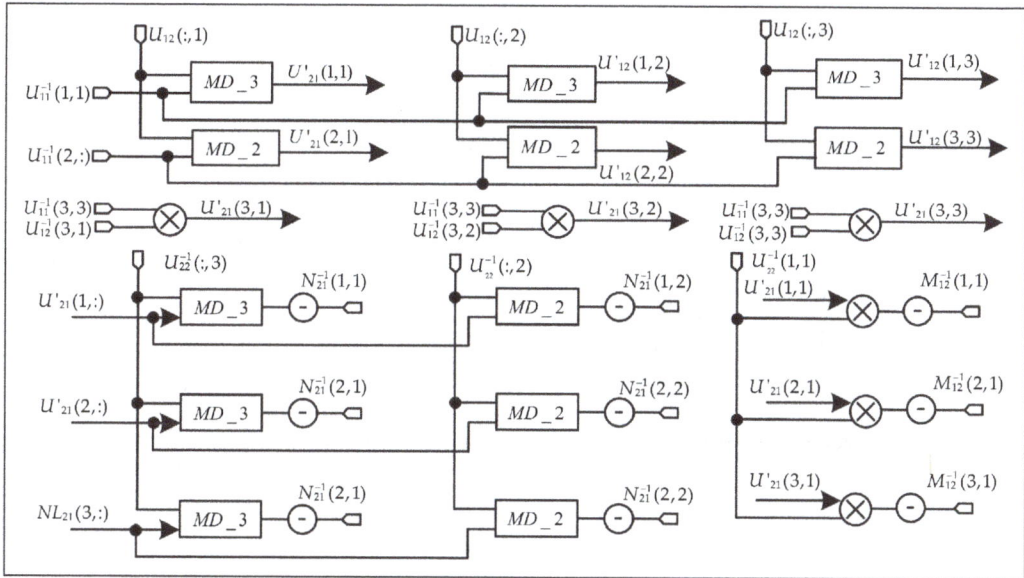

(b)

FIGURE 7.11   The implementation architecture for N_{21}^{-1} and M_{12}^{-1}: (a) for N_{21}^{-1}; (b) for M_{12}^{-1}.

1. **FPGA-based implementation of N_{21}^{-1} and M_{12}^{-1}**

   Figure 7.11 shows the implementation of N_{21}^{-1} and M_{12}^{-1}. It contains MD_3 module, MD_2 module and negative operation.

2. **FPGA-based implementation of $(A^TA)^{-1}$**

   The FPGA-based implementation for $(A^TA)^{-1}$ is shown in Figure 7.12. Figure 7.12(a) is an FPGA implementation of solving the $U_{11}^{-1}L_{11}^{-1} + M_{21}^{-1}N_{21}^{-1}$ matrix. Ten MD_3 modules, three MD_2 modules, two multipliers and nine adders are used to solve the $U_{11}^{-1}L_{11}^{-1} + M_{21}^{-1}N_{21}^{-1}$ matrix. Figure 7.12(b) shows the parallel computation for $M_{12}^{-1}L_{22}^{-1}$ matrix, three MD_3 modules

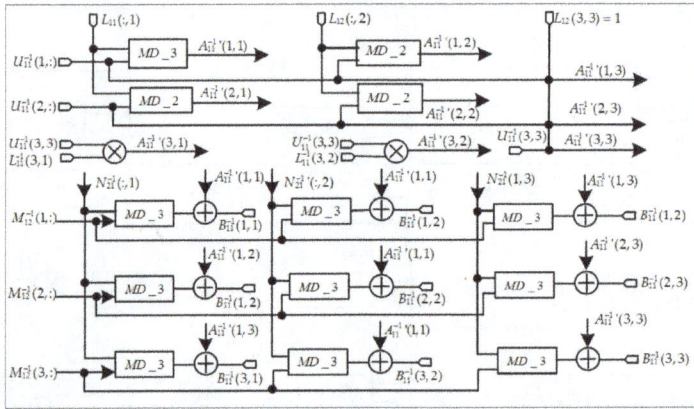

(a)

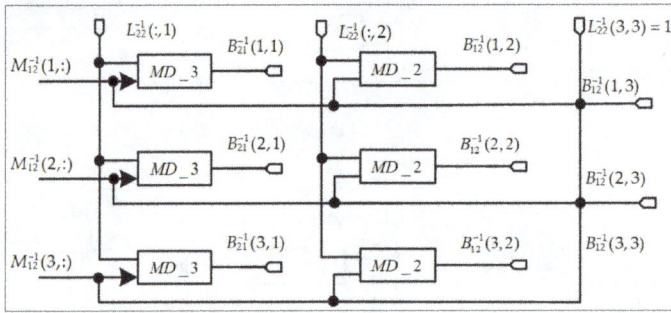

(b)

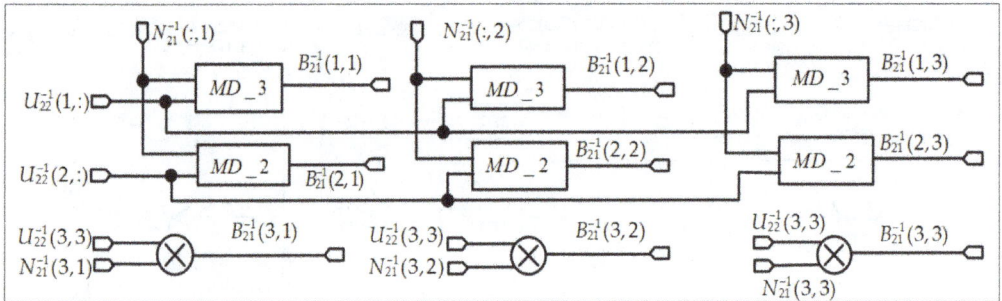

(c)

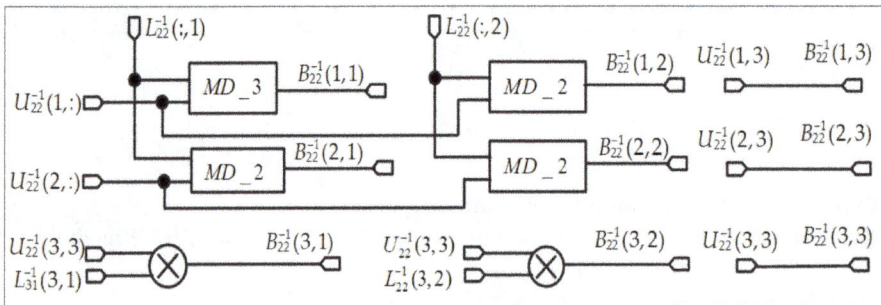

(d)

FIGURE 7.12 The FPGA-based implementation for $(A^T A)^{-1}$: (a) for calculating $U_{11}^{-1} L_{11}^{-1} + M_{21}^{-1} N_{21}^{-1}$ matrix, (b) for solving $M_{12}^{-1} L_{22}^{-1}$ matrix, (c) for calculating $U_{22}^{-1} N_{21}^{-1}$ matrix and (d) for computing $U_{22}^{-1} L_{22}^{-1}$ matrix.

and three MD_2 modules are used. The structure for solving the $U_{22}^{-1}N_{21}^{-1}$ matrix is shown in Figure 7.12(c), three MD_3 modules, three MD_2 modules and three multipliers are used. Figure 7.12(d) shows the implementation of the $U_{22}^{-1}L_{22}^{-1}$ matrix, one MD_3 module, two MD_2 modules and two multipliers are used. The Figure 7.12(a) thru (d) are parallel calculated. Finally, $(A^TA)^{-1}$ matrix is obtained and the elements are outputted at the same clock cycle.

7.3.1.4   FPGA-Based Implementation of *a* and *b* Matrices

From the ten GCPs data, L_x and L_y matrix can be obtained, since *a* and *b* matrices have the same structure, such as A^TL_x, $(A^TA)^{-1}(A^TL_x)$, A^TL_y and $(A^TA)^{-1}(A^TL_x)$. From Equation (7.34), six MD_10 modules are adopted. Six MD_6 modules are needed for solving Equation (7.35). Considering the limited resources of an FPGA, a serial framework is used to implement *a* (Figure 7.13).

The details of state transition graph (STG) are given as follows:

S_idle, idle state. When the rst signal is low, the system is in the reset state, all registers (R0, R1, ..., R5) and other signals are reset.

S_1 to S_6 are the six different state machines (SMs). Under the different SMs, different row values of the matrix $(A^TA)^{-1}$ with matrix A^TL_x are calculated in the MD_6 modules. The result of MD_6 is serial saved in the registers R5 to R0. When the six SMs are finished, the values of registers R5 to R0 are parallel outputted to the matrix *a*.

S_1, the first SM. When the rst signal goes high, the current state enters the first state when the enable signal is being received. The values of matrix $B^{-1}(1,:)=(A^TA)^{-1}(1,:)$ and A^TL_x are put into the MD_6 module for multiplication and addition. The result is saved in the register R5.

S_2, the second SM. When the S_1 state is complete, the current state enters the second state. The values $B^{-1}(2,:)=(A^TA)^{-1}(2,:)$ and A^TL_x begin to calculate. Then, the result is saved in the register R4.

With this given order above, when the S_6 SM (final State) is completed, the matrix *a* is derived. Under the same clock cycle, the values of the registers R5 to R0 (a_0-a_5) are parallel outputted. The current state returns to the first state. Additionally, each state should contain a certain delay time. The delay time should include MD_6 module operation time and serial storage time.

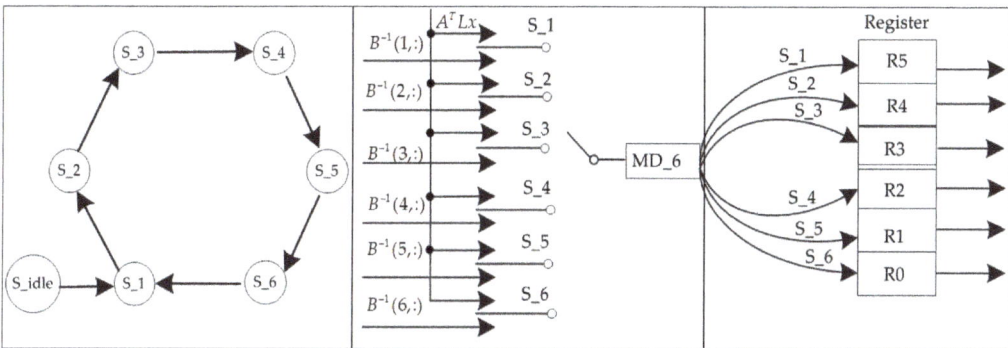

FIGURE 7.13   The processing of STG: (a) is a state machine, (b) is a serial framework and (c) is a serial input and parallel output.

$$A^T Lx = \begin{bmatrix} A_{11} & A_{21} & A_{31} & A_{41} & A_{51} & A_{61} & A_{71} & A_{81} & A_{91} & A_{a1} \\ A_{12} & A_{22} & A_{32} & A_{42} & A_{52} & A_{62} & A_{72} & A_{82} & A_{92} & A_{a2} \\ A_{13} & A_{23} & A_{33} & A_{43} & A_{53} & A_{63} & A_{73} & A_{83} & A_{93} & A_{a3} \\ A_{14} & A_{24} & A_{34} & A_{44} & A_{54} & A_{64} & A_{74} & A_{84} & A_{94} & A_{a4} \\ A_{15} & A_{25} & A_{35} & A_{45} & A_{56} & A_{65} & A_{75} & A_{85} & A_{95} & A_{a4} \\ A_{16} & A_{26} & A_{36} & A_{46} & A_{56} & A_{66} & A_{76} & A_{86} & A_{96} & A_{a6} \end{bmatrix} \begin{bmatrix} lx_1 \\ lx_2 \\ lx_3 \\ lx_4 \\ lx_5 \\ lx_6 \\ lx_7 \\ lx_8 \\ lx_9 \\ lx_a \end{bmatrix} = \begin{bmatrix} Lx1 \\ Lx2 \\ Lx3 \\ Lx4 \\ Lx5 \\ Lx6 \end{bmatrix} \qquad (7.34)$$

$$= \begin{bmatrix} A_{11}lx_1 + A_{21}lx_2 + A_{31}lx_3 + A_{41}lx_4 + A_{51}lx_5 + A_{61}lx_6 + A_{71}lx_7 + A_{81}lx_8 + A_{91}lx_9 + A_{a1}lx_a \\ A_{12}lx_1 + A_{22}lx_2 + A_{32}lx_3 + A_{42}lx_4 + A_{52}lx_5 + A_{62}lx_6 + A_{72}lx_7 + A_{82}lx_8 + A_{92}lx_9 + A_{a2}lx_a \\ A_{13}lx_1 + A_{23}lx_2 + A_{33}lx_3 + A_{43}lx_4 + A_{53}lx_5 + A_{63}lx_6 + A_{73}lx_7 + A_{83}lx_8 + A_{93}lx_9 + A_{a3}lx_a \\ A_{14}lx_1 + A_{24}lx_2 + A_{34}lx_3 + A_{44}lx_4 + A_{54}lx_5 + A_{64}lx_6 + A_{74}lx_7 + A_{84}lx_8 + A_{94}lx_9 + A_{a4}lx_a \\ A_{15}lx_1 + A_{25}lx_2 + A_{35}lx_3 + A_{45}lx_4 + A_{55}lx_5 + A_{65}lx_6 + A_{75}lx_7 + A_{85}lx_8 + A_{95}lx_9 + A_{a5}lx_a \\ A_{16}lx_1 + A_{26}lx_2 + A_{36}lx_3 + A_{46}lx_4 + A_{56}lx_5 + A_{66}lx_6 + A_{76}lx_7 + A_{86}lx_8 + A_{96}lx_9 + A_{a6}lx_a \end{bmatrix},$$

$$a = (A^T A)^{-1}(A^T L_x) = \begin{bmatrix} B_{11}^{-1} & B_{12}^{-1} \\ B_{21}^{-1} & B_{22}^{-1} \end{bmatrix} \begin{bmatrix} Lx1 \\ Lx2 \\ Lx3 \\ Lx4 \\ Lx5 \\ Lx6 \end{bmatrix} = \begin{bmatrix} a_0 \\ a_1 \\ a_2 \\ a_3 \\ a_4 \\ a_5 \end{bmatrix} \qquad (7.35)$$

7.3.2　FPGA-Based Implementation of Coordinate Transformation and Bilinear Interpolation

Figure 7.14 illustrates the flowchart for the coordinate transformation and bilinear interpolation method, which consists of the transformation from the output image coordinates to the ground coordinates, conversion the ground coordinate to the raw image coordinates and bilinear interpolation.

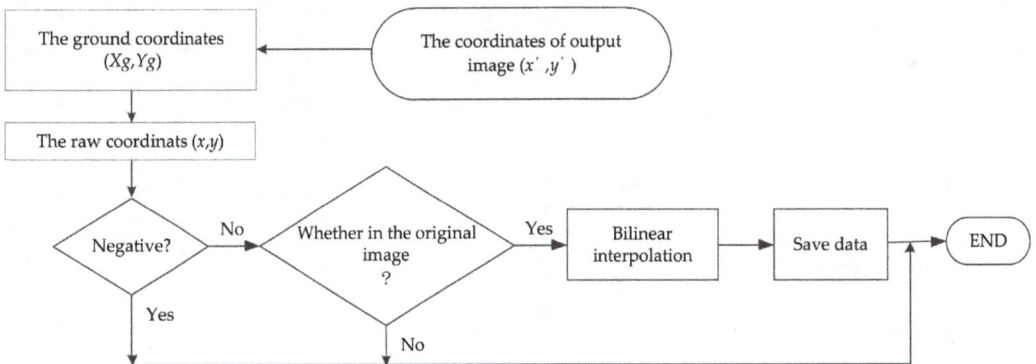

FIGURE 7.14　The flowchart of the coordinate transformation and bilinear interpolation.

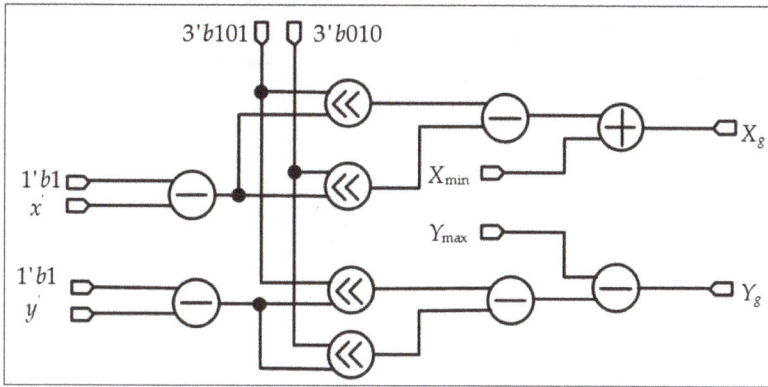

FIGURE 7.15　Parallel computation for Xg and Yg.

The implementation on the basis of FPGA for the coordinate transformation and bilinear interpolation algorithm are shown in Figure 7.15–7.19.

1. **FPGA-based implementation of X_g and Y_g**

   The purpose of this implementation is to obtain X_g and Y_g in Equation (7.12) in Section 7.2.1.2, as shown in Figure 7.15. In this part, 32-bit integer and shift register are used to reduce the resource utilization of an FPGA. To implement the shift operation, X_{GSD} and Y_{GSD} are approximated as 2^m or $2^m\text{-}2^n$ (m and n are integers). For instance, $X_{GSD} = X_{GSD} = 30$, which can be expressed as the form of $2^5\text{-}2$. In Figure 7.15, the symbol "$\ll$" in the circle represents the left shift operation.

2. **FPGA-based implementation of bilinear interpolation**

   When X_g and Y_g are calculated, the coordinates of x_row and y_column are obtained from the Equation (7.14) through (7.15). The bilinear interpolation algorithm is performed based on the x_row and y_column (Figure 7.16). As observed from Figure 7.16, the processing of bilinear interpolation is divided into three steps.

   First, the integer part i, j, the fractional part u, v and the weight part $1\text{-}u$ and $1\text{-}v$ of the floating-point (x_row, y_column) coordinates are computed, respectively.

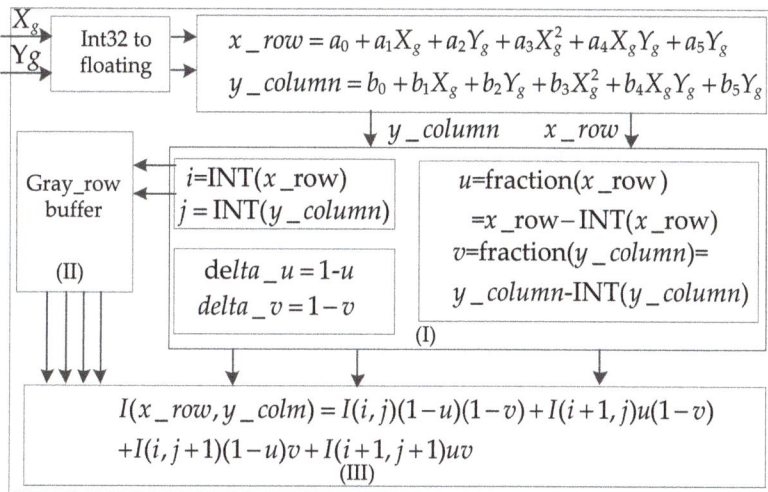

FIGURE 7.16　The diagram of bilinear interpolation.

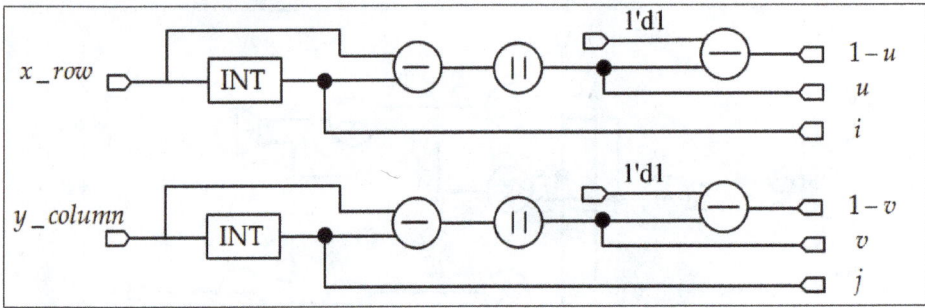

FIGURE 7.17   The parallel implementation method for *i, j, u, v,* 1-*u* and 1-*v*.

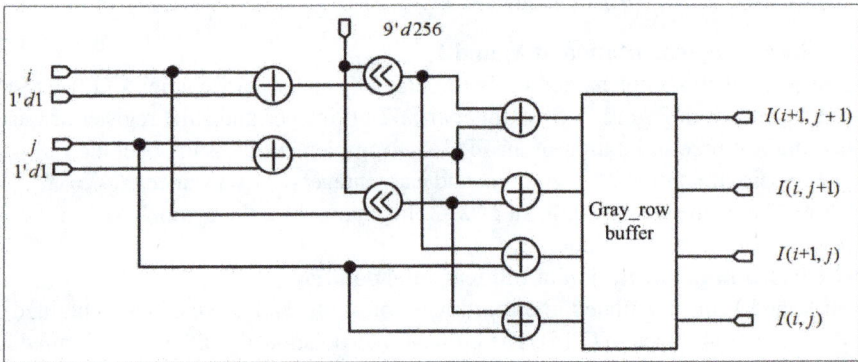

FIGURE 7.18   The architecture for parallel reading gray values.

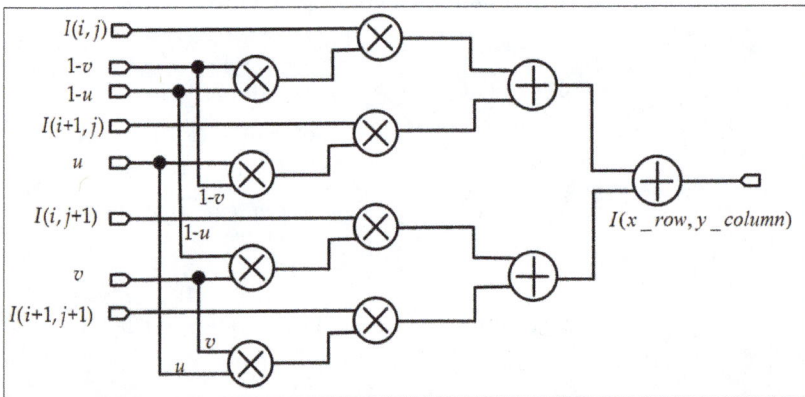

FIGURE 7.19   The parallel computation for bilinear interpolation.

Figure 7.17 shows the parallel implementation method for *i, j, u, v,* 1-*u* and 1-*v*. Four sub-tractors, two INT modules and two absolute modules ("||" in the circle) are used.

Second, it is mainly to calculate the address of four nearest neighbor pixels and read the corresponding gray values. The address of memory can be calculated from Equation (7.17) of Section 7.2.2.2. The processing of reading gray value is shown in Figure 7.18. To reduce the resources of an FPGA, the multiplication is converted into left shift operation.

Finally, after the values of *i, j, u, v,* 1-*u,* 1-*v, I(i, j), I(i+1, j), I(i, j+1)* and *I(i+1, j+1)* are computed, the gray value of the georeferenced image can be interpolation. With Equations (7.13) and (7.16), the implementation of bilinear interpolation algorithm is shown in Figure 7.19, where eight multipliers and three adders are used to compute the value of *I* (*x_row, y_column*).

7.4 VERIFICATIONS AND PERFORMANCE ANALYSIS

7.4.1 THE SOFTWARE AND HARDWARE ENVIRONMENT

The proposed method in this chapter is implemented on a Xilinx Virtex-7-XC7VX980t-ffg1930-1 FPGA that has 612,000 logic cells, 1,500 kB Block RAM, 1,224,000 Flip-Flops and 3,600 DSP slices. Additionally, the design tool is Vivado 2014.2, the simulation tool is ModelSim SE-64 10.4, and the hardware design language is Verilog HDL. To validate the proposed method, the georeferenced algorithm is also implemented by MATLAB R2014a, Visual Studio 2015 (C++) and ENVI 5.3 on a PC equipped with an Intel (R) Core i7-4790 CPU @ 3.6GHz and 8GB RAM, running Windows 7 (64-bit).

7.4.2 REMOTELY SENSED IMAGE DATA

To validate the method proposed in this chapter, two data sets are used to perform the georeferencing. The first data set is acquired from the ENVI example dataset, i.e., bldt_tm.img and bldt_tm.pts (Figure 7.20(a) and (b)). The second data set is obtained from the ERDAS example dataset, i.e., tmAtlanta.img and panAtanta.img (Figure 7.20 (b)). Both of the image size are 256×256 pixels^2. The other information of two datasets is shown in Table 7.2.

7.4.3 PROCESSING PERFORMANCE

7.4.3.1 Error Analysis

To quantitatively evaluate the accuracy of the georeferencing implemented using FPGA, the root mean squared error *(RMSE)* is used (Shlien 1979), i.e.,

$$RMSE_x = \sqrt{\frac{\sum_{k=1}^{n}(x_k' - x_k)^2}{n}}, \tag{7.36}$$

TABLE 7.2

The Two Data Sets

Image	Projection	Zone	Resolution (m)	Band	Wavelength (μm)
bldt_tm.img	UTM	13	30×30	3	0.63–0.69
tmAtlanta.img	State plan (NAD 27)	3676	30×30	4	0.76–0.90

(a)

(b)

FIGURE 7.20   The raw image. (a) The first raw image. (b) The second raw image.

$$RMSE_y = \sqrt{\frac{\sum_{k=1}^{n}(y'_k - y_k)^2}{n}}, \qquad (7.37)$$

$$RMSE = \sqrt{RMSE_x^2 + RMSE_y^2}, \qquad (7.38)$$

where x'_k and y'_k are coordinates of the georeferenced image which are computed by the proposed method; x_k and y_k are reference geodetic coordinates; and n is the number of check points (CPs).

To compute $RMSE$s, 100 CPs are selected (Figure 7.21).

With the experimental results, a few conclusions can be drawn up as follows:

1. From Equations (7.36)–(7.38), the $RMSE$s are computed based on FPGA, MATLAB, Visual Studio (C++) and ENVI, as shown in Table 7.3. It can be concluded that the $RMSE_x$, $RMSE_y$ and $RMSE$ of the georeferenced image of the first data set implemented using FPGA are 0.1441, 0.1672 and 0.2207 pixels, respectively. The accuracy

(a)

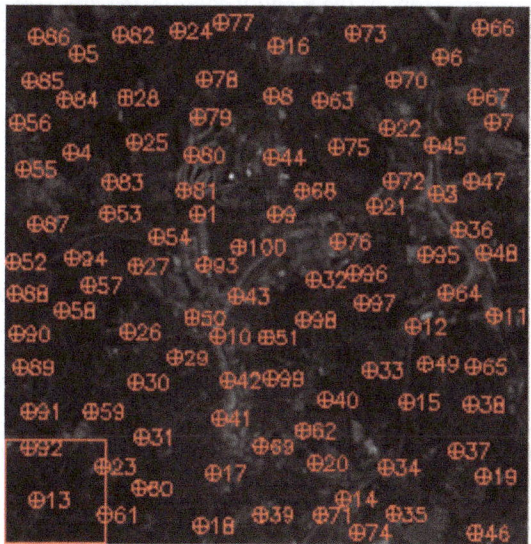

(b)

FIGURE 7.21   CP distribution in (a) the first image and (b) the second image.

TABLE 7.3

The Values of the RMSEs of the Georeferenced First Image

Platform	$RMSE_x$	$RMSE_y$	$RMSE$
FPGA	0.1441	0.1672	0.2207
MATLAB	0.1441	0.1672	0.2207
Visual Studio (C++)	0.1441	0.1672	0.2207
ENVI	0.1442	0.1699	0.2229

TABLE 7.4

Statistics Values of the Georeferenced First Image

Platform	Coordinates	Max Error (Pixel)	Min Error (Pixel)	Mean Error (Pixel)
FPGA	x	0.51	0.001	0.078
	y	0.41	0.003	0.095
MATLAB	x	0.51	0.001	0.078
	y	0.40	0.003	0.095
Visual Studio (C++)	x	0.51	0.001	0.078
	y	0.40	0.003	0.095
ENVI	x	0.53	0.001	0.083
	y	0.53	0.001	0.103

of the georeferenced using FPGA has the same values when implemented by MATLAB and Visual studio (C++), and is close to the values implemented using ENVI software. Additionally, other statistics, such as maximum, minimum and mean error are computed and listed in Table 7.4. It can be found that the proposed FPGA-based algorithm has a mean error of x and y coordinates at 0.0775 pixels and 0.0945 pixels, respectively; the minimum and the maximum errors of the x and y coordinates are 0.0008 and 0.0026 pixels, and 0.5113 and 0.4038 pixels, respectively. The accuracy of the various errors is calculated by the proposed algorithm has the same values than those based on MATLAB and Visual Studio (C++), and has a close accuracy than that based on ENVI software.

2. Table 7.5 lists the *RMSE*s of the georeferenced image of the second data set. The values of the $RMSE_x$, $RMSE_y$ and *RMSE* implemented by FPGA are 0.0965, 0.1268 and 0.1593 pixels, respectively. The *RMSE*s implemented using FPGA have the same accuracy than those based on MATLAB and Visual Studio (C++), and have a close accuracy implemented using ENVI software. But it can be considered acceptable for the absolute error is less than one pixel (Richards 1999). In addition, maximum, minimum and mean errors are computed and listed in Table 7.6. It can be found that the mean errors implemented using FPGA of the x and y coordinates are 0.0789 and 0.1144 pixels, respectively, and the minimum and maximum errors for the x and y coordinates are 0.0008 and 0.0046 pixels, and 0.2613 and 0.2081 pixels, respectively.

As observed form Tables 7.3–7.6, the accuracy of the georeferenced image when implemented using FPGA can reach the requirements because its *RMSE*s are less than one pixel (Schowengerdt 2007).

Figure 7.22(a)–(d) shows the georeferenced images of the first data set implemented by FPGA, MATLAB, Visual Studio (C++) and ENVI, respectively. Figure 7.23(a)–(d) shows georeferenced images of the second image implemented by FPGA, MATLAB, Visual Studio (C++) and ENVI, respectively.

TABLE 7.5

The Values of *RMSE*s of the Georeferenced Second Image

Platform	$RMSE_x$	$RMSE_y$	RMSE
FPGA	0.0965	0.1268	0.1593
MATLAB	0.0965	0.1268	0.1593
Visual Studio (C++)	0.0965	0.1268	0.1593
ENVI	0.0897	0.1009	0.1350

TABLE 7.6

Statistics Values of the Georeferenced Second Image

Platform	Coordinates	Max Error (Pixel)	Min Error (Pixel)	Mean Error (Pixel)
FPGA	x	0.2613	0.0008	0.0789
	y	0.2081	0.0046	0.1144
MATLAB	x	0.2613	0.0008	0.0789
	y	0.2081	0.0046	0.1144
Visual Studio (C++)	x	0.2613	0.0008	0.0789
	y	0.2081	0.0046	0.1144
ENVI	x	0.4039	0.0003	0.0588
	y	0.4039	0.0003	0.0673

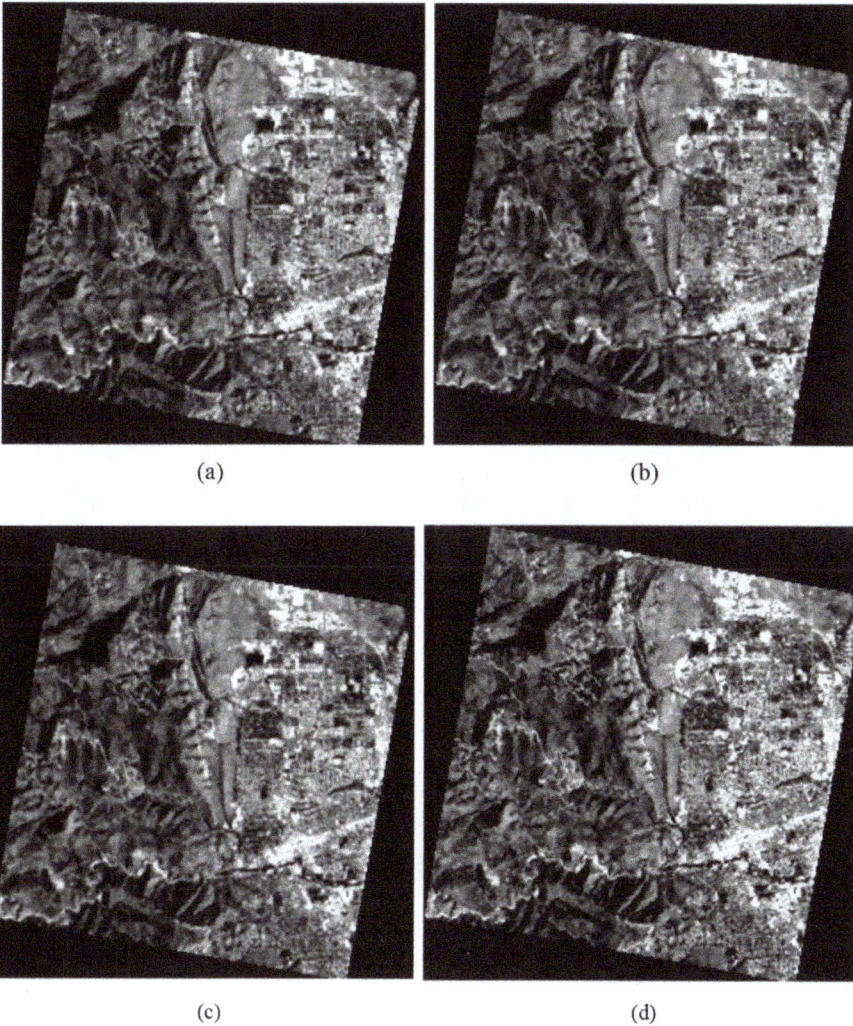

(a)   (b)

(c)   (d)

FIGURE 7.22   The georeferenced image of the first data set. (a) By FPGA. (b) By MATLAB. (c) By Visual Studio (C++). (d) By ENVI 5.3.

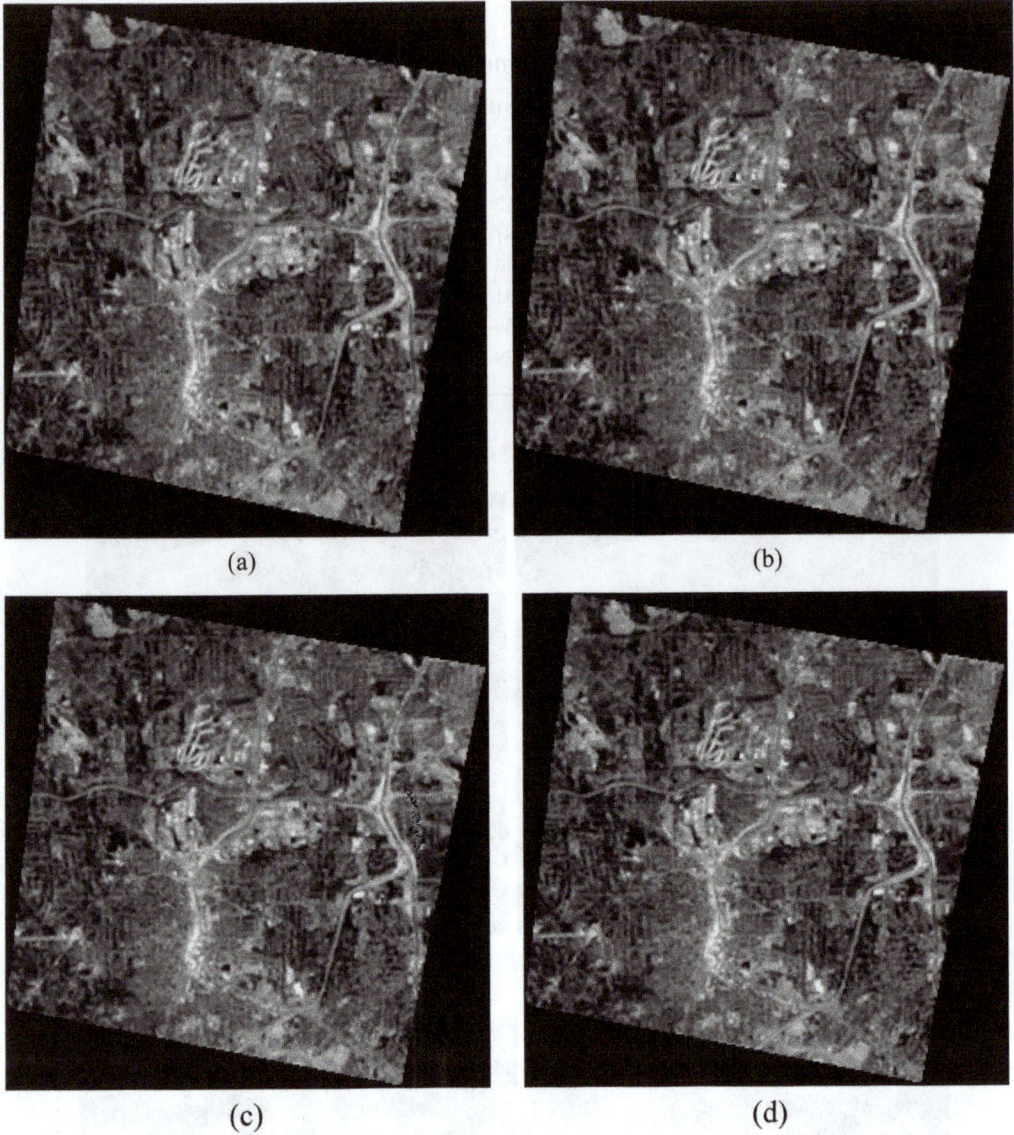

FIGURE 7.23 The georeferenced image of the second data set. (a) By FPGA. (b) By MATLAB. (c) By Visual Studio (C++). (d) By ENVI 5.3.

7.4.3.2 Gray Value Comparison

To verify the accuracy of gray value, the gray levels of georeferenced image are compared to those implemented using FPGA, MATLAB, Visual Studio (C++) and ENVI software. The georeferenced image implemented using FPGA as a referencing image, called "Ref-Img". The georeferenced image implemented using MATLAB, Visual Studio (C++) and ENVI are called "Img-MATLAB", "Img-C++" and "Img-ENVI", respectively. The gray differences between the Ref-Img and those georeferenced images implemented by MATLAB, Visual Studio (C++) and ENVI are obtained and shown in Figures 7.24 and 7.25.

As observed from Figure 7.24, the gray values of both Figures 7.24(a) and (b) are 0, which means the proposed method implemented using FPGA has the same accuracy with the implemented by MATLAB and Visual Studio (C++) (French et al. 2013; Shaffer 2018). Figure 7.24(c) indicates that

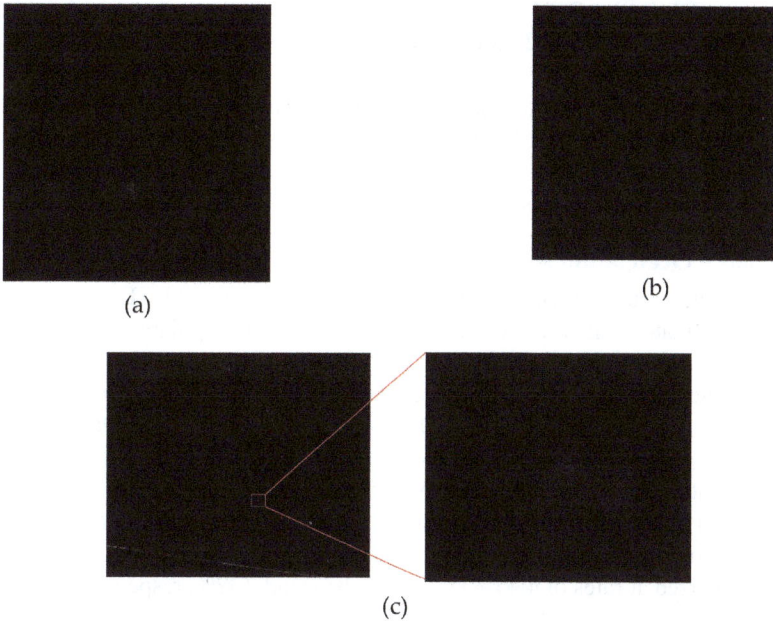

FIGURE 7.24   The differences between referencing image and other georeferenced images of the first image. (a) The differences between Ref-Img and Img-C++. (b) The differences between Ref-Img and Img-MATLAB. (c) The differences between Ref-Img and Img-ENVI.

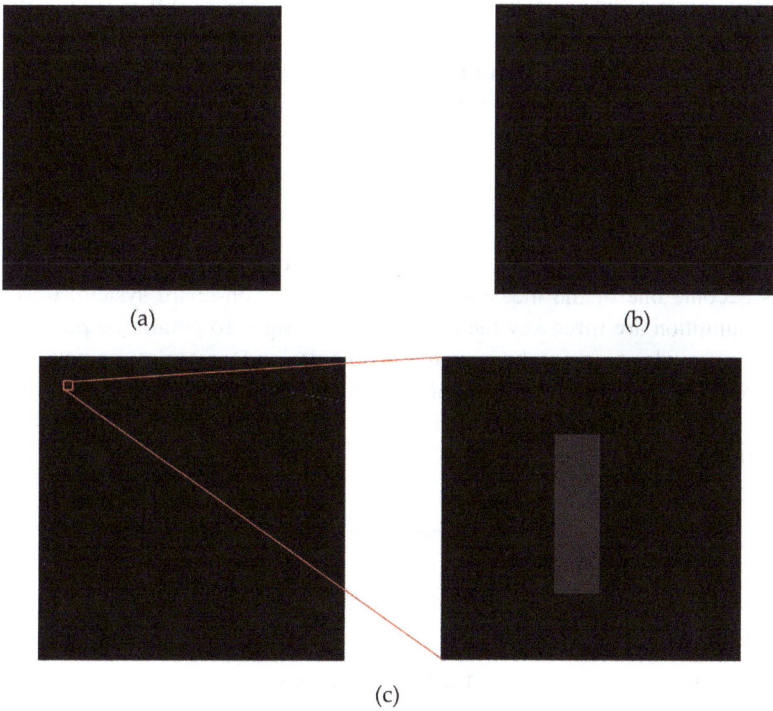

FIGURE 7.25   The differences between referenced image and other georeferenced images of the second image. (a) The differences between Ref-Img and Img-C++. (b) The differences between Ref-Img and Img-MATLAB. (c) The differences between Ref-Img and Img-ENVI.

the difference image between Ref-Img and Img-ENVI has many pixels are not same. The mean values of both Figures 7.24(a) and (b) are 0. The mean value of Figure 7.24(c) is 0.7143, which is less than 1.

As observed from Figure 7.25, the proposed method has the same accuracy as those implemented using MATLAB and Visual Studio (C++), and has a small portion of the gray values are different from the Img-ENVI. The mean values of Figures 7.25(a) and (b) are 0. The mean value of Figure 7.25(c) is 0.1265, which is less than 1.

7.4.3.3   Resource Occupation Analysis

The resource utilization ration, including the flip-flop (FF), look-up-table (LUT) and DSP48s of the FPGA for the coordinate transformation method and bilinear interpolation function are assessed, respectively.

In the coordinate transformation method, 250,656 LUTs, 499,268 registers and 388 DSP48s are utilized at rates of 40.96% (250,656/612,000=40.96%), 40.79% and 37.96%, respectively (see Table 7.7).

In the bilinear interpolation function, floating-point and 32-bit fixed-point mixed operations are adopted, which can reduce the resource consumption of an FPGA. Table 7.8 lists the resources occupied for the bilinear interpolation scheme; 27,218 LUTs, 45,823 registers, 456 RAM/FIFO and 267 DSP48s are utilized at rates of 4.45%, 3.74%, 30.40% and 7.42%, respectively.

7.4.3.4   Processing Speed Comparison

The processing speed is considered as one of the most important factors for implementation by an FPGA. Table 7.9 lists the processing speed implemented by FPGA, Visual Studio (C++) and MATLAB. The size of the first raw image is 256×256 pixels^2. After georeferencing, the size of georeferenced image is 281×281 pixels^2. The running time of the georeferencing method for the first image using FPGA, Visual Studio (C++) and MATLAB is 0.13, 1.06 and 1.12 s, respectively. The size of the second raw image is 256×256 pixels^2; after georeferencing, the image size is 285×277 pixels^2. The running time of the georeferencing method for the second raw image using FPGA, Visual Studio (C++) and MATLAB is 0.15, 1.21 and 1.26 s, respectively. To put it simply, the processing speed using FPGA is 8 times faster than that based on PC computer.

7.4.3.5   Power Consumption

With the development of technology and the improvement of system performance, low power consumption has become one of the measurement objectives of on-board system. Resources, speed and power consumption are three key factors in FPGA design. To obtain the power consumption, Vivado software provides a comprehensive methodologies and strategies for power consumption. As observed from Figure 7.26, the powers of the dynamic and the device are 0.280 W (43%) and 0.379 W (57%), respectively. the total on-chip power is 0.659 W, which is acceptable in on-board processing platform.

TABLE 7.7

The Logic Unit Utilization Ratio of the Coordinate Transformation Method

Parameter	Used	Available	Utilization Ratio (%)
Number of slice LUTs	250,656	612,000	40.96
Number of slice Registers	499,268	1,224,000	40.79
Number of DSP48s	388	3,600	37.96

TABLE 7.8

The Logic Unit Utilization Ratio of the Bilinear Interpolation Method

Parameters	Used	Available	Utilization Ratio (%)
Number of slice LUTs	27,218	612,000	4.45
Number of slice Registers	45,823	1,224,000	3.74
Number of block RAM/FIFO	456	1,500	30.40
Number of DSP48s	267	3,600	7.42

TABLE 7.9

The Consumption Speed

Raw Images	FPGA (s) F = 100MHz	Visual Studio (C++) (s)	MATLAB (s)	Size (Pixels^2)
bldt_tm.img	0.13	1.06	1.12	281×281
tmAtlanta.img	0.15	1.21	1.26	285×277

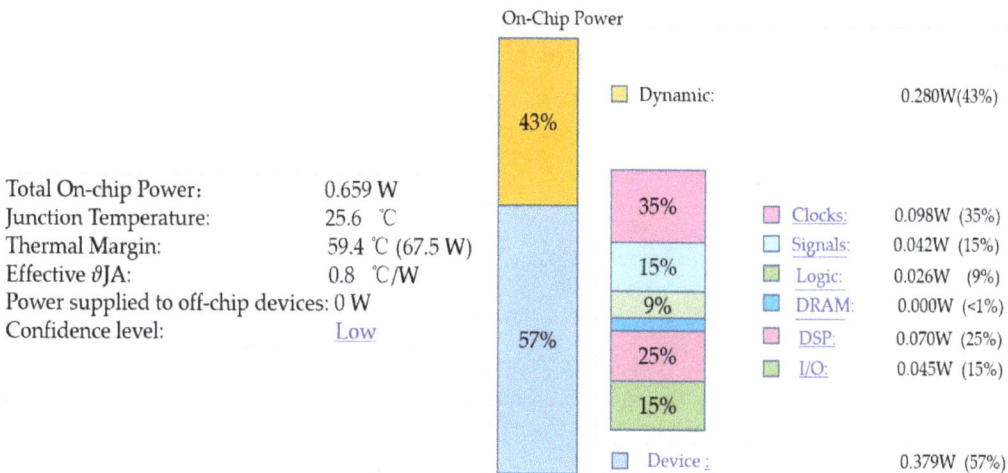

On-Chip Power

Total On-chip Power: 0.659 W
Junction Temperature: 25.6 ℃
Thermal Margin: 59.4 ℃ (67.5 W)
Effective ϑJA: 0.8 ℃/W
Power supplied to off-chip devices: 0 W
Confidence level: Low

Dynamic: 0.280W(43%)   43%

Clocks: 0.098W (35%)   35%
Signals: 0.042W (15%)   15%
Logic: 0.026W (9%)   9%
DRAM: 0.000W (<1%)
DSP: 0.070W (25%)   25%
I/O: 0.045W (15%)   15%

Device : 0.379W (57%)   57%

FIGURE 7.26 Power consumption.

7.5 CONCLUSIONS

This chapter describes on-board georeferencing using optimized second-order polynomial equation on the basis of FPGA, which consists of five modules: input data, coordinate transformation, bilinear interpolation and output data.

First, a comprehensive framework was developed to optimize the georeferencing algorithm based on an FPGA, which consists of (1) a floating-point block LU decomposition is used to inverse the matrix. Compared with the traditional LU decomposition, the multiplication and division operations are reduced by 1.02 and 1.25 times, respectively. The block LU decomposition method can reduce the complexity for inverting matrix and speed up the operation. (2) In order to reduce resource consumption of an FPGA, some strategies are adopted in programming, i.e., 32-bit integer and floating-point mixed operation and serial-parallel data communication.

Second, the performances of the proposed algorithm are evaluated by error analysis, gray value comparison, resource occupation analysis, processing speed comparison and power consumption. With the experimental results, it can be found that the accuracy of the FPGA-based georeferenced image are the same as those implemented using MATLAB and Visual Studio (C++). The processing speed using the proposed algorithm is 8 times faster than that based on PC computer does. Therefore, it can be concluded that the proposed georeferencing algorithm using FPGA with second-order polynomial model can achieve real-time operation.

REFERENCES

Bailey, D.G., *Design for Embedded Image Processing on FPGAs*. Wiley-IEEE Press, 2011, 275–305, ISBN: 9780470828519.

Bannari, A., Morin, D., Bénié, G.B., Bonn, F.J., A theoretical review of different mathematical models of geometric corrections applied to remote sensing images. *Remote Sensing Reviews*, 1995, 13: 27–47. doi: 10.1080/02757259509532295.

Chen, J., Ji, K., Shi, Z., Liu, W., Implementation of block algorithm for LU factorization. Los Angeles, CA, USA, 31 March–2 April 2009, 569–573.

Chen, J., Joang, T., Lu, W., Han, M., The geometric correction and accuracy assessment based on cartosat-1 satellite image. Image and signal processing (CISP). *2010 3rd International Congress on, IEEE*, Yantai, China, 16–18 October, 2010; pp. 1253–1257.

Chen, L.C., Teo, T.A., Liu, C.L., The geometrical comparisons of RSM and RFM for FORMOSAT-2 satellite images. *Photogrammetric Engineering and Remote Sensing*, 2006, 72: 573–579.

Dawood, A.S., Visser, S.J., Williams, J., A. Reconfigurable FPGAs for real time image processing in space. *In 2002 14th International Conference on Digital Signal Processing Proceedings*. DSP 2002 (Cat. No. 02TH8628), Greece, 1–3 July 2002; pp. 845–848.

Fang, L., Wang, M., Li, D., Pan, J., CPU/GPU near real-time preprocessing for ZY-3 satellite images: Relative radiometric correction, MTF compensation, and geocorrection. *ISPRS Journal of Photogrammetry and Remote Sensing*, 2014, 87: 229–240.

French, J.C., Balster, E.J., Turri, W.F., A 64-bit ortho-rectification algorithm using fixed-point arithmetic. *High-Performance Computing in Remote Sensing III, SPIE Remote Sensing*, Dresden, Germany, 2013: 8895.

Gill, T., Collett, L., Armston, J., Eustace, A., Danaher, T., Scarth, P., Flood, N., Phinn, S., Geometric correction and accuracy assessment of landsat-7 etm+ and landsat-5 tm imagery used for vegetation cover monitoring in Queensland, Australia from 1988 to 2007. *Surveyor*, 2010, 55: 273–287.

González, G., Bernabé, S., Mozos, D., Plaza, A., FPGA implementation of an algorithm for automatically detecting targets in remotely sensed hyperspectral images. *IEEE Journal of Selected Topics in Applied Earth Observations and Remote Sensing*. 2016, 9: 4334–4343.

González, C., Mozos, D., Resano, J., Plaza, A., FPGA implementation of the n-FINDR algorithm for remotely sensed hyperspectral image analysis. *IEEE Transactions on Geoscience and Remote Sensing*, 2012, 50: 374–388.

Govindu, G., Prasanna, V.K., Daga, V., Gangadharpalli, S., Sridhar, V., Efficient floating-point based block LU decomposition on FPGAs. *International Conference on Engineering of Reconfigurable Systems and Algorithms*, Ersa'04, Las Vegas, Nevada, USA DBLP, 21–24 June 2004; pp. 276–279.

Gribbon, K.T., Bailey, D.G., A novel approach to real-time bilinear interpolation. *IEEE International Workshop on Electronic Design, Test and Applications*, IEEE Computer Society, Perth, WA, Australia, 28–30 January 2004; pp. 126–131.

Huang, J., *FPGA-Based Optimization and Hardware Implementation of P-H Method for Satellite Relative Attitude and Absolute Attitude Solution*. Tianjin: Tianjin University, 2019, 1.

Huang, J., Zhou, G., Zhang, D., Zhang, G., Zhang, R., Baysal, O., An FPGA-based implementation of corner detection and matching with outlier rejection. *International Journal of Remote Sensing*, August, 2018a, 39(23), 1–20.

Huang, J., Zhou, G., Zhou, X. A new FPGA architecture of fast and brief algorithm for on-board corner detection and matching. *Sensors*. 2018b, 18: 1014, doi:10.3390/s18041014.

Jensen, J.R., Lulla, K. Introductory digital image processing-a remote sensing perspective. *Environmental and Engineering Geoscience*, 2007, 13: 89–90. doi: 10.2113/gseegeosci.13.1.89.

Joyce, K.E., Belliss, S.E., Samsonov, S.V., Mcneill, S.J., Glassey, P.J., A review of the status of satellite remote sensing and image processing techniques for mapping natural hazards and disasters. *Progress in Physical Geography*, 2009, 33: 183–207. doi:10.1177/0309133309339563.

Kartal, H., Sertel, E., Alganci, U. Comparative analysis of different geometric correction methods for very high resolution Pleiades images. *Recent Advances in Space Technologies (RAST), 2017 8th International Conference on IEEE*, Istanbul, Turkey, 19–22 June 2017; pp. 171–175.

Lee, C.A., Gasster, S.D., Plaza, A., Chang, C.I., Huang, B. Recent developments in high performance computing for remote sensing: A review. *IEEE Journal of Selected Topics in Applied Earth Observations & Remote Sensing*, 2011, 4: 508–527.

Liu, D., Zhou, G., Huang, J., Zhang, R., Shu, L., Zhou, X., Xin, C.S., On-board georeferencing using FPGA-based optimized second-order polynomial equation. *Remote Sensing*, 2019, 11(2): 124. doi: 10.3390/rs11020124.

Liu, D., Zhou, G., Zhang, Dianjun, et al. Ground control point automatic extraction for spaceborne georeferencing based on FPGA, *IEEE Journal of Selected Topics in Applied Earth Observations and Remote Sensing*, 2020, 13: 3350–3366. doi:10.1109/JSTARS.2020.2998838.

Liu, D., Research on Parallel Algorithms Using FPGA for Geometric Correction and Registration of Remotely Sensed Imagery. *Dissertation, Tianjin University*, May 2022.

Long, T., Jiao, W., He, G., Zhang, Z. A fast and reliable matching method for automated georeferencing of remotely-sensed imagery. *Remote Sensing*, 2016, 8: 56.

Lopez, S., Vladimirova, T., Gonzalez, G., Resano, J., Mozos, D. The promise of reconfigurable romputing for hyperspectral imaging onboard systems: A review and trends. *Proceedings of the IEEE*, 2013, 101: 698–722. doi:10.1109/JPROC.2012.2231391.

López-Fandiño, J., Barrius, P.Q., Heras, D.B., Argüello, F. Efficient ELM-based techniques for the classification of hyperspectral remote sensing images on commodity GPUs. *IEEE Journal of Selected Topics in Applied Earth Observations & Remote Sensing*, 2015, 8: 2884–2893.

Lu, J., Zhang, B., Gong, Z., Li, E., Liu, H., The remote-sensing image fusion based on GPU, *The International Archives of the Photogrammetry, Remote Sensing and Spatial Information Sciences*, Beijing, China, 23 October 2008, 1233–1238.

Ma, Y., Chen, L., Liu, P., Lu, K., Parallel programing templates for remote sensing image processing on GPU architectures: Design and implementation. *Computing*, 2016, 98: 7–33.

Pakartipangi, W., Darlis, D., Syihabuddin, B., Wijanto, H., Analysis of camera array on board data handling using FPGA for nano-satellite application. *International Conference on Telecommunication Systems Services and Applications*. IEEE, Bandung, Indonesia, 25–26 November 2015, pp. 1–6.

Plaza, A., Du, Q., Chang, Y.L., King, R.L. High performance computing for hyperspectral remote sensing. *IEEE Journal of Selected Topics in Applied Earth Observations and Remote Sensing*, 2011, 4: 528–544, doi: 10.1109/JSTARS.2010.2095495.

Plaza, A., Valencia, D., Plaza, J., Martinez, P., Commodity cluster-based parallel processing of hyperspectral imagery. *Journal of Parallel and Distributed Computing*. 2006, 66: 345–358.

Qi, B., Shi, H., Zhuang, Y., Chen, H., Chen, L., On-board, real-time preprocessing system for optical remote-sensing imagery. *Sensors*, 2018, 18: 1328. doi: 10.3390/s18051328.

Raffa, M., Mercogliano, P., Galdi, C., Georeferencing raster maps using vector data: A meteorological application. *Metrology for Aerospace*, IEEE, Florence, Italy, 22–23 June 2016, pp. 102–107.

Reguera-Salgado, J., Calvino-Cancela, M., Martin-Herrero, J. GPU geocorrection for airborne pushbroom imagers. *IEEE Transactions on Geoscience and Remote Sensing*, 2012, 50: 4409–4419.

Richards, J.A., *Remote Sensing Digital Image Analysis*, Germany/Springer: Berlin/Heidelberg, 1999, 39. ISBN: 978-3-642-30062-2.978-3-642-30062-2.

Sanyal, J., Lu, X.X. Application of remote sensing in flood management with special reference to monsoon Asia: A review. *Natural Hazards*, 2004, 33: 283–301.

Savoy, F.M., Dev, S., Lee, Y.H., Winkler, S., Geo-referencing and stereo calibration of ground-based whole sky imagers using the sun trajectory, *Geoscience and Remote Sensing Symposium*, IEEE, Beijing, China, 10–15 July 2016, pp. 7473–7476.

Schowengerdt, R.A. Chapter 7-correction and calibration. *Remote Sensing*, 3rd ed. Academic Press: Cambridge, MA, USA, 2007, P. 285–XXII, ISBN: 978-0-12-369407-2.

Shaffer, D.A., *An FPGA implementation of large-scale image ortho-rectification*. Doctoral dissertation, University of Dayton, 2018, dayton1523624621509277.

Shlien, S., Geometric correction, registration, and resampling of Landsat imagery. *Canadian Journal of Remote Sensing*, 1979, 5: 74–89.

Swann, R., Hawkins, D., Westwellroper, A., Johnstone, W., The potential for automated mapping from geocoded digital image data. *Photogrammetric Engineering and Remote Sensing*, 1988, 54: 187–193.

Thomas, O., Trym, V.H., Ingebrigt, W, Ingebrigt, W., Real-time georeferencing for an airborne hyperspectral imaging system. *Algorithms and Technologies for Multispectral, Hyperspectral, and Ultraspectral Imagery XVII*, 2011, 8048: 80480S.

Toutin, T., Geometric processing of remote sensing images: Models, algorithms and methods. *International Journal of Remote Sensing*, 2004, 25: 1893–1924.

Tralli, D.M., Blom, R. G., Zlotnicki, V., Donnella, A., Evans, D.L., Satellite remote sensing of earthquake, volcano, flood, landslide and coastal inundation hazards. *ISPRS Journal of Photogrammetry and Remote Sensing*, 2005, 59: 185–198.

Van der Jeught, S., Buytaert, J.A.N., Dirckx, J. J., Real-time geometric lens distortion correction using a graphics processing unit, *Optical Engineering*, 2012, 51(2), 027002. https://doi.org/10.1117/1.OE.51.2.027002.

Wang, T., Zhang, G., Li, D., Tang, X., Pan, H., Zhu, X., Chen, C., Geometric accuracy validation for ZY-3 satellite imagery. *IEEE Geoscience and Remote Sensing Letters*, 2014, 11: 1168–1171.

Williams, J.A., Dawood, A.S., Visser, S.J., FPGA-based cloud detection for real-time onboard remote sensing. *In Field-Programmable Technology, 2002. (FPT). Proceedings*. Hong Kong, China, China, 16–18 December, 2002, pp. 110–116.

Yu, G., Vladimirova, T., Sweeting, M.N., Image compression systems on board satellites. *Acta Astronautica*. 2009, 64: 988–1005, doi: 10.1016/j.actaastro.2008.12.006.

Zhang, Y., Kerle, N., Satellite remote sensing for near-real time data collection. *Geospatial Information Technology for Emergency Response*. CRC Press, 2008, 91–118. ISSN: 9780203928813

Zhou, G., Geo-referencing of video flow from small low-cost civilian UAV. *IEEE Transactions on Automation Engineering and Science*, 2010, 7: 156–166.

Zhou, G., *Urban High-Resolution Remote Sensing: Algorithms and Modelling*, Taylor & Francis/CRC Press, 2020, ISBN: 978-03-67-857509, 465 pages.

Zhou, G., Near real-time orthorectificatoin and nosaic of small UAV-based video flow for time-critical event response. *IEEE Transactions on Geoscience and Remote Sensing*, 2009, 47: 739–747.

Zhou, G., Baysal, O., Kaye, J., Habib, S., Wang, C., Concept design of future intelligent earth observing satellites. *International Journal of Remote Sensing*, 2004, 25: 2667–2685.

Zhou, G., Chen, W., Kelmelis, J.A., Zhang, D., A comprehensive study on urban true ortho-rectification. *IEEE Transactions on Geoscience and Remote Sensing*, 2005, 43: 2138–2147.

Zhou, G., Huang, J., Shu, L., An FPGA-based p-h method on-board solution for satellite relative altitude. *Geomatics and Information Science of Wuhan University*, 2018, 43: 1–9.

Zhou, G., Li, R., Accuracy evaluation of ground points from IKONOS high-resolution satellite imagery, *Photogrammetric Engineering and Remote Sensing*, 2000, 66: 1103–1112.

Zhou, G., Yue, T., Shi, Y., Zhang, R., Huang, J., Second-order polynomial equation-based block adjustment for ortho-rectification of DISP imagery. *Remote Sensing*, 2016, 8: 680.

Zhou, G., Zhang, R., Liu, N., Huang, J., Zhou, X. On-board ortho-rectification for images based on an FPGA. *Remote Sensing*. 2017, 9: 874, doi: 10.3390/rs9090874.

Zhu, H., Cao, Y., Zhou, Z., Gong, M., Parallel multi-temporal remote sensing image change detection on GPU. *IEEE, International Parallel and Distributed Processing Symposium Workshops & PhD Forum*, IEEE Computer Society, Shanghai, China, 21–25 May 2012, pp. 1898–1904.

Ziboon, A.R.T., Mohammed, I.H., Accuracy assessment of 2D and 3D geometric correction models for different topography in Iraq. *Engineering and Technology Journal, Part (A) Engineering*. 2013, 31: 2076–2085. ISSN: 16816900 24120758.

8 On-Board Image Ortho-Rectification Using Collinearity

8.1 INTRODUCTION

Given technological development, remotely sensed images can be acquired quickly and easily. However, the speed of image processing cannot catch up with the speed of obtaining remote sensing images because of the limitations of image processing technology. Conventionally, to process the acquired images (such as mosaic, fusion and ortho-rectification), they need to be sent back to the ground performance center. Moreover, many traditional image processing systems, such as ENVI and ERDAS IMAGINE, are serial instruction systems based on personal computers (PC). Thus, these image processing systems hardly meet the demand in response of time-critical disasters, making the abundant image resources underutilized (Zhang 2019, Zhang et al. 2018).

Ortho-rectification is an essential step in the remotely sensed image processing, which aims to remove the geometric distortions and obtain the mapping-based geographic coordinates of the image. It is important for and the basis of the subsequent image processing and applications. The traditional ortho-rectification methods correct images and remove distortion pixel-by-pixel using a PC-based platform on the basis of a digital elevation model (DEM). It is difficult to achieve a demand for (near) real-time performance because the processing unit is a pixel and there is a great amount of image data. On the other hand, since the algorithm complexity of ortho-rectification is very high, serial instruction processing systems take much time to perform the ortho-rectification algorithms. Thus, how to improve the speed of the ortho-rectification process has become an urgent issue when applied in on-board processing of a spacecraft.

Due to the limitation of the speed of serial instruction processing, many parallel processing methods for image processing have been proposed, such as (Warpenburg and Siegel 1982; Wittenbrink and Somani 1995; Sylvain and Serge 1999; Halle 2001; Dai and Yang 2011; Pan et al. 2016; Jiang and Nooshabadi 2016). Pan et al. (2016) presented a fast motion estimation method to reduce the encoding complexity of the H.265/HEVC encoder implemented by Intel Xeon CPU E5-1620 v2 and random access memory (RAM). Jiang and Nooshabadi (2016) proposed a scalable massively parallel fast research algorithm to reduce the computational cost of motion estimation and disparity estimation using a central processing unit (CPU)/graphical processing unit (GPU). In these methods, the GPU and CPU are combined to process images. Although the ground parallel processing system had improved the speed of image processing, the RS images needed to be sent back to the ground processing centers. Within the entire process, much time is still wasted. Additionally, most of the parallel processing methods are based on the multiple task operating system of the GPU, which cannot essentially solve the problem of a serial instruction method (Zhang 2019).

An effective solution for the (near) real-time processing of image ortho-rectification is to perform the ortho-rectification on hardware. In recent decades, the field programmable gate array (FPGA) has been widely used in the image processing such as imaging compression (Escamilla-Hernández et al. 2008; Kate 2012), filtering (Pal et al. 2016; Wang et al. 2016; Zhang et al. 2016), edge detection (Ontiveros-Robles et al. 2017; 2016), real-time processing of video images (Li et al. 2016a, 2016b) and motion estimation (Botella et al. 2010; González et al. 2012; 2013) to make real-time processing come true. González et al. (2012; 2013) optimized matching-based motion estimation

DOI: 10.1201/9781003319634-8

algorithms using an Altera custom instruction-based paradigm and a combination of synchronous dynamic random access memory (SDRAM) and on-chip memory in Nios II processors, and presented a low-cost system. Botella et al. (2010) proposed an architecture for a neuromorphic robust optical flow based on FPGA, which was applied in a difficult environment. In addition to a very-high-speed integrated circuit hardware description language (VHDL) and Verilog HDL, OpenCL is usually used to design an FPGA (Rodriguez-Donate et al. 2015; Waidyasooriya et al. 2016). Waidyasooriya et al. (2016) proposed an FPGA architecture for three-dimensional (3D) finite difference time domain (FDTD) acceleration applying OpenCL, which solved the problem of designing time. Rodriguez-Donate et al. (2015) evaluated the use of a convolution operator in signal processing disciplines focused on FPGA evaluation under different optimizations with respect to thread and memory level exploitation, in which OpenCL was used. In the RS community, although there is also some research on applying FPGA to the real-time processing of RS images, it is still not enough to meet the requirement in practice. For example, Thomas et al. (2008) and Kalomiros and Lygouras (2008) proposed an image processing system by combining software and hardware that can improve the speed of image correction and mosaicing. David and Don (2010) presented a processing method whereby the computing process is migrated to an FPGA, which aimed at solving the problem when the number of pixels in an image was huge and the transformation calculation of the floating-point matrix was complex. Through applying the characteristics of FPGA parallel computing, the proposed method by David and Don (2010) improved the speed of image correction. Winfried (2001) designed an on-board bispectral infrared detection (BIRD) system based on the neural network processor NI1000, a digital signal processor (DSP) and an FPGA. The system can perform on-board radiometric correction, geometric correction and texture extraction. Malik et al. (2014) built a quick process hardware platform using an FPGA. The hardware platform could process 390 frames of 640×480 images per second. Tomasi et al. (2012) researched a stereo vision algorithm based on an FPGA to perform the correction of video graphics array (VGA) images (57 fps). Pierre et al. (Greisen et al. 2011) applied the pipeline method of an FPGA to correct the color of stereo video images. Kumar and Sridharan (2007) realized the real-time correction of images using an FPGA under a dynamic environment.

To our understanding, a hardware system for image correction is mainly about the field of real-time correction of video images, stereo-pair real-time correction, etc. There are few studies on ortho-rectification of remotely sensed images (Zhang 2019; Zhang et al. 2018). Thus, this chapter develops a hardware platform based on an FPGA for RS image ortho-rectification. Through decomposing the ortho-rectification algorithm, several basic algorithms of image processing can be obtained, reducing the complexity of the algorithm and reaching the purpose of (near) real-time ortho-rectification. The proposed FPGA-based ortho-rectification platform integrates three modules: a memory module, a coordinate transformation module (including the transformation from geodetic coordinates to photo coordinates and the transformation from photo coordinates to scanning coordinates), and an interpolation module.

8.2 ORTHO-RECTIFICATION USING FPGA

8.2.1 A Brief Review of the Ortho-Rectification Algorithm

Many ortho-rectification models have been proposed over the past few decades. According to the type of image, terrain of the covered area and geomorphic features, an appropriate model can be chosen to ortho-rectify the remotely sensed images. Generally, ortho-rectification models contain a rigorous correction model based on the collinearity condition equation, a rational function model, or an affine-transformation-based correction model. In this chapter, the collinearity condition equation is used to implement remotely sensed image ortho-rectification on the proposed FPGA platform.

The ortho-rectification method can be classified as the direct method and indirect method. The direct method applies the image coordinates of the original image to compute the coordinates of the ortho-photo, and the indirect method applies the image coordinates of the ortho-photo to compute the image coordinates of the original image. In this chapter, the indirect method is used. The processes of the indirect method based on the collinearity condition equation include the following:

1. **The determination of the geodetic coordinates of the pixels in the ortho-photo**
   Let (X_{g0}, Y_{g0}) be the geodetic coordinates of a marginal point on the left bottom of the ortho-photo; (I, J) be the column and row coordinates of an arbitrary pixel in the ortho-photo; Δx and Δy be the sample intervals of columns and rows, respectively; M be the scale denominator of the ortho-photo; and (X_g, Y_g) be the geodetic coordinates of pixel G. The geodetic coordinates of pixel G can be obtained by

$$\begin{cases} X_g = X_{g0} + M \bullet (I + 0.5) \bullet \Delta x \\ Y_g = Y_{g0} + M \bullet (J + 0.5) \bullet \Delta y. \end{cases} \tag{8.1}$$

   After determining the geodetic coordinates of pixels in the ortho-photo, the elevation (i.e., Z_g) of each pixel can be acquired through interpolating the digital surface model (DSM).

2. **The transformation from geodetic coordinates to photo coordinates**
   At the time of imaging, the ground point, the center of projection and the photo point are on a line. According to the collinearity relationship among them, the photo coordinates of the ground point can be obtained by:

$$\begin{cases} u = x_0 - f\dfrac{a_1(X_g - X_s) + b_1(Y_g - Y_s) + c_1(Z_g - Z_s)}{a_3(X_g - X_s) + b_3(Y_g - Y_s) + c_3(Z_g - Z_s)} \\[3mm] v = y_0 - f\dfrac{a_2(X_g - X_s) + b_2(Y_g - Y_s) + c_2(Z_g - Z_s)}{a_3(X_g - X_s) + b_3(Y_g - Y_s) + c_3(Z_g - Z_s)} \end{cases} \tag{8.2}$$

where (u, v) are the photo coordinates of the ground point; x_0, y_0 and f are the interior orientation elements; X_s, Y_s and Z_s are the exterior orientation elements; and a_h, b_h and c_h ($h = 1, 2, 3$) are the elements of the rotation matrix R that can be obtained by Equation (8.3).

$$\begin{aligned} R &= \begin{pmatrix} a_1 & b_1 & c_1 \\ a_2 & b_2 & c_2 \\ a_3 & b_3 & c_3 \end{pmatrix} \\ &= \begin{pmatrix} \cos\omega\cos\kappa - \sin\varphi\sin\omega\sin\kappa & \cos\omega\sin\kappa & \sin\varphi\cos\kappa + \cos\varphi\sin\omega\sin\kappa \\ -\cos\varphi\sin\kappa - \sin\omega\sin\varphi\cos\kappa & \cos\omega\cos\kappa & -\sin\varphi\sin\kappa + \cos\varphi\sin\omega\cos\kappa \\ -\sin\varphi\cos\omega & -\sin\omega & \cos\varphi\cos\omega \end{pmatrix} \end{aligned} \tag{8.3}$$

where φ, ω and κ are three rotational angles along the x-, y- and z-axes in coordinate transformation, respectively.

3. **The transformation from photo coordinates to scanning coordinates**
   Because there is affine deformation between the photo coordinate system and the scanning coordinate system, the following affine transformation is used to get the scanning coordinates for the pixels:

$$
\begin{bmatrix} i' \\ j' \end{bmatrix} = \begin{bmatrix} m_1 & n_1 \\ m_2 & n_2 \end{bmatrix} \begin{bmatrix} u \\ v \end{bmatrix} + \begin{bmatrix} k'_1 \\ k'_2 \end{bmatrix} + \begin{bmatrix} i'_0 \\ j'_0 \end{bmatrix}
$$

$$
= \begin{bmatrix} m_1 & n_1 \\ m_2 & n_2 \end{bmatrix} \begin{bmatrix} u \\ v \end{bmatrix} + \begin{bmatrix} k_1 \\ k_2 \end{bmatrix}
$$

(8.4)

where m_t, n_t and k'_t ($t = 1, 2$) are the coefficients of affine transformation; (i'_0, j'_0) are the scanning coordinates of the principle point; specially, $k_1 = k'_1 + i'_0$ and $k_2 = k'_2 + j'_0$; and (i', j') are the scanning coordinates of an arbitrary pixel.

4. **Gray-scale bilinear interpolation**
   The gray-scale of pixels in the ortho-photo can be determined according to the acquired scanning coordinates. Because the obtained scanning coordinates may not only be in the center of a pixel, a gray-scale interpolation process is required. In this chapter, the bilinear interpolation method is applied.

$$
\begin{aligned}
f(i+p, j+q) = (1-p)(1-q)f(i,j) + (1-p)qf(i,j+1) \\
+ p(1-q)f(i+1,j) + pqf(i+1,j+1)
\end{aligned}
$$

(8.5)

where i and j are nonnegative integers; p and q are in the range of $(0, 1)$; and $f(i, j)$ are gray values.

8.2.2 FPGA-BASED IMPLEMENTATION FOR ORTHO-RECTIFICATION

The parallel processing of an FPGA is a hot topic in the high-performance rapid calculation community. The calculation speed of an FPGA is affected by multi-factors (such as the

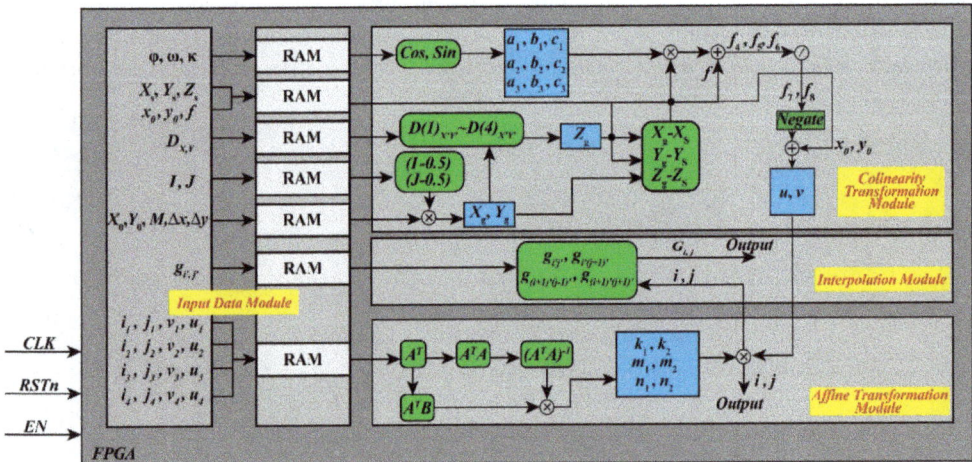

FIGURE 8.1 The proposed field programmable gate array (FPGA)-based architecture for the ortho-rectification of remotely sensed (RS) images. RAM, random access memory.

amount of logical resource in the chosen FPGA and the optimal design of algorithms). Through analyzing the structure of an ortho-rectification algorithm and optimizing it, an FPGA-based hardware architecture for ortho-rectification was designed in Figure 8.1, which consists of three modules: (i) an input data module; (ii) a coordinate transformation module that includes a collinearity equation transformation module (CETM) and an affine transformation module (ATM); and (iii) an interpolation module (IM). The details of the modules are described as follows.

1. The original data and parameters are stored in the RAM of the input data module. These original data and parameters are sent to the CETM, ATM and IM in the same clock cycle, when the enable signal is being received.
2. The coefficients of the collinearity conditional equation, geodetic coordinates of the ortho-photo and photo coordinates are calculated by the CETM. In the same clock cycle, the acquired photo coordinates are sent to the ATM.
3. The coefficients of affine transformation are calculated in the ATM and then the coefficients and photo coordinates are combined to calculate the scanning coordinates, which are sent to the IM and output in the same clock cycle.
4. The gray-scale of the ortho-photo is obtained by scanning the coordinates and cached gray-scale of the original image in the IM. In the same clock cycle, the obtained gray-scale of the ortho-photo is output to the external memory.

8.2.2.1 FPGA-Based Implementation for a Two-Row Buffer

As is well known, to perform bilinear interpolation, several neighborhoods of a pixel are needed. However, an FPGA is not like software for storing the whole image in the internal storage and reading the value of the image pixel according to the index. Thus, according to the size of the neighborhood, several rows of image data should be cached in advance. Moreover, the cache should be built to store the required image data. Many research studies have made efforts to design the structure of a buffer, such as (Hsiao et al. 2010; Cao et al. 2012; Kazmi et al. 2019; Hu and Zhu 2010). Especially, the structure of a state machine (Hu and Zhu 2010) is a useful method for solving the issue of buffers.

According to Hu and Zhu (2010), a two-row buffer based on an FPGA is proposed. As shown in Figure 8.2, four states, i.e., IDEL, the beginning of all, the end of all and beginning of the line, are used for initialization and cache cleaning. In the "transing" state, according to the control signal and the address of the reading and writing, the corresponding operations are

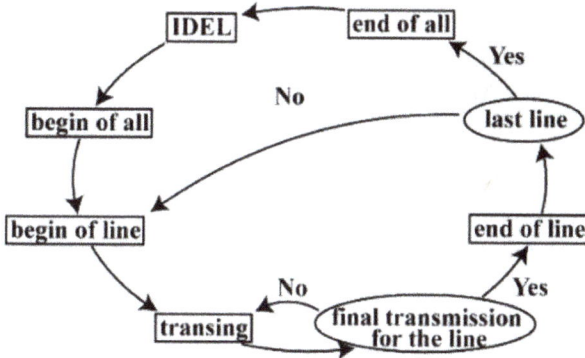

FIGURE 8.2 State diagram of cache for image input (Hu and Zhu 2010).

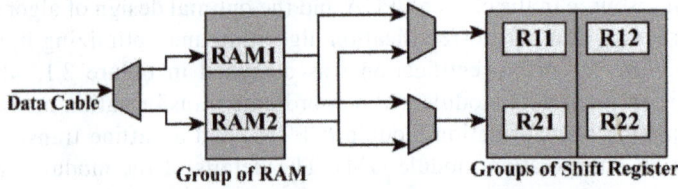

FIGURE 8.3   The model of the two-row buffer.

chosen, including storing the image date in the dual RAM from the external storage, waiting for image data to be taken from the dual RAM and turning to the "end of line" state. The number of dual RAM modules, i.e., n_{RAM}, can be determined by the number of rows of neighbors, i.e., n_{nb}, where n_{RAM} is equal to n_{nb}. To implement bilinear interpolation, 2×2 neighbors are needed. Thus, as shown in Figure 8.3, two dual RAMs are used as the cache media for the first level cache of image input. When two dual RAMs are used to buffer data, a multiplexer is exploited to determine which dual RAM should store the data. For 2×2 neighbors, two groups of cycle shift registers are used to store the data from neighboring pixels of the input image in the second level cache. For each group consisting of end-to-end registers, two registers, noted as R_{mn} ($m, n = 1, 2$), are contained. The data sent from the dual RAMs first are input selectors that ensure the row order is the same as the row order of the input image, and then the output of the selectors is sent to the groups of the cycle shift register. When storing the data from the dual RAMs into R_{m1} each time, the data stored in R_{mn} ($m = 1, 2, n = 1$) is assigned to $R_{m(n+1)}$ ($m = 1, 2, n = 1$), while the data stored in R_{m4} ($m = 1, 2$) are thrown out.

8.2.2.2   FPGA-Based Implementation for Coordinate Transformation

According to Section 8.2.1, the ortho-rectification method based on the collinearity condition equation needs two coordinate transformations. As shown from Equations (8.1) – (8.4), these equations involve compound operations. However, the hardware implementation for an algorithm is based on the most basic logic operations. Thus, the ortho-rectification method based on the collinearity equation must be divided into several simple add and multiplication operations, corresponding to the hardware components of the adder and multiplier, respectively. The details of the FPGA-based implementation for coordinate transformation modules are described as follows.

1. **FPGA-based implementation for calculating geodetic coordinates X_g, Y_g and Z_g**
   To obtain the geodetic coordinates X_g, Y_g and Z_g at part (1) of Section 8.2.1 using an FPGA chip, a parallel computation architecture is presented in Figure 8.4. As shown in Figure 8.4, four adders and four multipliers are used to compute X_g and Y_g in parallel by I, X_0, Δx, M, J, Δy and Y_0, respectively. After obtaining X_g and Y_g, they are sent to the interpolation module for calculating Z_g. To ensure that X_g, Y_g and Z_g are output in the same clock cycle, the delay units are utilized for computing processing. The details of the interpolation module are presented in the following Section 8.2.2.3, because its computing process is the same as gray interpolation.

2. **FPGA-based implementation for the transformation from geodetic coordinates to photo coordinates**
   For an FPGA-based implementation in parallel computing for the transformation from geodetic coordinates to photo coordinates, a modification of Equation (8.2) is needed. The intermediate variables produced in the computing process can be divided into three levels. The goal of the first level is to implement the margin calculation of the geodetic

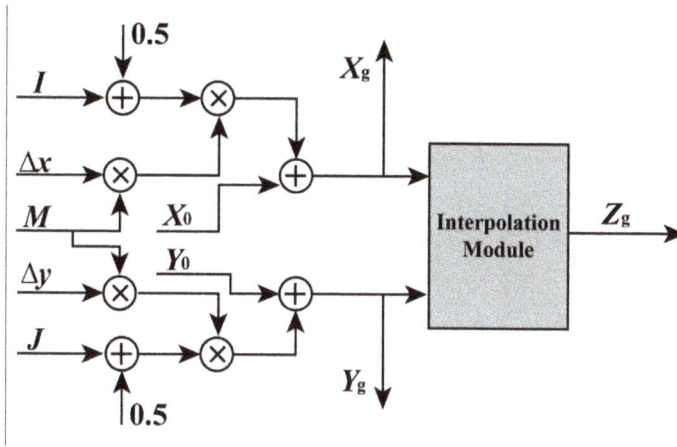

FIGURE 8.4 FPGA-based parallel computation architecture for calculating geodetic coordinates X_g, Y_g and Z_g.

coordinates between any point and the principal point of the photograph. The first level contains:

$$f_1 = X_g - X_s; \qquad f_2 = Y_g - Y_s; \qquad f_3 = Z_g - Z_s. \tag{8.6}$$

The second level (see Equation (8.7)) is to calculate the products between the transformation coefficients and intermediate variables f_1, f_2 and f_3.

$$f_4 = a_1 f_1 + b_1 f_2 + c_1 f_3; \qquad f_5 = a_2 f_1 + b_2 f_2 + c_2 f_3; \qquad f_6 = a_3 f_1 + b_3 f_2 + c_3 f_3 \tag{8.7}$$

The computation of the third level (see Equation (8.8)) is based on the first and the second levels to obtain the photo coordinates,

$$f_7 = f_4 / f_6; \qquad f_8 = f_5 / f_6. \tag{8.8}$$

To compute f_4, f_5 and f_6 in Equation (8.7), the elements of the rotation matrix R (i.e., a_i, b_i and c_i, i = 1, 2, 3) should be first calculated based on Equation (8.3). To compute the rotation matrix R using an FPGA chip, a parallel computation module is presented in Figure 8.5(a), in which a_i, b_i and c_i (i = 1, 2, 3) are calculated by the *sin* and *cos* functions of the three rotational angles φ, ω and κ. Through using the CORDIC IP core, the sin and cos functions can be implemented by an FPGA. To ensure that a_i, b_i and c_i (i = 1, 2, 3) are obtained in the same clock cycle, the delay units should be exploited in the computing process, where there are twelve multipliers and four adders.

To implement the transformation from geodetic to photo coordinates, i.e., Equations (8.6) – (8.8) using an FPGA chip, a parallel computation module is presented in Figure 8.5(b). As shown in Figure 8.5(b), the initial data (i.e., x_0, y_0, f, X_S, Y_S and Z_S) stored in the input data module, the rotation matrix calculated by using the *sin* and *cos* functions and rotation angles and the geodetic coordinates (X_g, Y_g and Z_g) calculated by Figure 8.4 are used to compute the photo coordinates (u, v). In the computing process, eleven multipliers, eight adders and two dividers are employed. Moreover, the delay units are also utilized to ensure that the outputs (u, v) are sent into the next module in the same clock cycle.

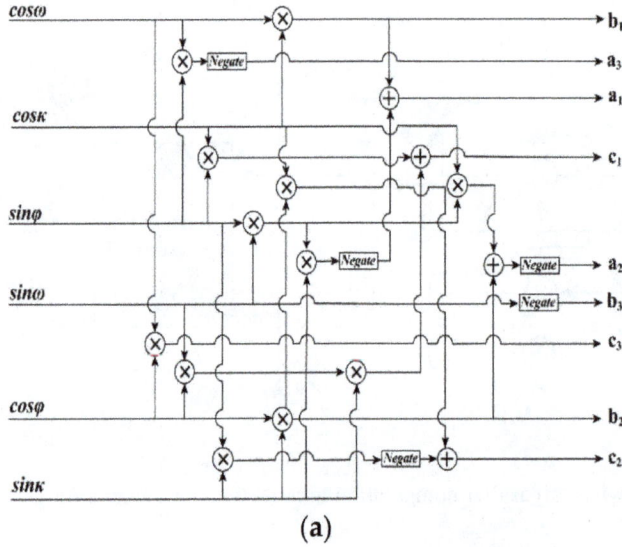

(a)

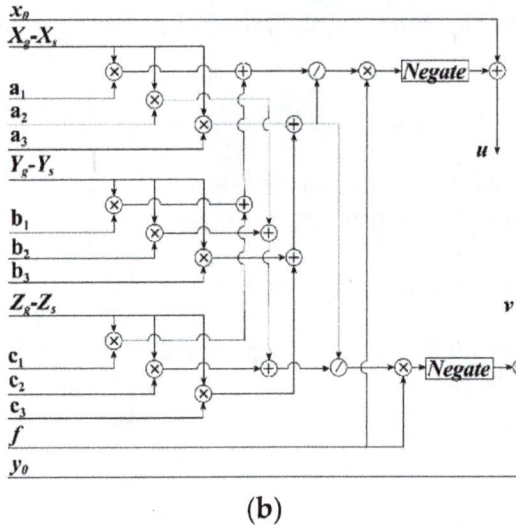

(b)

FIGURE 8.5   (a) FPGA-based computation for elements of rotation matrix R; (b) The FPGA implementation architecture for coordinate transformation from geodetic to photo coordinates.

3. **FPGA-based implementation for the transformation from photo to scanning coordinates**

   To realize the transformation from photo to scanning coordinates using an FPGA chip, Equation (8.4) is divided into two levels. The first level is:

$$g_1 = A_1 B; \qquad g_2 = A_2 B. \tag{8.9}$$

Additionally, the second level includes:

$$g_3 = g_1 + k_1; \qquad g_4 = g_2 + k_2 \tag{8.10}$$

$$i = g_3 + \ddot{i}_0; \qquad j = g_4 + \ddot{j}_0 \tag{8.11}$$

where $A_1 = (m_1, n_1)$ and $A_2 = (m_2, n_2)$ are the affine transformation coefficients, and $B = (u, v)^T$ are the photo coordinates. The transformation coefficients, i.e., m_s, n_s and k_s ($s = 1, 2$), should be calculated first based on the following least squares method.

To compute the coefficients of affine transformation, four control points are needed. Let H be the matrix of the control point scanning coordinates, Q be the matrix of the photo coordinates of the known fiducial points and C be the matrix of the affine transformation coefficients:

$$H = \begin{bmatrix} i_1 & j_1 & i_2 & j_2 & i_3 & j_3 & i_4 & j_4 \end{bmatrix}^T \tag{8.12}$$

$$Q = \begin{bmatrix} 1 & 0 & u_1 & 0 & v_1 & 0 \\ 0 & 1 & 0 & u_1 & 0 & v_1 \\ 1 & 0 & u_2 & 0 & v_2 & 0 \\ 0 & 1 & 0 & u_2 & 0 & v_2 \\ 1 & 0 & u_3 & 0 & v_3 & 0 \\ 0 & 1 & 0 & u_3 & 0 & v_3 \\ 1 & 0 & u_4 & 0 & v_4 & 0 \\ 0 & 1 & 0 & u_4 & 0 & v_4 \end{bmatrix} \tag{8.13}$$

$$C = \begin{bmatrix} k_1 & k_2 & m_1 & m_2 & n_1 & n_2 \end{bmatrix}^T \tag{8.14}$$

$$H = QC. \tag{8.15}$$

According to the least squares method, Equation (8.15) can be solved by:

$$C = (Q^T Q)^{-1} (Q^T H). \tag{8.16}$$

- **FPGA-based implementation for $Q^T Q$ and $Q^T H$**

  To implement the matrix multiplication $Q^T Q$ based on an FPGA chip, it can be rewritten using an optimum method. As shown in Equation (8.17), the elements of the matrix contain three formats, i.e., "$a_1 + b_1 + c_1 + d_1$", "$a_1^2 + b_1^2 + c_1^2 + d_1^2$" and "$a_1 a_2 + b_1 b_2 + c_1 c_2 + d_1 d_2$". Moreover, the matrix of $G = Q^T Q$ is symmetric, i.e., the elements of the lower triangular matrix are the same as the elements of upper triangular matrix. To save the resources of the FPGA chip, the upper triangular matrix is computed only in parallel based on the FPGA architecture of Figure 8.6. In the presented architecture, twelve multipliers and fifteen adders are used in the parallel computing process. The delay units are applied to ensure that the results are output into next module in the same clock cycle.

  Because the format of matrix E's elements is similar to the format of matrix G (see Equations (8.17), (8.18)), the FPGA-based architecture for the parallel computing process of $Q^T H$ can be referenced from the architecture of b_{13} and b_{35} in Figure 8.6. The details of the parallel computation for $Q^T H$ based on an FPGA are not repeated here.

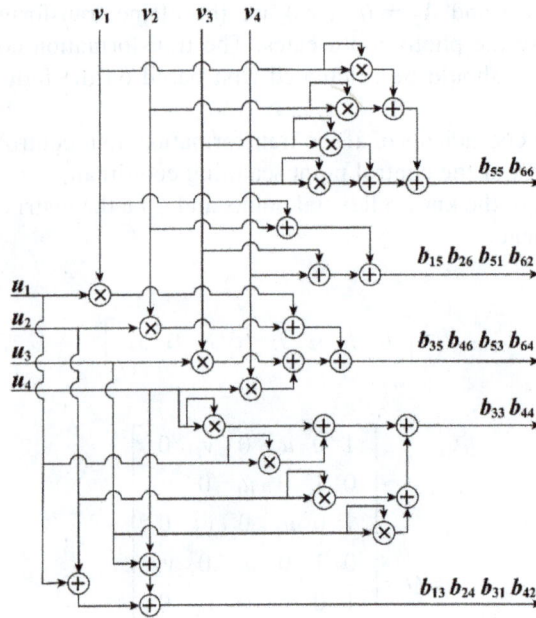

FIGURE 8.6 The FPGA implementation architecture of Q^TQ.

$$
G = Q^T Q = \begin{bmatrix}
1+1+1+1 & 0 & u_1+u_2+u_3+u_4 \\
0 & 1+1+1+1 & 0 \\
u_1+u_2+u_3+u_4 & 0 & u_1^2+u_2^2+u_3^2+u_4^2 \\
0 & u_1+u_2+u_3+u_4 & 0 \\
v_1+v_2+v_3+v_4 & 0 & u_1v_1+u_2v_2+u_3v_3+u_4v_4 \\
0 & v_1+v_2+v_3+v_4 & 0
\end{bmatrix}
$$

$$
\begin{bmatrix}
0 & v_1+v_2+v_3+v_4 & 0 \\
u_1+u_2+u_3+u_4 & 0 & v_1+v_2+v_3+v_4 \\
0 & u_1v_1+u_2v_2+u_3v_3+u_4v_4 & 0 \\
u_1^2+u_2^2+u_3^2+u_4^2 & 0 & u_1v_1+u_2v_2+u_3v_3+u_4v_4 \\
0 & v_1^2+v_2^2+v_3^2+v_4^2 & 0 \\
u_1v_1+u_2v_2+u_3v_3+u_4v_4 & 0 & v_1^2+v_2^2+v_3^2+v_4^2
\end{bmatrix}
\tag{8.17}
$$

$$
E = Q^T H = \begin{bmatrix} i_1+i_2+i_3+i_4 & j_1+j_2+j_3+j_4 & i_1u_1+i_2u_2+i_3u_3+i_4u_4 \\ j_1u_1+j_2u_2+j_3u_3+j_4u_4 & i_1v_1+i_2v_2+i_3v_3+i_4v_4 & j_1v_1+j_2v_2+j_3v_3+j_4v_4 \end{bmatrix}^T \tag{8.18}
$$

- **FPGA-based implementation for G^{-1}**

  To implement the inversion of matrix G (i.e., G^{-1}) based on the FPGA chip, it can be divided into two parts: (i) the implementation for decomposing matrix G using the LDL^T method and (ii) the implementation of G^{-1}. The details of the implementation are described as follows.

To implement the decomposing of matrix G, the LDL^T method is used to modify the matrix G as the following Equation (8.19):

$$G = \begin{bmatrix} g_{11} & g_{12} & \cdots & g_{1n} \\ g_{21} & g_{22} & \cdots & g_{2n} \\ \vdots & \cdots & \cdots & \vdots \\ g_{n1} & \cdots & \cdots & g_{nn} \end{bmatrix} = \begin{bmatrix} 1 & & & \\ l_{21} & 1 & & \\ \vdots & \cdots & \ddots & \\ l_{n1} & \cdots & l_{n(n-1)} & 1 \end{bmatrix} \begin{bmatrix} d_{11} & & & \\ & d_{22} & & \\ & & \ddots & \\ & & & d_{nn} \end{bmatrix} \begin{bmatrix} 1 & & & \\ l_{21} & 1 & & \\ \vdots & \cdots & \ddots & \\ l_{n1} & \cdots & l_{n(n-1)} & 1 \end{bmatrix}^T = LDL^T$$

(8.19)

where matrix L is a lower triangular matrix, matrix D is a diagonal matrix and L^T is a transposed matrix of L. The elements of matrix L and matrix D, i.e., l_{ij} and d_{ii}, can be solved by Equation (8.20):

$$\begin{cases} d_{ii} = g_{ii} - \sum_{k=1}^{i-1} l_{ik} d_{kk} l_{ik} & \text{(a)} \\ l_{ij} = (g_{ij} - \sum_{k=1}^{j-1} l_{ik} d_{kk} l_{jk}) / d_{jj} & \text{(b)} \end{cases}, \quad \begin{matrix} j = 1,2,\cdots,n \\ i = j+1, j+2,\cdots,n. \end{matrix}$$

(8.20)

To reduce the times of multiplication, an intermediate variable $u_{ij} = l_{ij} d_{jj}$ is introduced. Therefore, Equation (8.20) can be modified as Equation (8.21):

$$\begin{cases} d_{ii} = g_{ii} - \sum_{k=1}^{i-1} u_{ik} l_{ik} & \text{(a)} \\ u_{ij} = g_{ij} - \sum_{k=1}^{j-1} u_{ik} l_{jk} & \text{(b)} \\ l_{ij} = u_{ij} / d_{ii} & \text{(c)} \end{cases}, \quad \begin{matrix} j = 1,2,\cdots,n \\ i = j+1, j+2,\cdots,n. \end{matrix}$$

(8.21)

According to the characteristics of Equation (8.21), the FPGA-based architecture for calculating l_{ij} and d_{ii} is shown in Figure 8.7. In the LDL^T method, the d_{ii} are first calculated based on Equation (8.21); subsequently, the elements of the same column of matrix L are calculated in parallel. Moreover, Equation (8.21) shows that the elements of the later column of matrix L depend on the elements of the former column. As shown in Figure 8.7, five multipliers and twenty-five adders are used to calculate l_{ij} and d_{ii}. In the computing processing, delay units are applied to ensure that the results are output into the next process in the same clock cycle.

After completing the computation of l_{ij} and d_{ii}, the inversion of matrix G can be calculated on the basis of l_{ij} and d_{ii}. According to Equation (8.19), the inversion of matrix G can be rewritten as Equation (8.22):

$$G^{-1} = (LDL^T)^{-1} = (L^{-1})^T D^{-1} L^{-1}.$$

(8.22)

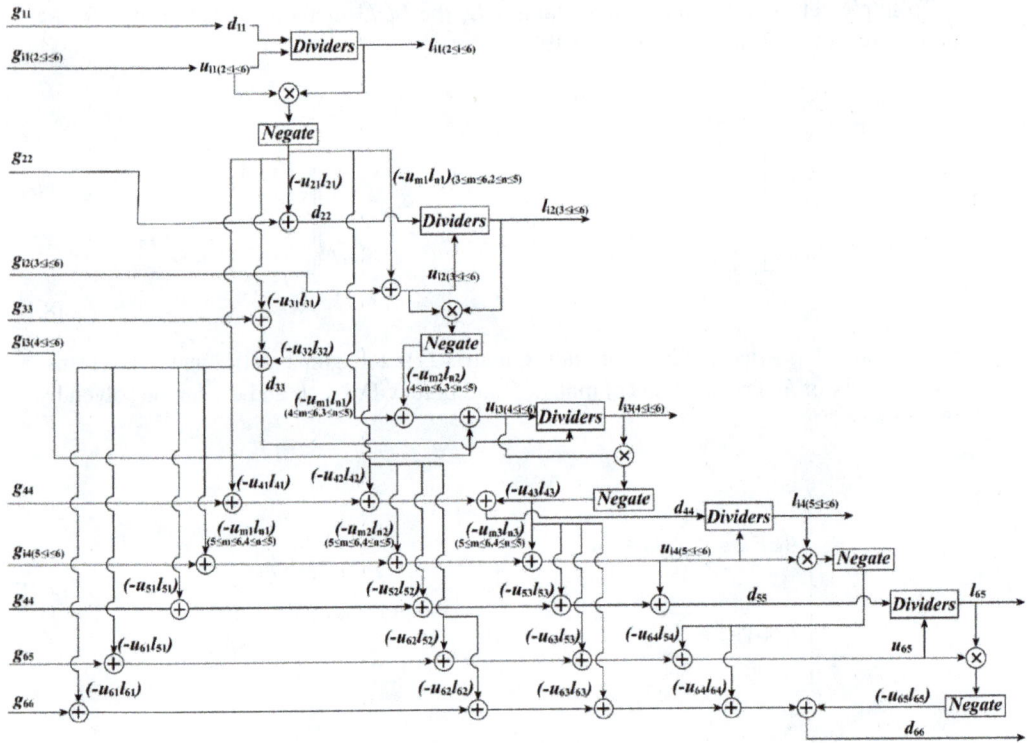

FIGURE 8.7   The FPGA-based architecture for calculating l_{ij} and d_{ii}.

To implement Equation (8.22) based on the FPGA chip, an FPGA-based architecture is presented in Figure 8.8. In the presented architecture, five parts are contained. The details of the presented architecture are described as follows.

In the (i) part, MUX is applied to construct the column elements of matrix G.

In the (ii) part, the LDL^T method is used to calculate the elements of matrix L and matrix D, i.e., l_{ij} and d_{ii}.

In the (iii) part, the inversions of matrix D and matrix L are calculated in parallel. Moreover, in this part, the first MUX is used to construct the vector of L, and L^{-1} is calculated using a systolic array architecture that is applied for the fast inversion of dense matrices (El-Amawy 1989). The second MUX of this part is utilized to construct the vector of $(L^{-1})^T$. Through calculating the reciprocal of the elements of the diagonal matrix D, the D^{-1} can be obtained and output. For D^{-1}, a row vector is constructed consisting of the elements of matrix D.

In the (iv) part, the transposed matrix of L^{-1} is multiplied by D^{-1}, denoted as $P = (L^{-1})^T D^{-1}$, and delay units are used for delaying the output of L^{-1}.

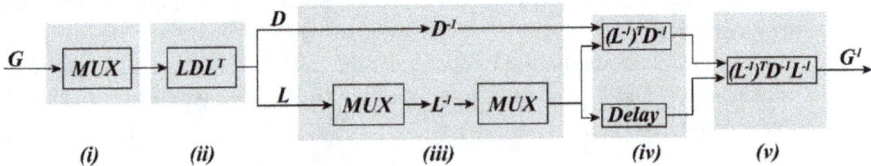

FIGURE 8.8   The FPGA-based architecture for the inversion of G.

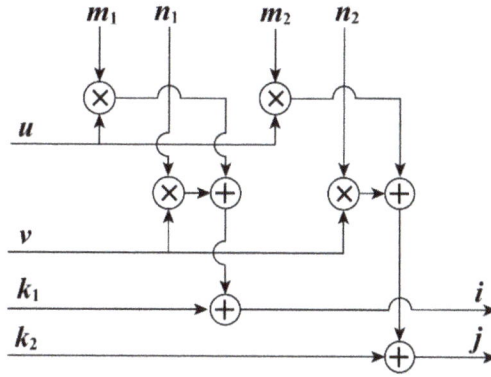

FIGURE 8.9 The FPGA-based architecture for the transformation from photo coordinates to scanning coordinates.

In the (v) part, through multiplying P and L^{-1}, the inversion of matrix G, i.e., G^{-1}, is obtained and output.

- **FPGA-based implementation for the transformation from photo to scanning coordinates.**
  After obtaining the coefficients of the affine transformation based on the FPGA chip, the scanning coordinates can be calculated with these coefficients, and the photo coordinates based on the FPGA-based architecture are presented in Figure 8.9. As shown in Figure 8.9, four multipliers and four adders are used.

8.2.2.3 FPGA-Based Implementation for Bilinear Interpolation

In the whole process of ortho-rectification, the interpolation process is needed in two stages, i.e., the interpolation for geodetic coordinate Z_g and gray-scale. In Section 8.2.2.2 (1), after obtaining the geodetic coordinates X_g and Y_g, Z_g can be obtained using X_g and Y_g to interpolate the DSM. In a similar way, after acquiring the scanning coordinates i and j, the gray-scale of the ortho-photo can be acquired using i and j to interpolate the gray-scale of the original image. Because these two interpolation processes are similar, the FPGA-based architecture for interpolation is shared between them.

Considering the interpolation effect, the algorithm's complexity, and the resources of the FPGA, the bilinear interpolation method is used to implement the interpolation for Z_g and gray-scale. However, as shown in Equation (8.5), the original bilinear interpolation method has eight times of multiplication, three times of adding and two times of subtraction. Since the multiplication will take up many resources, Equation (8.5) is rewritten as Equation (8.23), which contains only three times of multiplication, three times of adding and three times of subtraction (Zhang 2019, Zhou and Zhang 2017).

$$\begin{cases} r_{11} = r_1 + q \bullet (r_2 - r_1) \\ r_{12} = r_3 + q \bullet (r_4 - r_3) \\ r_{out} = r_{11} + p \bullet (r_{12} - r_{11}) \end{cases} \tag{8.23}$$

where r represents the value of the DSM or gray-scale of the original image, and $p = |i\text{-}INT(i)|$ and $q = |j\text{-}INT(j)|$ are intermediate variables. The FPGA implementation architecture of the bilinear interpolation algorithm is shown in Figure 8.10. The two-row buffer, in Figure 8.10, is an independent function module presented in Figure 8.3. The two-row buffer is packaged as an independent subsystem so that it can decrease the utilization of the buffer module. In this architecture for the bilinear interpolation algorithm, four multipliers and eight adders are utilized.

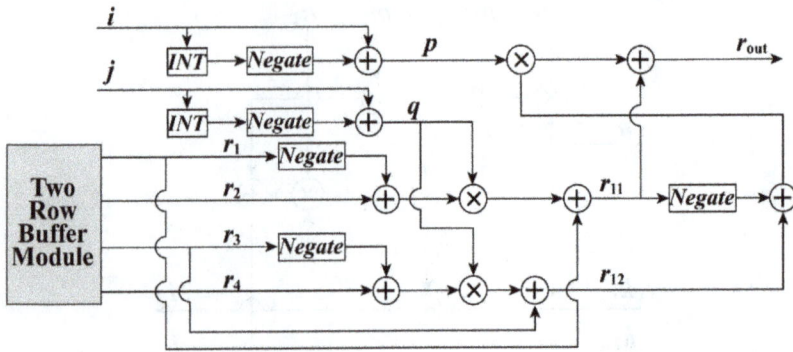

FIGURE 8.10   The FPGA implementation architecture for the bilinear interpolation algorithm.

8.3   EXPERIMENT

8.3.1   THE SOFTWARE AND HARDWARE ENVIRONMENT

The hardware platform used in this chapter is the AC701 Evaluation Kit of Artix-7 series produced by the Xilinx company. The version of the FPGA is Xilinx Artix-7 XC7A200T FBG676ACX1349 D4658436A ZC. The design tool is ISE 4.7 and System Generator. The simulation tool is ModelSim SE10.1a. As shown in Figure 8.11, the FPGA Evaluation Kit uses the UART and JTAG ports to connect with the computer. The power port provides 250 V. The LCD panel and the screen show the results at the same time. To validate the proposed method, the ortho-rectification algorithm is also implemented using MATLAB 2015a on PC with a Windows 7 (64-bit) operation system, which is equipped with an Intel(R) Core(TM) i7-4790 CPU @ 3.6GHz and 8 GB RAM.

8.3.2   DATA SETS

To validate the proposed system based on an FPGA, two test data sets are used to perform the ortho-rectification. The first study area is located in central Denver, Colorado. The exploited aerial image (17,054 × 17,054), see Figure 8.12(a), was collected on 17 April 2000 using an RC 30 aerial camera,

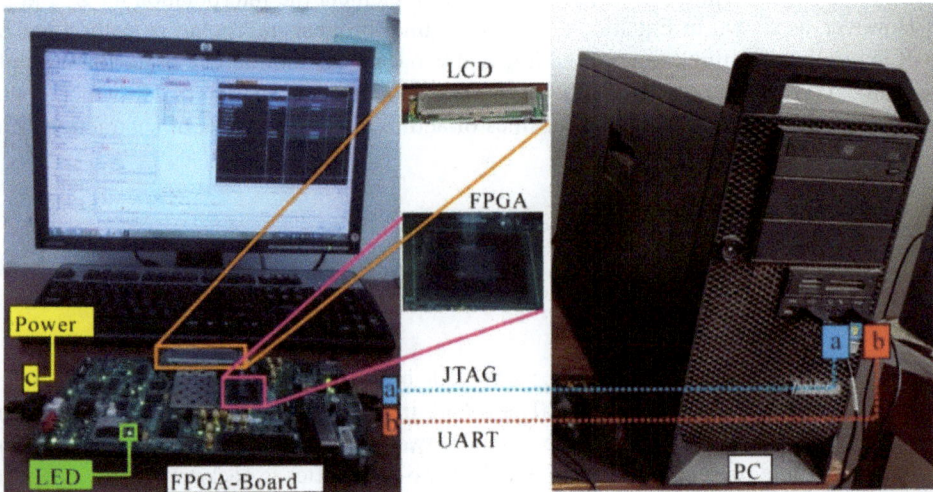

FIGURE 8.11   System diagram.

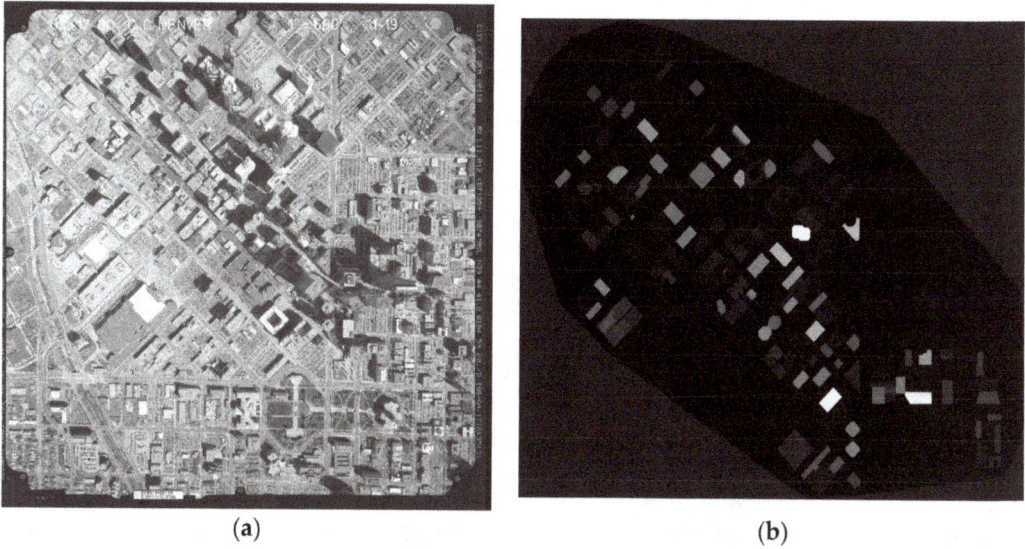

FIGURE 8.12 (a) The original aerial image covering the first study area; (b) digital surface model (DSM) covering the first study area. (Zhou et al. (2005).)

which is the same as Zhou et al. (2005). The focal length is 153.022 mm and the flying height is 1650 m above the mean ground elevation of the imaged area. The second data set is acquired from an ERDAS IMAGINE example dataset, i.e., ps_napp.img (2294 × 2294) and ps_dem.img (see Figure 8.13). The known parameters, provided by the vendors, are listed in Table 8.1.

As described in part (2) of Section 8.2.2.2, to obtain the scanning coordinates, the affine transformation coefficients must be solved. To this end, four fiducial points for each study area are used

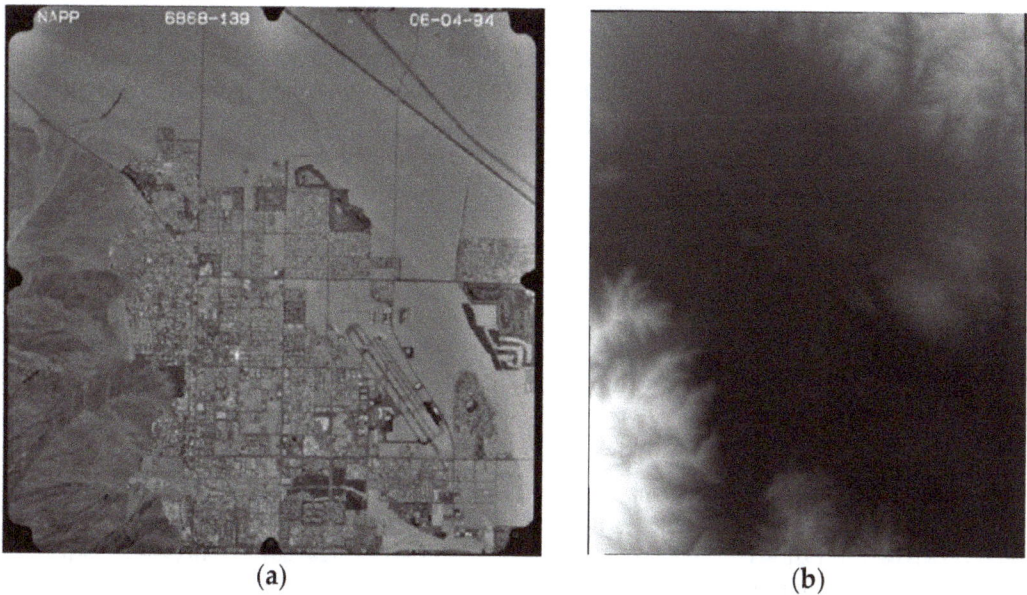

FIGURE 8.13 (a) The original aerial image covering the second study area; (b) digital elevation model (DEM) covering the second study area. (From ERDAS IMAGINE 9.2.)

TABLE 8.1

The Known Parameters for the Data Sets of the Two Study Areas

Known Parameters	First Study Area	Second Study Area
x_0	0.002	−0.004
y_0	−0.004	0.000
f (mm)	153.022	152.8204
X_S (m)	3,143,040.5560	543,427.1886
Y_S (m)	1,696,520.9258	3,744,740.3247
Z_S (m)	9072.2729	6743.2730
ω (rad)	−0.02985539	0.63985182
φ (rad)	−0.00160606	−0.65999005
κ (rad)	−1.55385318	0.86709830

to acquire the affine transformation coefficients according to Equation (8.16), and they are shown in Table 8.2.

After the above necessary parameters are acquired, they are taken as the constants and input to the proposed FPGA-based ortho-rectification system. The ortho-rectified results (ortho-photo) using the proposed FPGA-based method are shown in Figures 8.14(b) and 8.15(b). To validate the rectification's accuracy and speed, ortho-rectification for the same data sets was also implemented using the PC-based platform. The ortho-rectification results using the PC-based software are shown in Figures 8.14(a) and 8.15(a).

8.4   DISCUSSION

8.4.1   VISUAL CHECK

To validate the rectified accuracy, the ortho-photo results ortho-rectified by a PC-based platform are taken as the references. In each of the study areas, three sub-areas are chosen and zoomed into to visually check the accuracy (see Figures 8.16 and 8.17). As observed from Figures 8.16 and 8.17, the ortho-photos ortho-rectified by the proposed method expose one pixel's difference when compared to the results from the PC-based platform. Through the visual check, it can be concluded that the proposed method can meet the demand of ortho-rectification in practice.

TABLE 8.2

Four Fiducial Points for Each Study Area

#	The First Study Area				The Second Study Area			
	I	j	u	v	i	j	u	v
FP$_1$	683.403	881.001	−196.100	191.150	87.500	88.501	−106.000	106.000
FP$_2$	15,820.521	835.103	182.325	192.300	2208.499	83.501	105.999	105.994
FP$_3$	15,868.602	15,970.971	183.525	−186.075	2213.503	2204.504	105.998	−105.999
FP$_4$	730.452	16,019.980	−194.925	−187.300	92.500	2209.501	−106.008	−105.998

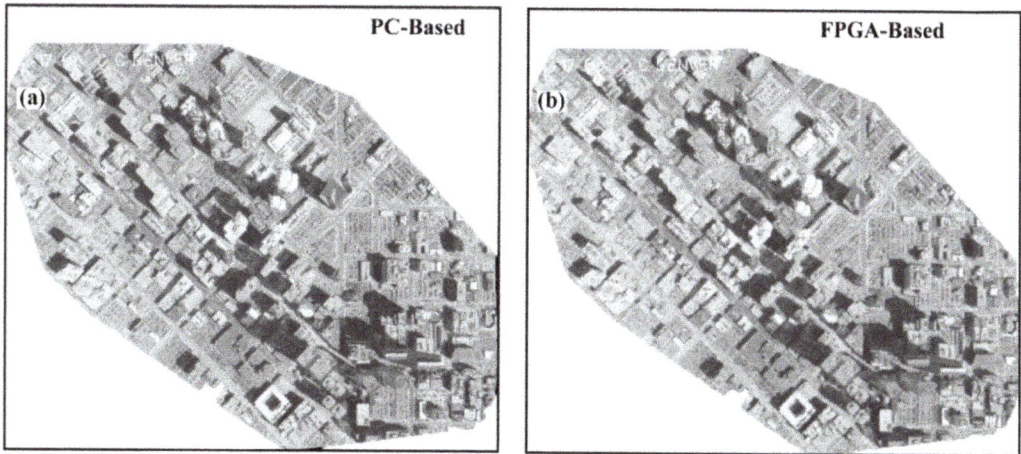

FIGURE 8.14 The ortho-photo ortho-rectified by (a) the personal computer (PC)-based platform; (b) and the proposed method in the first study area.

8.4.2 ERROR ANALYSIS

To further quantitatively evaluate the ortho-photo accuracy obtained by the proposed method, the root-mean-square error (RMSE) (Shi and Shaker 2003; Reinartz et al. 2006) is applied to quantitatively analyze the rectification error of the proposed method. The RMSEs of the planimetric coordinates along the x- and y-axes, and distance (φ_X, φ_Y and φ_S), are computed by, respectively

$$\varphi_Y = \sqrt{\frac{\sum_{k=1}^{n}(Y_k' - Y_k)^2}{n-1}} \tag{8.24}$$

FIGURE 8.15 The ortho-photo ortho-rectified by (a) the PC-based platform; (b) and the proposed method in the second study area.

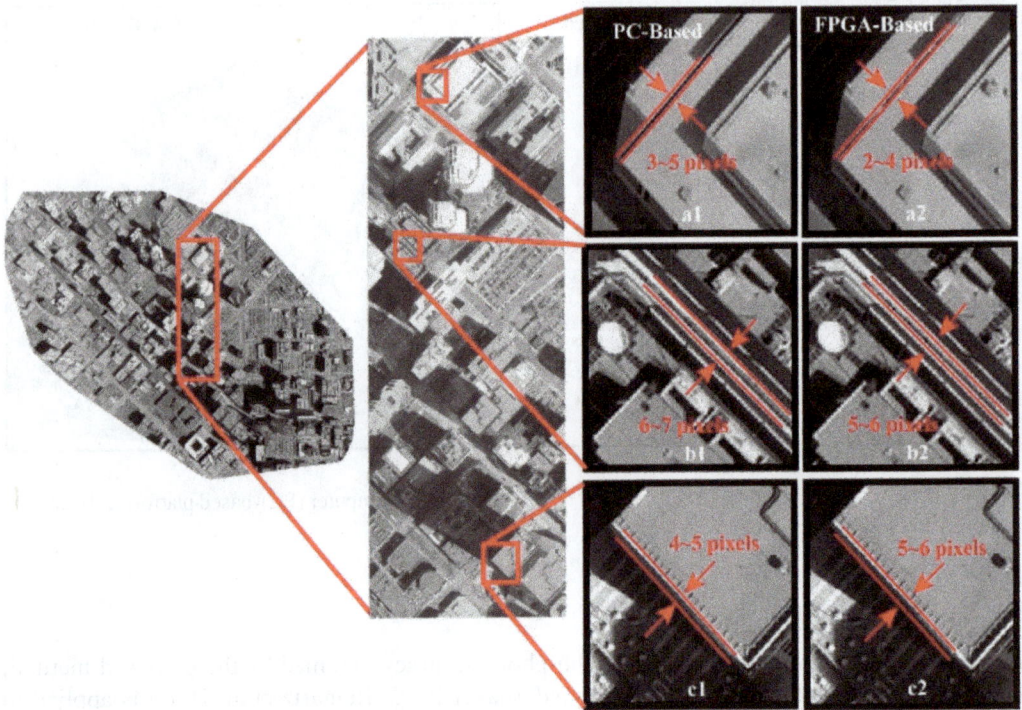

FIGURE 8.16 Visual check analysis for the ortho-rectified results in the three subareas of the first study area.

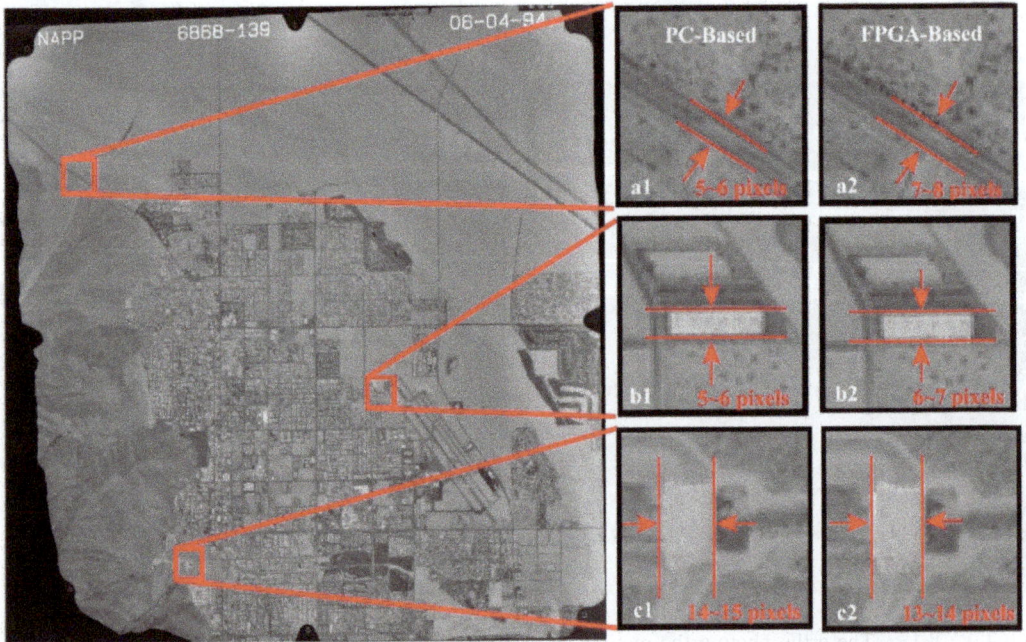

FIGURE 8.17 Visual check analysis for the ortho-rectified results in the three subareas of the second study area.

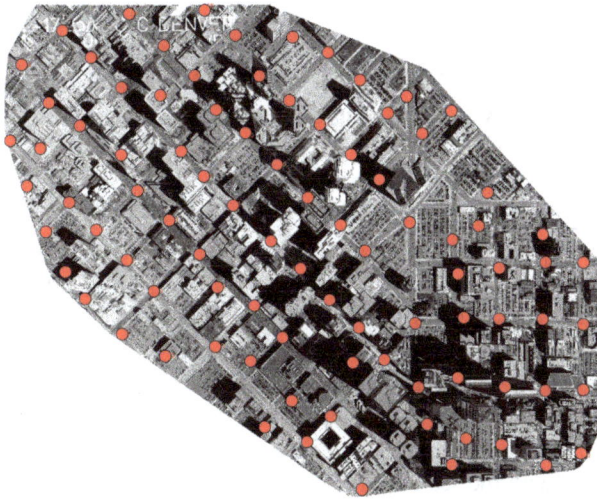

FIGURE 8.18   The distribution of the 90 check points labeled as red in the first study area.

where X'_k and Y'_k are the geodetic coordinates rectified by the proposed method; X_k and Y_k are the reference geodetic coordinates; and n is the number of check points.

To this end, ninety check points (see Figure 8.18) for the first study area were selected to validate the accuracy achieved by the proposed method. The differences of coordinates obtained between the proposed method and the PC-based platform are shown in Figure 8.19. According to Equation (8.21) and Figure 8.19, the RMSEs of φ_X, φ_Y and $\varphi_{s\,s}$ are 1.09, 1.61 and 1.93 m, respectively. In addition, other statistics, such as maximum value, minimum value, standard deviation, and mean

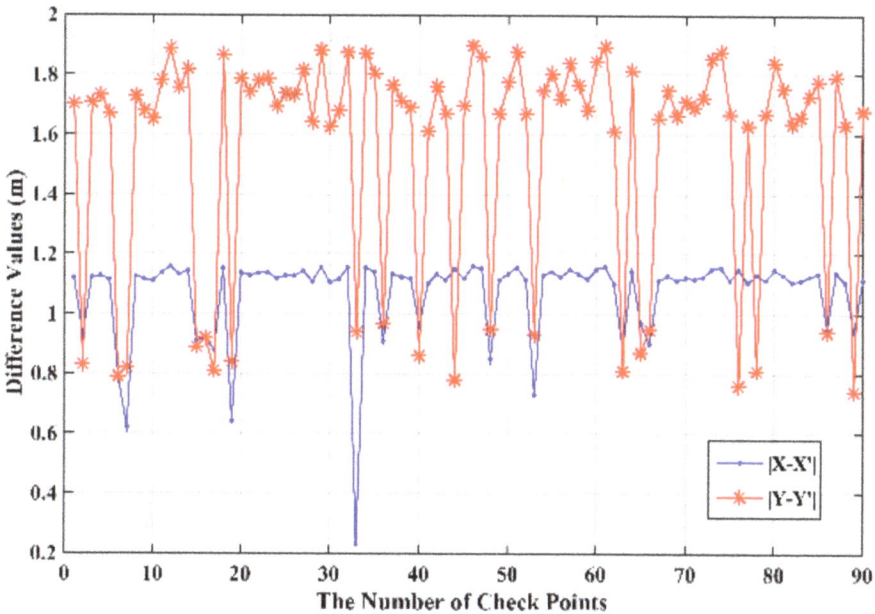

FIGURE 8.19   Different statistics analysis for ortho-photos obtained by our method and the PC-based platform.

TABLE 8.3

Statistics of Difference Value of Geodetic Coordinate

#	Max	Min	Mean	Standard Deviation
X coordinates	1.16 m	0.23 m	1.07 m	0.14 m
Y coordinates	1.89 m	0.74 m	1.55 m	0.38 m

of difference value are also computed and shown in Table 8.3. As shown in Table 8.3, the standard deviations of X and Y are very small. From Table 8.3, it can be found that the maximum error of the X and Y coordinates are 1.16 and 1.89 m, respectively, and the standard deviation of the X and Y coordinates are 0.14 and 0.38 m, respectively. According to Zhang et al. (2006), the ultimate purpose of RS image rectification is to produce thematic maps from the rectified images. Whether the rectified images can satisfy the cartographic requirement of thematic maps depends on the scale of thematic maps. Because the minimum resolving distance on any map is only 0.1 mm, the tolerable errors on the ground distance vary with the scales of thematic maps. The tolerable error would be equivalent of 10 m on the ground if the scale of thematic map is 1:100,000. Because the rectification error obtained by the proposed method ranges from ~1 m to ~2 m, thus, the correction accuracy level of the proposed FPGA-based platform is suitable for compiling 1: 10,000 to 1: 20,000 thematic maps.

However, it is also noted that differences between the proposed method and the PC-based platform in the X and Y coordinates still exist. The difference may be caused by the algorithms implemented through the FPGA hardware, such as fix-point computation, which propagate and accumulate. In addition, the proposed FPGA-based platform only applies two octaves, while the PC-based platform applied at least eight octaves.

Moreover, another method (i.e., receiver operating characteristics curve, ROC curve) is used to evaluate the error of the proposed method. The ROC curve is useful for organizing classifiers and visualizing the performance of classifiers. The detailed information of the ROC curve can be found in (Fawcett 2006). In an ROC graph, the vertical axis represents the true positive rate *(TPR)* acquired by Equation (8.22) and the abscissa axis is the false positive rate *(FPR)* obtained by Equation (8.23). Let the difference of X-coordinate and Y-coordinates, which are less than 1 m, be of a positive class, and the others be of a negative class. Then, the differences are sorted by descending order. Finally, three differences of X-coordinates (the same as Y-coordinates) are used as a group to calculate the *TPR* and *FPR*. The ROC curves of the X-coordinates and Y-coordinates are shown in Figure 8.20(a) and (b), respectively. As shown in Figure 8.20, there are 17 X-coordinates and 20 Y-coordinates that

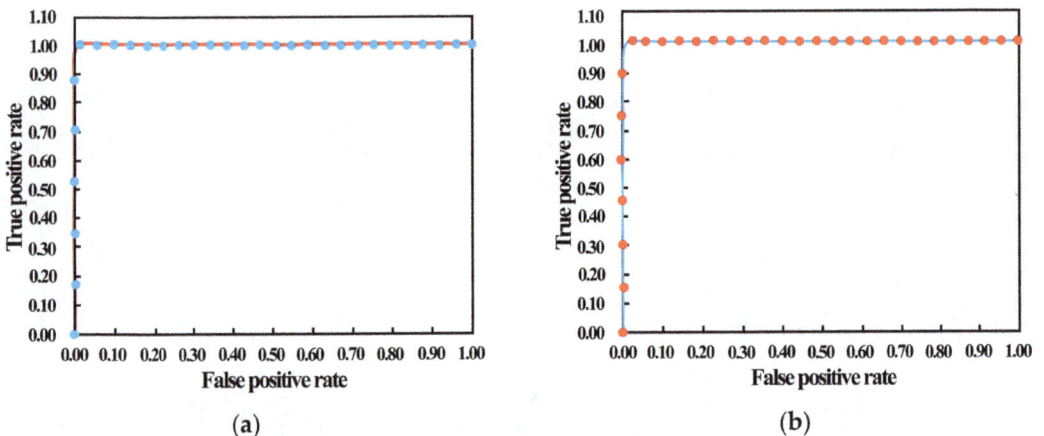

(a)   (b)

FIGURE 8.20   The receiver operating characteristics (ROC) curve analysis through the difference of the X-coordinates (a) and Y-coordinates (b).

are less than 1 m. The ROC curve of the X-coordinates point at (0.013, 1) produces the highest TPR. Additionally, the ROC curve of the Y-coordinates point at (0.014, 1) produces the highest *TPR*.

$$TPR = \frac{Positives\ correctly\ classified}{Total\ positives} \tag{8.25}$$

$$FPR = \frac{Negatives\ incorrectly\ classified}{Total\ Negatives} \tag{8.26}$$

8.4.3 PROCESSING SPEED COMPARISON

One of the most importance factors on on-board ortho-rectification is the processing speed. To evaluate and compare the speed of the proposed method and the PC-based platform, a normalized metric, i.e., throughput representing the capacity of processing pixels per second, is used. For the proposed FPGA-based platform, the throughput is 11,182.3 kilopixels per second and the whole time of the ortho-rectification processing is 26.01 s for the first study area. However, for the PC platform, the throughput is 2582.9 kilopixels per second and the total time of the ortho-rectification processing is 112.6 s for the same image. That means the processing speed by the proposed FPGA-based platform is approximately 4.3 times faster than that by the PC-based platform.

8.4.4 RESOURCE CONSUMPTION

In addition, this chapter takes the utilization ratio of each type of resource, such as input buffers (IBUF), output buffers (OBUF) and signal processors (DSP48E) to assess the proposed method.

First, the utilization ratios for the resources of the coordinate transformation module and the bilinear interpolation module are analyzed, independently. For implementing the coordinate transformation module of the proposed FPGA-based platform, the main hardware consists of 192 IBUF, 870 OBUF and 78 DSP48E, as well as a few adding units (ADDSUB), multiplier units (MULT) and lookup tables (LUT). The results of the utilization ratios of the logic unit for implementing the coordinate transformation module are shown in Table 8.4.

For the bilinear interpolation module of the proposed FPGA-based platform, 129 slice resources, 291 IOs and 75,779 LUTs are used. The utilization ratio of the IOs can reach 72% (291/400 ≈ 0.72). Additionally, the utilization ratio of LUT is 56% (75,779/134,600 ≈ 0.56) (see Table 8.5).

Generally, if the utilization ratio of a resource reaches 60–80%, it shows that the selected device can meet the requirements of the design scheme. If the utilization ratio of a resource is too low,

TABLE 8.4
The Utilization Ratios of Logic Units for the Coordinate Transformation Module (LUT, Look-up Table; FF, Flip Flop)

#	Name of Logic Unit	Utilization Ratio (%)
Logic unit resource	Register	34
Distribution of logic unit	Flip Flop	12
	LUT	27
	LUT-FF Pairs	58
	Control Sets	2
Input and output (IO)	IOs	78
	IOBs	54

TABLE 8.5

The Utilization Ratios of Logic Units for the Bilinear Interpolation Module

#	Name of Logic Unit	Utilization Ratio (%)
Use ratio of logic unit	Register	24
Distribution of logic unit	Flip Flop	17
	LUT	56
	LUT-FF Pairs	64
	Control Sets	7
Input and output (IO)	IOs	72
	IOBs	65

it demonstrates that the selected device is wasted for implementing the design scheme. As shown in Tables 8.4 and 8.5, the utilization ratios of register used in the coordinate transformation module and the bilinear interpolation module are only 34% and 24%, respectively. The utilization ratio of the register is relatively low in both models. The utilization ratios LUT applied in these two models are 27% and 56%, respectively. The utilization ratio of LUT used in the bilinear interpolation module is about twice higher than the utilization ratio of LUT applied in the coordinate transformation module. The reason is that the bilinear interpolation module needs to store the gray values of neighbors. Additionally, the utilization ratios of the Flip Flop and control sets are low in both models. Moreover, the utilization ratios of LUT–FF pairs, IOs and IOBs are relatively high in both models. In summary, according to the above comprehensive utilization ratios for resources, it can be demonstrated that the resources of the selected FPGA can meet the design requirement of the proposed FPGA-based ortho-rectification method.

8.5 CONCLUSIONS

In this chapter, an FPGA-based ortho-rectification method, which is intended to perform the ortho-rectification process on-board spacecraft, is proposed for accelerating the speed of the ortho-rectification of remotely sensed images. The proposed FPGA-based ortho-rectification platform consists of a memory module, a coordinate transformation module (including the transformation from geodetic to photo coordinates and the transformation from photo to scanning coordinates) and a gray-scale interpolation module based on a bilinear interpolation algorithm.

To validate the ortho-rectification's accuracy, an ortho-photo ortho-rectified by a PC-based platform is taken as a reference. Two study areas, including three subareas, are chosen to validate the proposed method. The RMSE, associated with maximum, minimum, standard deviation, and mean of the X and Y coordinates' differences are used. The experimental results demonstrated that the maximum errors of the X and Y coordinates are 1.16 and 1.89 m, respectively, and the standard deviations of the X and Y coordinates are 0.14 and 0.38 m, respectively. The RMSEs of the planimetric coordinates along the X- and Y-axes (φX, φY) and the distance φS are 1.09, 1.61 and 1.93 m, respectively. Thereby, it can be concluded from these quantitative analyses that the proposed method can meet the demand of ortho-rectification in practice.

In addition, through analyzing the processing speed of ortho-rectification, it can be found that the processing speed by the proposed FPGA-based platform is approximately 4.3 times faster than that by the PC-based platform. In terms of the resource consumptions, it can be found that the bilinear interpolation module of the proposed method utilizes 129 slice resources and 291 IOs, whose utilization ratio of the IOs can reach 72%, and the LUT achieves 56%.

REFERENCES

Botella, G., Garcia, A., Rodriguez-Alvarez, M., Ros, E., Meyer-Baese, U., Molina, M.C., Robust bioinspired architecture for optical-flow computation. *IEEE Transaction VLSI Systems* 2010, 18: 616–629.

Cao, T.P., Elton, D., Deng, G., Fast buffering for FPGA implementation of vision-based object recognition systems. *Journal Real-Time Image Processing,* 2012, 7: 173–183.

Dai, C., Yang, J., Research on ortho-rectification of remote sensing images using GPU-CPU cooperative processing. *In Proceedings of the International Symposium on Image and Data Fusion*, Tengchong, Yunnan, China, 9–11 August 2011, pp. 1–4.

David, K., Don, G., Real-time ortho-rectification by FPGA-based hardware acceleration. *In Remote Sensing, International Society for Optics and Photonics*, SPIE Remote Sensing, 2010, Toulouse, France, 2010, pp. 78300Y–78300Y-7.

El-Amawy, A., A systolic architecture for fast dense matrix inversion. *IEEE Transactions Computers*, 1989, 38: 449–455.

Escamilla-Hernández, E., Kravchenko, V., Ponomaryov, V., Robles-Camarillo, D., Ramos, L.E., Real time signal compression in radar using FPGA. *Científica*, 2008, 12: 131–138.

Fawcett, T., An introduction to ROC analysis. *Pattern Recognition Letters*, 2006, 27: 861–874.

González, D., Botella, G., García, C., Prieto, M., Tirado, F., Acceleration of block-matching algorithms using a custom instruction-based paradigm on a NIOS II microprocessor. *EURASIP Journal on Advances in Signal Processing*, 2013, 2013: 118.

González, D., Botella, G., Meyer-Baese, U., García, C., Sanz, C., Prieto-Matías, M., A low-cost matching motion estimation sensor based on the Nios II microprocessor, *Sensors*, 2012, 12: 13126.

Greisen, P., Heinzle, S., Gross, M., Burg, A.P., An FPGA-based processing pipeline for high-definition stereo video. *EURASIP J. Image Video Processing*, 2011, 2011: 1–13.

Halle, W., Thematic data processing on board the satellite BIRD. In Proceedings of the SPIE 4540, *Sensors, Systems, and Next-Generation Satellites V, Toulouse, France*, 17–20 September 2001.

Hsiao, P.Y., Lu, C.L., Fu, L.C., Multilayered image processing for multiscale Harris corner detection in digital realization. *IEEE Transactions on Industrial Electronics*, 2010, 57: 1799–1805.

Hu, X., Zhu, Y., Research on FPGA based image input buffer design mechanism. *Microcomputer Information*, 2010, 2: 5–6+25.

Jiang, C., Nooshabadi, S., A scalable massively parallel motion and disparity estimation scheme for multiview video coding. *IEEE Transactions. Circuits Systems Video Technology*, 2016, 26: 346–359.

Kalomiros, J.A., Lygouras, J., Design and evaluation of a hardware/software FPGA-based system for fast image processing. *Microprocessors & Microsystems*, 2008, 32: 95–106.

Kate, D., Hardware implementation of the Huffman encoder for data compression using altera DE2 board. *International Journal of Advances in Engineering*, 2012, 2: 11–15.

Kazmi, M., Aziz, A., Akhtar, P., An efficient and compact row buffer architecture on FPGA for real-time neighbourhood image processing. *Real-Time Image Processing*, 2019, 16: 1845–1858.

Kumar, P.R., Sridharan, K., VLSI-efficient scheme and FPGA realization for robotic mapping in a dynamic environment. *IEEE Transactions on Very Large-Scale Integration Systems*, 2007, 15: 118–123.

Li, H.H., Liu, S., Piao, Y., Snow removal of video image based on FPGA. *In Proceedings of the 5th International Conference on Electrical Engineering and Automatic Control*, Weihai, China, 16–18 October 2015, Huang, B., Yao, Y., Eds., Springer: Berlin/Heidelberg, Germany, 2016a, pp. 207–215.

Li, H., Xiang, F., Sun, L., *Based on the FPGA Video Image Enhancement System Implementation*. DEStech Transactions on Computer Science, 2016b.

Malik, A.W., Thornberg, B., Imran, M., Lawal, N., Hardware architecture for real-time computation of image component feature descriptors on a FPGA. *International Journal of Distributed Sensor Networks*, 2014, 2014: 14.

Ontiveros-Robles, E., Vázquez, J.G., Castro, J.R., Castillo, O., A FPGA-based hardware architecture approach for real-time fuzzy edge detection. *Studies in Computational Intelligence*, 2017, 667: 519–540.

Ontiveros-Robles, E., Gonzalez-Vazquez, J.L., Castro, J.R., Castillo, O., A hardware architecture for real-time edge detection based on interval type-2 fuzzy logic. *In Proceedings of the IEEE International Conference on Fuzzy Systems, Vancouver, BC, Canada*, 24–29 July 2016.

Pal, C., Kotal, A., Samanta, A., Chakrabarti, A., Ghosh, R., An efficient FPGA implementation of optimized anisotropic diffusion filtering of images. *International Journal of Reconfigurable Computing*, 2016, 2016: 1.

Pan, Z., Lei, J., Zhang, Y., Sun, X., Kwong, S., Fast motion estimation based on content property for low-complexity h.265/HEVC encoder. *IEEE Transactions Broadcast*, 2016, 62: 675–684.

Reinartz, P., Müller, R., Lehner, M., Schroeder, M., Accuracy analysis for DSM and orthoimages derived from SPOT HRS stereo data using direct georeferencing. *ISPRS Journal of Photogramm Remote Sensing*, 2006, 60: 160–169.

Rodriguez-Donate, C., Botella, G., Garcia, C., Cabal-Yepez, E., Prieto-Matias, M., Early experiences with OpenCL on FPGAs: Convolution case study. *In Proceedings of the 2015 IEEE 23rd Annual International Symposium on Field-Programmable Custom Computing Machines*, Vancouver, BC, Canada, 2–6 May 2015, pp. 235–235.

Shi, W., Shaker, A., Analysis of terrain elevation effects on IKONOS imagery rectification accuracy by using non-rigorous models. *Photogrammetric Engineering & Remote Sensing*, 2003, 69: 1359–1366.

Sylvain, C.V., Serge, M., A load-balanced algorithm for parallel digital image warping. *International Journal of Pattern Recognition Artificial Intelligence*, 1999, 13: 445–463.

Thomas, U., Rosenbaum, D., Kurz, F., Suri, S., Reinartz, P., A new software/hardware architecture for real time image processing of wide area airborne camera images. *Journal of Real-Time Image Processing*, 2008, 4: 229–244.

Tomasi, M., Vanegas, M., Barranco, F., Diaz, J., Ros, E., Real-time architecture for a robust multi-scale stereo engine on FPGA. *IEEE Transactions on Very Large-Scale Integration Systems*, 2012, 20: 2208–2219.

Waidyasooriya, H., Hariyama, M., Ohtera, Y., FPGA architecture for 3-D FDTD acceleration using open CL. *In Proceedings of the 2016 Progress in Electromagnetic Research Symposium (PIERS)*, Shanghai, China, 8–11 August 2016, pp. 4719–4719.

Wang, E., Yang, F., Tong, G., Qu, P., Pang, T., Particle filtering approach for gnss receiver autonomous integrity monitoring and FPGA implementation. *TELKOMNIKA (Telecommunication Computing Electronics and Control)*, 2016, 14: 1321–1328.

Warpenburg, M.R., Siegel, L.J., SIMD image resampling. *IEEE Transactions Computer*, 1982, 31: 934–942.

Winfried, H., Thematic data processing on board the satellite BIRD, *Proceeding of SPIE*, 2001, 4540: 412–419.

Wittenbrink, C.M., Somani, A.K. 2D and 3D optimal parallel image warping. *Journal of Parallel & Distributed Computing*, 1995, 25: 197–208.

Zhang, R., GA-RLS-RFM ortho-rectification algorithm of satellite images and its hardware implementation using FPGA. Dissertation, Tianjin University, Tianjin China, August 2019.

Zhang, Y., Zhang, D., Gu, Y., Tao, F., Impact of GCP distribution on the rectification accuracy of Landsat TM imagery in a coastal zone. *ACTA Oceanol, Acta Oceanologica Sinica*, 2006, 25: 14.

Zhang, R., Zhou, G., Zhang, G, Zhou, X., Zhou. RPC-based orthorectification for satellite images using FPGA, *Sensors*, 2018, 18: 2511. doi:10.3390/s18082511.

Zhang, C., Liang, T., Mok, P.K. T., Yu, W., FPGA implementation of the coupled filtering method. *In Proceedings of the 2016 IEEE International Conference on Bioinformatics and Biomedicine (BIBM)*, Shenzhen, China, 15–18 December 2016, pp. 435–442.

Zhou, G., Chen, W., Kelmelis, J.A., Zhang, D., A comprehensive study on urban true ortho-rectification. *IEEE Transactions Geoscience Remote Sensing*, 2005, 43: 2138–2147.

Zhou, G., Zhang, R., et al., Real-time ortho-rectification for remote sensing images, *Int. J. of Remote Sensing*, 2019, 40(5–6): 2451–2465. doi:10.1080/01431161.2018.1488296.

Zhou, G., Zhang, R.; et al. On-board ortho-rectification for images based on an FPGA. *Remote Sens*, 2017, 9: 874.

9 On-Board Image Ortho-Rectification Using RPC Model

9.1 INTRODUCTION

Ortho-rectification is a process that orthorectifies an image onto its upright planimetry position in a given geodetic coordinate system and removes the displacement caused by terrain and distortions (Zhou et al. 2002; Zhou 2009; French and Balster 2014). Ortho-rectification is a prerequisite for remotely sensed image applications in such as land resource investigation, disaster monitoring, forestry inventory, environmental changes analysis, etc. The orthorectified image, is called digital orthophoto map (DOM), or digital orthoimage map (DOM). It not only contains the geometric information of a symbol map but also has the features of the image gray. However, in the past 20 years, studies on ortho-rectification have been based almost entirely on ground-image processing systems(Zhou et al. 2002; Zhou 2009; Aguilar et al. 2013; Marsetič et al. 2015; Habib et al. 2017), which is unable to meet the demand with respect to time-critical disasters. Examples are Zhou et al. (2002) presented a rigorous model for geometric ortho-rectification of declassified intelligence satellite photography (DISP) imagery. Zhou (2009) proposed a near real-time ortho-rectification method for mosaic of video flow acquired by an unmanned aerial vehicle (UAV). Aguilar et al. (2013) used rigorous model and rational function model to orthorectify GeoEye-1 and WorldView-2 images and assessed the geometric accuracy of the orthophoto. The results showed that the best horizontal geo-positioning accuracies were acquired by using third order rational functions with vendor's RPCs (Rational Polynomial Coefficients in Rational Function Model – RFM) data. Marsetič et al. (2015) presented an automatic processing chain for ortho-rectification of optical pushbroom sensors. Habib et al. (2017) proposed an approach using generated orthophotos from frame camera to improve the ortho-rectification of hyperspectral pushbroom scanner imagery. Thus, it is important to determine how to improve the speed of the ortho-rectification process when used in the on-board processing of a spacecraft.

With the increasing demands in (near) real-time remotely sensed imagery applications for applications such as military deployments, quick response to terrorist attacks and disaster rescue (e.g., flooding monitoring), the on-board implementation of ortho-rectification has attracted much research worldwide in recent years. For example, researchers have proposed multiple parallel-processing methods and employed hardware acceleration such as the approach by Warpenburg and Siegel (1982), who performed resampling in a single instruction stream-multiple data stream environment. Wittenbrink and Somani (1993) presented optimal concur-rent-read-exclusive-write and exclusive-read-exclusive-write parallel-random-access-machine algorithms for spatial image warping. Liu et al. (2010) proposed a parallel algorithm that is focused on massive remotely sensed ortho-rectification. Zhou et al. (2005) first presented the concept of "on-board geometric correction," but details pertaining to its on-board implementation were not given. Zhou (2009) proposed a method for a real-time mosaic of video flow acquired by a small low-cost unmanned aerial vehicle. However, the method was implemented based on software, which was a serial instruction system, and would affect the real-time processing efficiency (Zhou et al. 2004). Dai and Yang (2011) proposed a fast graphic processing unit (GPU) – central processing unit (CPU) cooperative processing algorithm that is based on computer unified device architecture for the ortho-rectification of remotely sensed images. Reguera-Salgado et al. (2012) proposed a method for the real-time geocorrection of images acquired by airborne push-broom sensors using the hardware acceleration and

DOI: 10.1201/9781003319634-9

parallel-computing characteristics of modern GPUs. Quan et al. (2016) presented an optical aerial image ortho-rectification parallel algorithm that employs GPU acceleration. These ground-based parallel-processing systems have increased to an extent the processing speed for remotely sensed image ortho-rectification. However, the remotely sensed images still need to be sent back to the ground-based processing centers. This process is time consuming. In addition, most of parallel-processing methods are based on the multiple task operating system of the GPU, which cannot essentially solve the problem of responding to time-critical events.

To implement on-board ortho-rectification in (near) real-time, an efficient approach is to apply field-programmable gate array (FPGA) chip because the FPGA chips offer a highly flexible design, scalable circuits and a high efficiency in data processing for its pipeline structure and fine-grained parallelism. In recent decades, researchers have widely used FPGA for image processing applications. Examples are Halle et al. (2000), who proposed an on-board image data processing system based on the neural network processor NI100, digital signal processors and FPGA. Eadie et al. (2003) investigated the use of FPGA for the correction of geometric image distortion. Kumar and Sridharan (2007) realized the real-time correction of images using an FPGA under a dynamic environment. Escamilla-Hernández et al. (2008) and Kate (2012) used an FPGA to implement data compression. Tomasi et al. (2012) proposed a stereo vision algorithm using an FPGA to perform the correction of video graphics array images (57 fps). Pal et al. (2016), Wang et al. (2016) and Zhang et al. (2016) applied FPGAs to accelerate the image data and signal filtering processes. Ontiveros-Robles et al. (2016; 2017) proposed FPGA-based hardware architectures for real-time edge detection using fuzzy logic algorithm. Li et al. (2016a, 2016b) utilized FPGAs to realize the real-time processing of video images to remove snow and fog. Huangand Zhou (2017) proposed an FPGA-based method for the on-board detection and matching of the feature points. Huang et al. (2018) presented a new FPGA architecture of a fast and brief algorithm for on-board corner detection and matching.

This chapter describes a FPGA-based method for the on-board implementation of ortho-rectification. The proposed method can be divided into three modules, i.e., reading parameters module, coordinates transformation module and interpolation module (Zhou et al. 2018).

9.2   RPC-BASED ORTHO-RECTIFICATION MODEL USING FPGA CHIP

To deal with various types of images, many geometric imaging models are presented. One of the most widely used models is the rational polynomial coefficient (RPC) model, which is a general imaging model that is independent of the satellite sensor and platform. Unlike rigorous physical models that are based on the collinear equation, the RPC model does not require knowledge of the interior orientation parameters (IOPs) and exterior orientation parameters (EOPs), which are sometimes not provided by vendors. The RPC model has been therefore widely applied to orthorectify satellite images (Wang and Ellis 2005; Hoja et al. 2008; Zhang et al. 2012; Yang and Zhu 2013; Yang et al. 2017). The details of RPC ortho-rectification method can be reference to Fraser et al. (2002) and Grodecki and Dial (2003). In this section, the RPC algorithm implemented using FPGA chip is described below.

9.2.1   THE IMPROVEMENT OF RPC ALGORITHM WITH CONSIDERING ON-BOARD IMPLEMENTATION

Ferrer et al. (2005) and Balster et al. (2011) presented FP processing method that is able to accelerate the calculation. The basic idea of the FP processing is: To make the transformation between a fixed-point variable and a floating-point variable, multiplication by a constant can be used to maintain the precision. When the constant is set to a power of 2, the multiplication can be seen as a single bit shift, i.e.,

$$F = 2^\tau F'$$

$$(9.1)$$

TABLE 9.1

Scale Factors and Integer Variables

Variable Name	Scale Factor	Integer Variable Name
a'_i, b'_i, c'_i and d'_i (i=1 to 20)	τ_1	a_i, b_i, c_i and d_i (i=1 to 20)
Lon', Lat', Hei'	τ_2	Lon, Lat, Hei
Lat'_{off}, Lat'_{scale}, Lon'_{off}, Lon'_{scale}, H'_{off}, H'_{scale}	τ_3	Lat_{off}, Lat_{scale}, Lon_{off}, Lon_{scale}, H_{off}, H_{scale}
$Line'_{off}$, $Line'_{scale}$, $Samp'_{off}$, $Samp'_{scale}$	τ_3	$Line_{off}$, $Line_{scale}$, $Samp_{off}$, $Samp_{scale}$

---

where F' is a floating-point variable, F is a fixed-point variable and τ is a scale factor, which affects the binary accuracy of the resulting integer representation. A larger scale factor will produce a higher degree of the binary accuracy.

With the FP processing method, FP-RPC algorithm is presented in this Section, where all of the variables and constants are transformed to integers using Equation (9.1). Table 9.1 gives the integer variables and their scale factors. As observed from Table 9.1, a'_i, b'_i, c'_i and d'_i (i = 1 to 20) are multinomial coefficients. Generally, the values of b'_1 and d'_1 are 1. Lon', Lat' and Hei' are geodetic coordinates, which represent the longitude, latitude and height, respectively. Lat'_{off}, Lat'_{scale}, Lon'_{off}, Lon'_{scale}, H'_{off}, H'_{scale}, $Line'_{off}$, $Line'_{scale}$, $Samp'_{off}$ and $Samp'_{scale}$ are the parameters for normalization. $Samp'$ and $Line'$ represent the image coordinates, sample and line.

According to Fraser et al. (2002), Grodecki and Dial (2003) and Equation (9.1), the normalized coordinates are converted into integers by

$$L = 2^{\tau_2} \frac{2^{-\tau_2} Lon - 2^{-\tau_3} Lon_{off}}{2^{-\tau_3} Lon_{scale}} \qquad P = 2^{\tau_2} \frac{2^{-\tau_2} Lat - 2^{-\tau_3} Lat_{off}}{2^{-\tau_3} Lat_{scale}} \qquad (9.2)$$

$$H = 2^{\tau_2} \frac{2^{-\tau_2} Hei - 2^{-\tau_3} H_{off}}{2^{-\tau_3} H_{scale}}$$

$$X = 2^{\tau_2} \frac{2^{-\tau_2} Samp - 2^{-\tau_3} Samp_{off}}{2^{-\tau_3} Samp_{scale}} \qquad Y = 2^{\tau_2} \frac{2^{-\tau_2} Line - 2^{-\tau_3} Line_{off}}{2^{-\tau_3} Line_{scale}} \qquad (9.3)$$

Moreover, the polynomials are converted into integers by

$$2^{-\tau_2} \mathbf{ND}_{LS} = (2^{-\tau_1} \mathbf{C})(2^{-\tau_2} \mathbf{N}^{T}) \qquad (9.4)$$

$$\mathbf{ND}_{LS} = 2^{-\tau_1} \mathbf{C} \mathbf{N}^{T} \qquad (9.5)$$

where $\mathbf{ND}_{LS} = \begin{bmatrix} Num_L & Den_L & Num_S & Den_S \end{bmatrix}^{T}$, $\mathbf{C} = \begin{bmatrix} a_1 & a_2 & ... & a_{19} & a_{20} \\ b_1 & b_2 & ... & b_{19} & b_{20} \\ c_1 & c_2 & ... & c_{19} & c_{20} \\ d_1 & d_2 & ... & d_{19} & d_{20} \end{bmatrix}$,

$\mathbf{N} = \begin{bmatrix} 1 & L & P & H & LP & LH & PH & LL & PP & HH & PLH & LLL & LPP & LHH & LLP & PPP & PHH & LLH & PPH & HHH \end{bmatrix}$.

In addition, the normalized image coordinates (X', Y') are converted into integers by

$$2^{-\tau_2} Y = \frac{2^{-\tau_2} Num_L}{2^{-\tau_2} Den_L}, \quad 2^{-\tau_2} X = \frac{2^{-\tau_2} Num_S}{2^{-\tau_2} Den'_S} \tag{9.6}$$

$$Y = 2^{\tau_2} \frac{Num_L}{Den_L}, \quad X = 2^{\tau_2} \frac{Num_S}{Den_S} \tag{9.7}$$

Finally, the image coordinates $(Samp', Line')$ are converted into integers by

$$\begin{cases} 2^{-\tau_3} Samp = (2^{-\tau_2} X)(2^{-\tau_3} Samp_{scale}) + 2^{-\tau_3} Samp_{off} \\ 2^{-\tau_3} Line = (2^{-\tau_2} Y)(2^{-\tau_3} Line_{scale}) + 2^{-\tau_3} Line_{off} \end{cases} \tag{9.8}$$

$$\begin{cases} Samp = (2^{-\tau_2} X) Samp_{scale} + Samp_{off} \\ Line = (2^{-\tau_2} Y) Line_{scale} + Line_{off} \end{cases} \tag{9.9}$$

9.2.2 PARALLEL COMPUTATION OF ORTHO-RECTIFICATION USING FPGA

Many factors impact the computation speed when an FPGA is adopted, such as the optimal design of algorithms and the logical resource of the utilized FPGA. By analyzing the structure of the FP-RPC algorithm and optimizing it, an FPGA-based architecture for FP-RPC-based ortho-rectification is designed, and shown in Figure 9.1. As described from Equations (9.2)–(9.9), their structures are

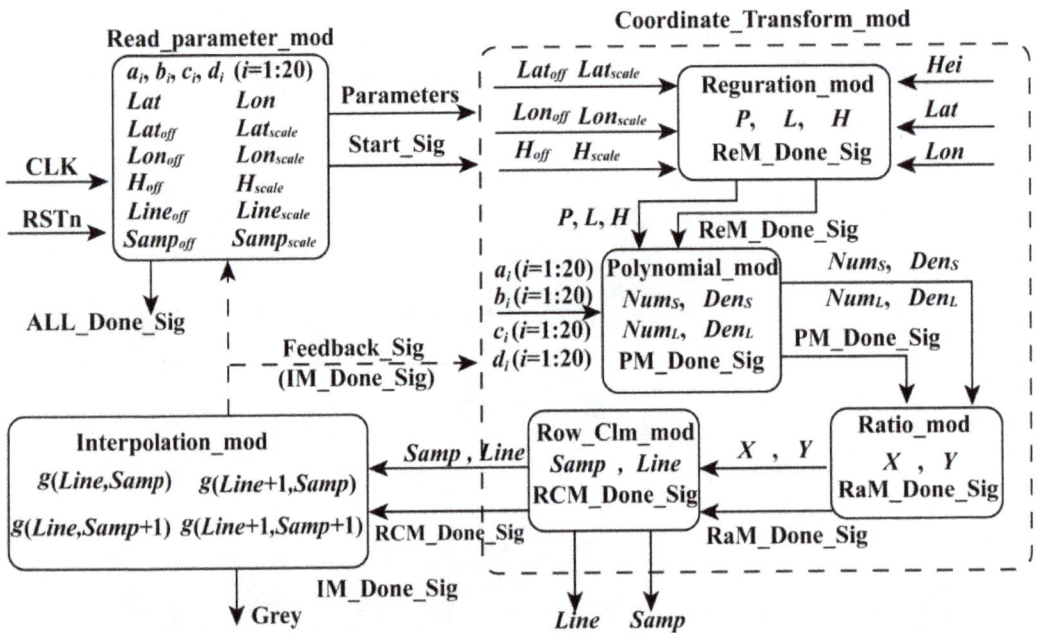

FIGURE 9.1 An FPGA-based architecture for FP-RPC-based ortho-rectification.

similar. It is convenient for FPGAs to be implemented in parallel. As shown in Figure 9.1, the FPGA-based FP-RPC module can be divided into three submodules:

1. Read_parameter_mod (RPM), which is used to send parameters to other modules,
2. Coordinate_Transform_mod (CTM), which is applied to transform geodetic coordinates to image coordinates and
3. Interpolation_mod (IM), which is utilized to perform bilinear interpolation.

The details of these modules are given as follows (Zhang et al. 2018; Zhang 2019).

1. For RPM, the constants, including the coefficients (a_i, b_i, c_i, d_i, i = 1 to 20) of polynomials and the parameters for normalization (Lat_{off}, Lat_{scale}, Lon_{off}, Lon_{scale}, H_{off}, H_{scale}, $Line_{off}$, $Line_{scale}$, $Samp_{off}$ and $Samp_{scale}$), can be obtained using the method in Shi and Shaker (2003). When the enable signal is being received, the geodetic coordinates (Lon, Lat, Hei) stored in the RAMs are read and sent to Coordinate_Transform_mod (CTM) with the constants and the start signal (Start_Sig) in the same clock cycle.
2. When the Start_Sig, the constants, and the geodetic coordinates are being received in the CTM, the normalized coordinates (P, L, H) are first calculated in the regularization module (Regulation_mod, ReM); then, the normalized coordinates and the done signal of ReM (ReM_Done_Sig) are sent to the polynomial module (Polynomial_mod, PM) with a_i, b_i, c_i and d_i (i = 1 to 20) in the same clock cycle to compute the numerators and denominators (Num_L, Num_S, Den_L and Den_S) of Equation (9.7). Subsequently, when the done signal (PM_Done_Sig) of PM, Num_L, Num_S, Den_L and Den_S are being received, the normalized coordinates (X, Y) of the image coordinates are calculated. Finally, when the normalized coordinates (X, Y) and the done signal (RaM_Done_Sig) of the ratio module (Ratio_mod, RaM) are being received, the image coordinates ($Samp$, $Line$) and the done signal (RCM_Done_Sig) of the image coordinate calculation module (Row_Clm_mod, RCM) are acquired and sent to the interpolation module (Interpolation_mod, IM) in the same clock cycle.
3. When the image coordinates ($Samp$, $Line$) and RaM_Done_Sig are being received in IM, the gray of pixel ($Samp$, $Line$) is obtained by interpolating and the done signal (IM_Done_Sig) of IM is produced.
4. When the posedge clk of the signal, ALL_Done_Sig is being detected, the processing is finished (Zhou et al. 2017; Zhang et al. 2018).

9.2.2.1 Read Parameter Module

To ensure that the constants, geodetic coordinates and the start signal (Start_Sig) are sent in the same clock cycle, a parallel module (i.e., RPM) is designed (see Figure 9.2). In the RPM, the constants are assigned corresponding values, while the geodetic coordinates are stored in RAM. With such a design, all values are expressed using a fixed point of 32 bits to ensure computational accuracy.

In the RPM, the geodetic coordinates are sent to the next module according to the order of the column. First, the address of RAM is initialized as 0. When the enable signal is detected, the first group of geodetic coordinates (Lat_0, Lon_0, Hei_0) is read from the RAM and sent to the next module with the constants and the Start_Sig in the same clock cycle. Starting from the second group of geodetic coordinates, the rules for reading and sending geodetic coordinates are changed. In other words, after the second group of geodetic coordinates (Lat_1, Lon_1, Hei_1), the geodetic coordinates will be read and sent unless the enable signal and the feedback signal (Feedback_Sig), which are sent by the interpolation module, are detected at the same time. After the final group of geodetic coordinates are read and sent, if the Feedback_Sig is received, the done signal (ALL_Done_Sig)

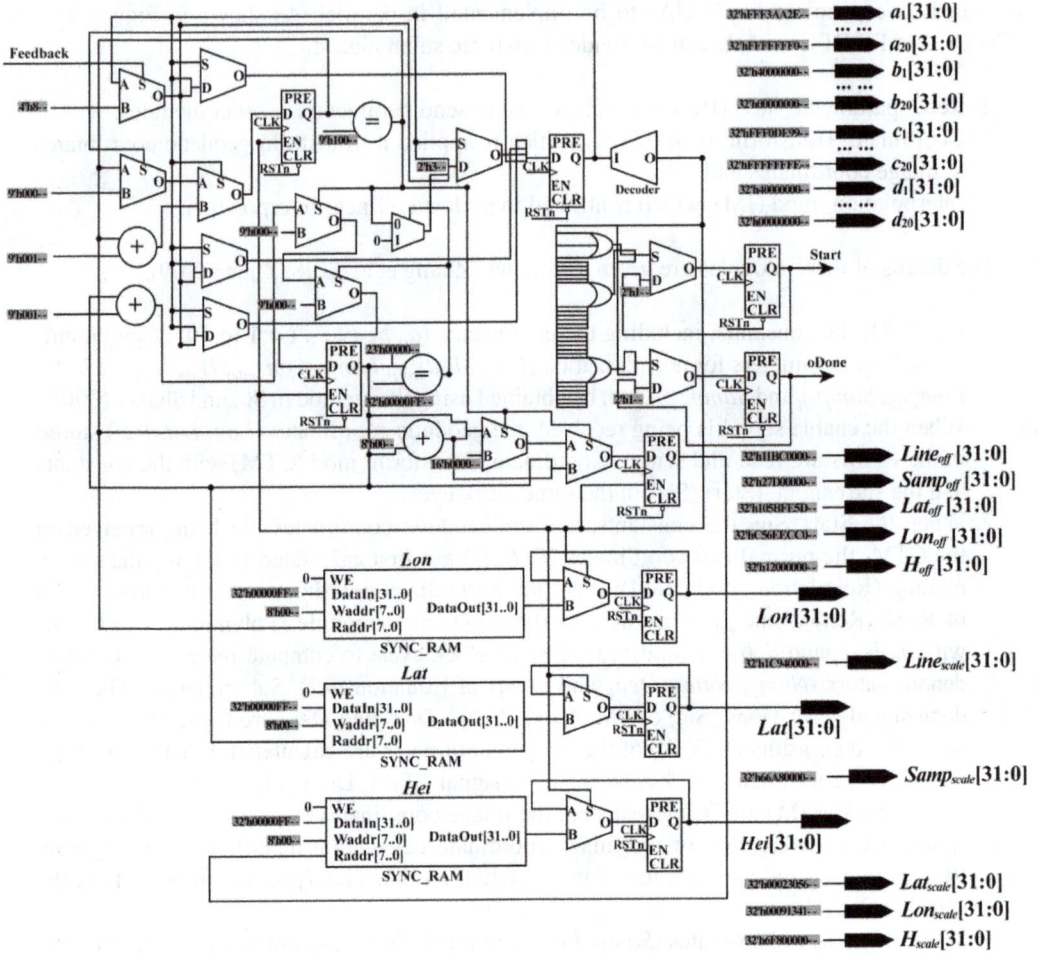

FIGURE 9.2　Architecture of the read parameter module.

of ortho-rectification is produced. When the ALL_Done_Sig is detected, the process of ortho-rectification is stopped (Zhang et al. 2018; Zhang 2019).

9.2.2.2　Coordinate Transformation Module

As shown in Figure 9.1, for the CTM, the inputs contain the constants, the geodetic coordinates and the Start_Sig; the outputs include image coordinates and the done signal of this module. The CTM can be divided into four submodules, namely ReM, PM, RaM and RCM. The details regarding the four submodules are described as follows.

- **Regulation Module**

    With Section 9.2.1, the geodetic coordinates (*Lat*, *Lon*, *Hei*) should be first transformed as the normalized coordinates (*L*, *P*, *H*) based on Equation (9.2), because this operation can minimize the introduction of errors during the computation of the numerical stability of equations (Halle et al. 2000). Moreover, Equation (9.2) is uniform, in other words, it is appropriate for implementation using FPGA. To obtain the normalized coordinates (*L*, *P*, *H*) of the geodetic coordinates (*Lat*, *Lon*, *Hei*) using an FPGA chip, a parallel

Regularization Module

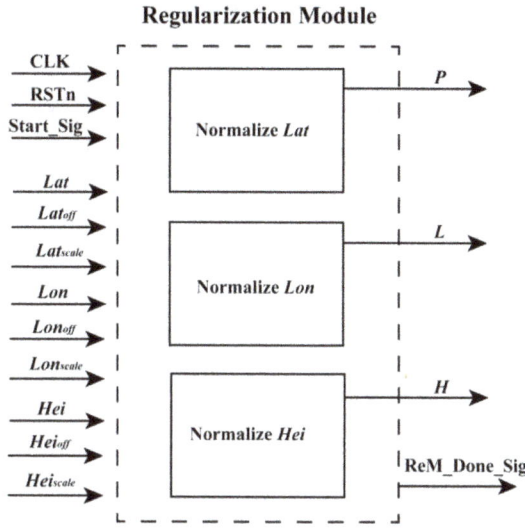

FIGURE 9.3   Schematic diagram of the ReM.

computation architecture is presented in Figure 9.3, where the structures of "Normalize *Lat*," "Normalize *Lon*" and "Normalize *Hei*" are similar. The architecture of "Normalize *Lat*" is depicted in Figure 9.4.

As observed from Figure 9.4, during the computation process, one divider, 10 adders, 10 flipflops and 16 multiplexer units are mainly used to normalize the *Lat*. With this design, the relationship between "Normalize *Lat*," "Normalize *Lon*" and "Normalize *Hei*" is parallel. The normalized coordinates (*L*, *P*, *H*) are obtained in the same clock cycle as the done signal.

Normalize *Latitude*

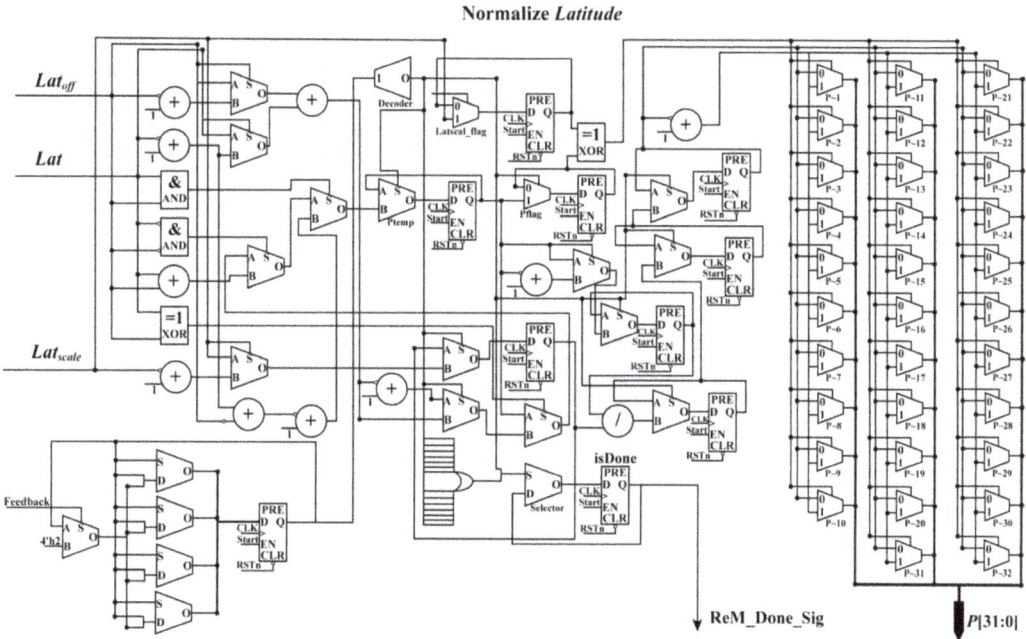

FIGURE 9.4   Schematic diagram of normalizing *Latitude*.

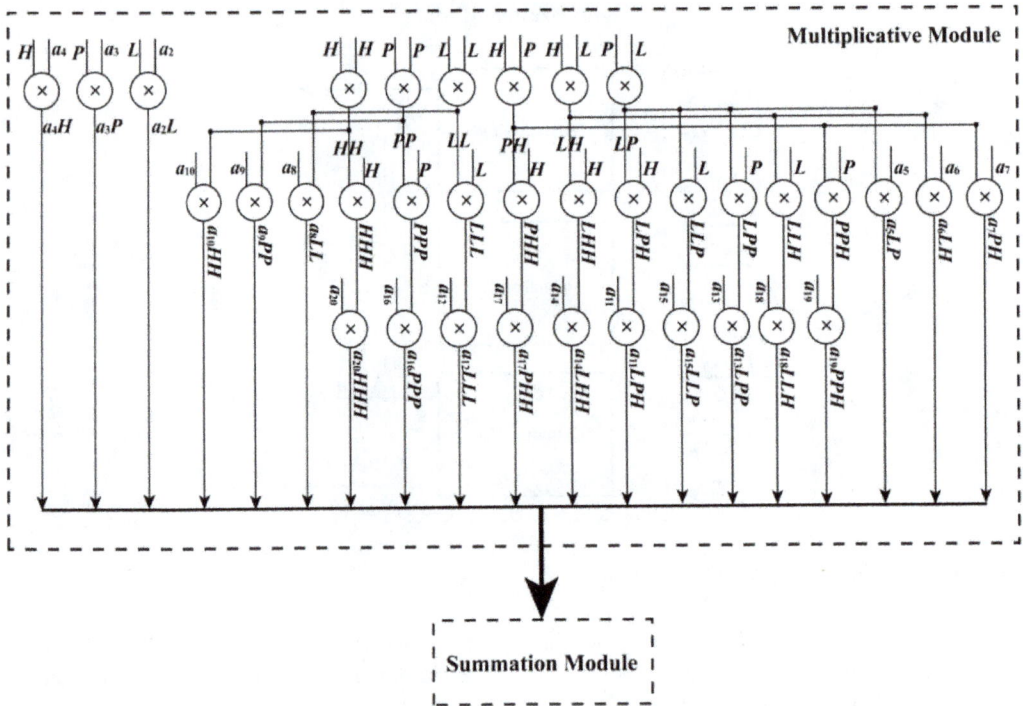

FIGURE 9.5 Architecture of the polynomial module.

- **Polynomial Module**

    When the ReM_Done_Sig and (L, P, H) are being received by the PM module, the PM module starts to work. In addition, Equations (9.4) and (9.5) have a uniform form, which are appropriate for the implementation of an FPGA chip in parallel. In the equations, variables such as LH, LP and PH are shared. To implement these polynomials in parallel using an FPGA chip, a parallel computation architecture is proposed in Figures 9.5 and 9.6, where the PM module is divided into two parts: one is used to perform multiplication and the other is applied to manipulate addition. When performing addition, some of the special

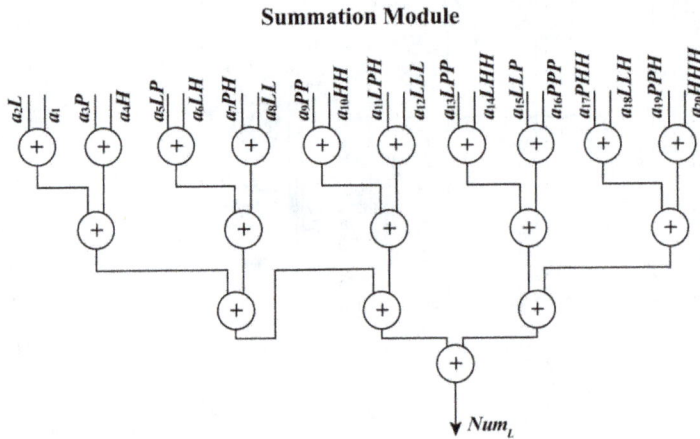

FIGURE 9.6 Architecture of the summation module.

operations about the positive and negative sets of data should be considered. Thus, for the additions in Figure 9.6, each of them is extended to the similar form (Zhang et al. 2018). Taking the addition between a_3P and a_4H as an example in Figure 9.7, three types of situations are considered: (i) a_3P and a_4H are both positive; (ii) a_3P and a_4H are both negative; and (iii) a_3P and a_4H have an opposite sign. The details for an extended addition are shown in Figure 9.7.

To implement each polynomial, 35 multipliers are utilized in the multiplication, and 19 extended additions are used. In each of the extended addition, 3 flipflops, 4 selectors,

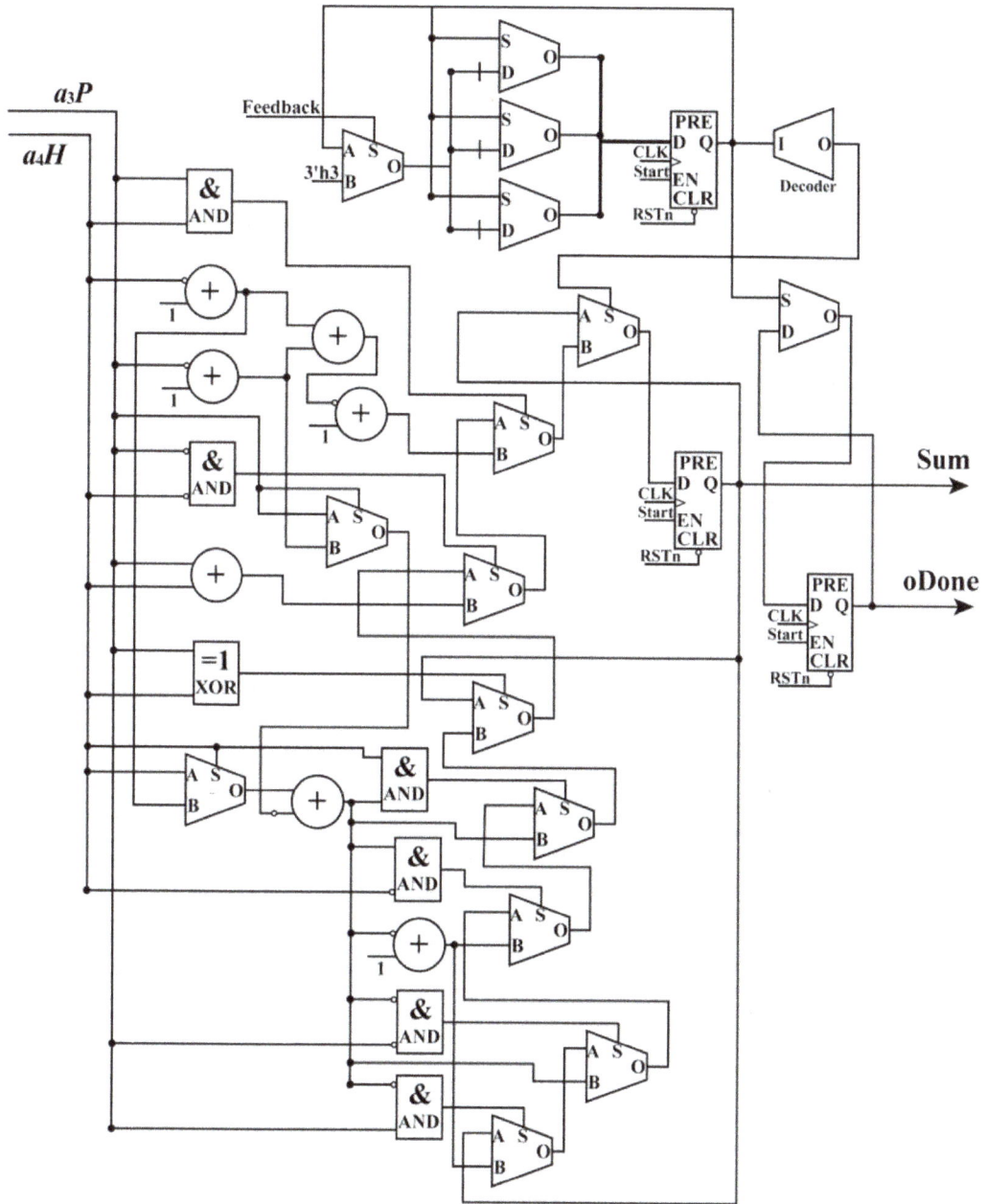

FIGURE 9.7 Details showing an example of the extended addition.

7 adders and 11 multiplexers are applied. After processing the PM module, four sums, i.e., Num_L, Num_S, Den_L and Den_S are obtained with the done signal of the PM module, PM_Done_Sig, in the same clock cycle.

- **Ratio Module**

    When the PM_Done_Sig, Num_L, Num_S, Den_L and Den_S are being received, the RaM module starts to calculate the normalized coordinates (X, Y) of image coordinates. As shown in Equation (9.6), the forms for the two equations are the same. This means that it is convenient to calculate X and Y in parallel using an FPGA chip. As shown in Figure 9.8, a parallel-computing architecture that is used to calculate X is presented. In the same way, the Y coordinate can be obtained.

    To obtain the X (or Y) coordinate, 1 divider, 3 adders, 6 multiplexers, 6 flipflops (two flipflops are public) and 32 selectors are applied. After the processing of the RaM module,

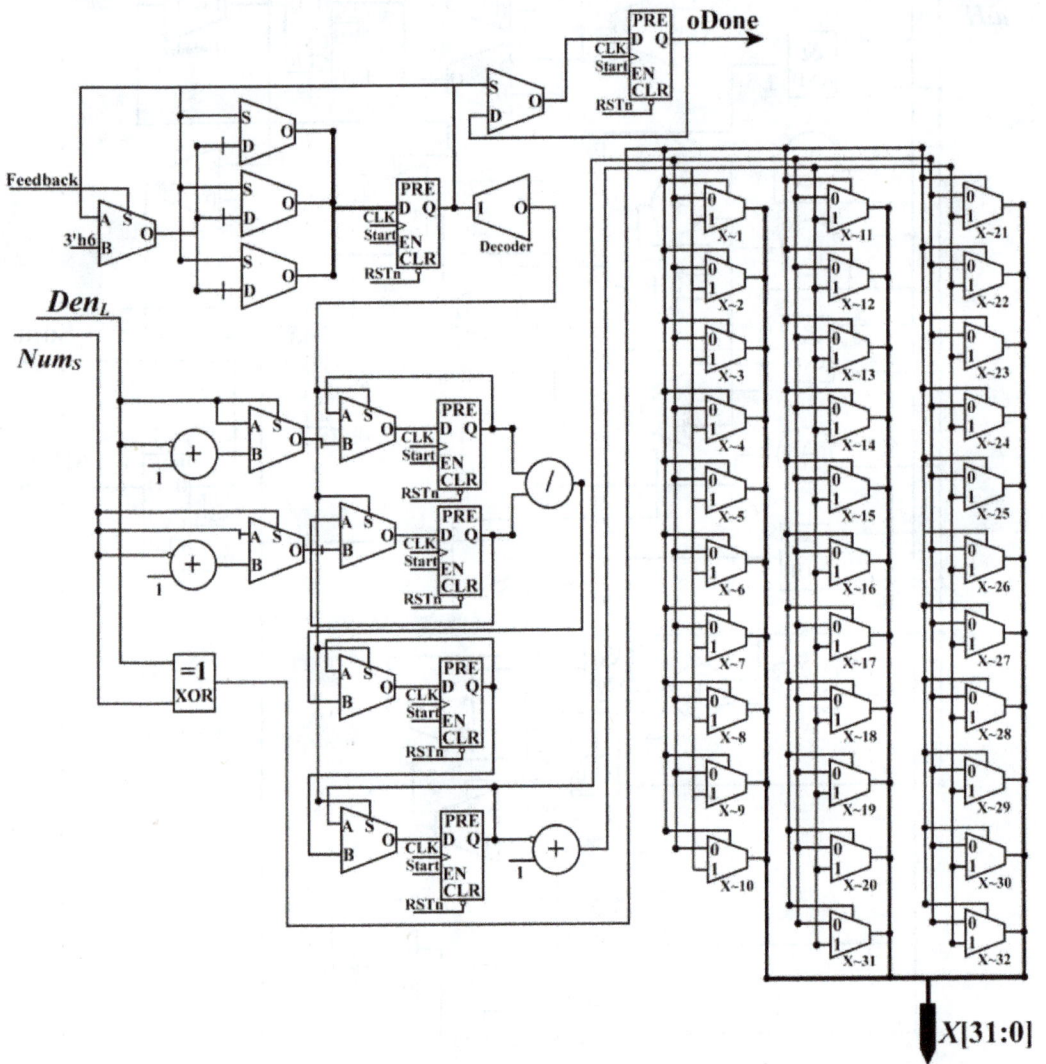

FIGURE 9.8 Architecture of the ratio module.

the X coordinate and Y coordinate are acquired with the done signal, RaM_Done_Sig, in the same clock cycle (Zhou et al. 2017; Zhang 2019).

- **Image Coordinate Calculation Module**

When the RaM_Done_Sig, *X* and *Y* coordinates are being detected, the RCM module starts to calculate the image coordinates (*Samp, Line*), i.e., column and row indexes. As shown in Equation (9.9), the equations give the relationship between the normalized coordinates (*X, Y*) and image coordinates (*Samp, Line*).

In addition, Equation (9.9) has a uniform form, which is helpful for implementation using an FPGA. To calculate the image coordinates (*Samp, Line*) in parallel, a parallel-computing hardware architecture is designed. Because the forms of the equations in Equation (9.9) are similar, only the architecture used for calculating the *Samp* coordinate is given. As shown in Figure 9.9, 1 multiplier, 4 flipflops (two of them are shared when

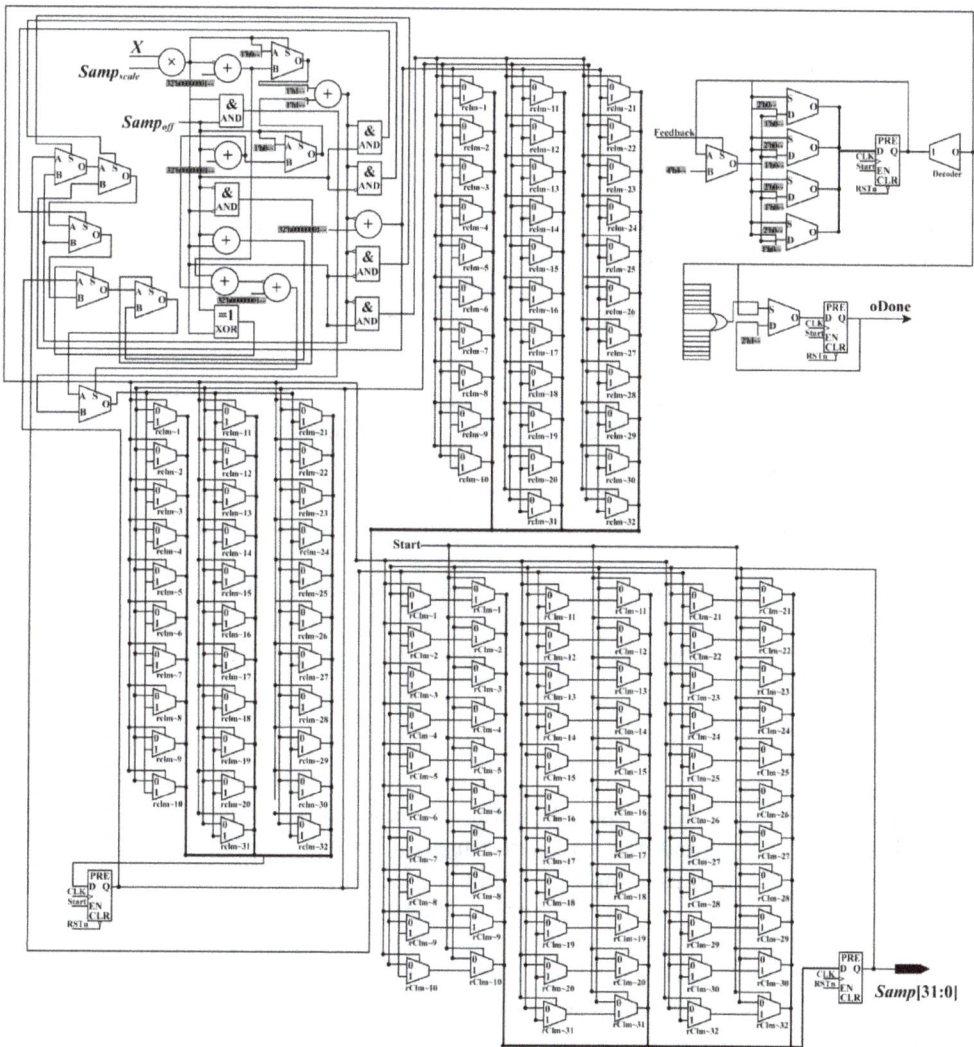

FIGURE 9.9   Architecture used for calculating *Samp* coordinate.

calculating *Line* coordinate), 5 selectors (MUX) are shared when calculating *Line* coordinate, using 7 adders and 136 multiplexers (MUX21).

After the processing of the RCM module, the image coordinates, that is, the column and row indexes (*Samp*, *Line*), and the done signal (RCM_Done_Sig) are obtained in the same clock cycle. Up to this point, the whole processing of the coordinate transformation is done. The obtained image coordinates are sent to the interpolation module to interpolate the grayscale.

9.2.2.3 Interpolation Module

It is almost impossible for the obtained column and row are located at the center of a pixel, it is therefore necessary to adopt the interpolation method to obtain the grays in the obtained column and row. Considering the interpolation effect, the complexity of interpolation algorithm and the resources of an FPGA, the bilinear interpolation method is selected to implement the gray interpolation. The details of the bilinear interpolation algorithm can be referenced to Section 7.2. The gray interpolation can mathematically be expressed by

$$\begin{aligned} g(i+p, j+q) = (1-p)(1-q)g(i, j) + (1-p)qg(i, j+1) \\ + p(1-q)g(i+1, j) + pqg(i+1, j+1) \end{aligned}$$

$$(9.10)$$

where i and j are nonnegative integers; the intermediates $p=|i\text{-int}(i)|$ and $q=|j\text{-int}(j)|$ are within the range of $(0, 1)$; and $g(i, j)$ represents gray values.

To implement the gray bilinear interpolation algorithm in parallel using an FPGA chip, a parallel computation architecture was designed in Figure 9.10, which contains four submodules:

1. The subtract_mod, which is used to obtain the integer part (*iLine* and *iSamp*) and fractional part (p and q) of *Line* and *Samp* indexes, and to calculate the subtraction (1_p and 1_q) in Equation (9.10).
2. The get_gray_addr_mod is applied to obtain the address of gray in RAM.
3. The multiplication part, which is utilized to calculate the multiplications in Equation (9.10).
4. The calculate_sum_mod, which is used to compute the sum in Equation (9.10). After the processing of the calculate_sum_mod, the results of interpolation in (*Samp*, *Line*) are obtained. The details of subtract_mod, get_gray_addr_mod and calculate_sum_mod are described as follows.

- **Subtract_mod**

  As shown in Equation (9.10), the gray values of four neighbors around the acquired column and row are required. Thus, the acquired column and row should be pre-processed to obtained the integer and fractional part, which are used to calculate (1-q) and (1-p). To implement the function using an FPGA chip, a parallel-computing architecture is proposed, named "*subtract_mod*". In the subtract_mod, the methods used to acquire *iSamp*, q and 1_q are similar to those for obtaining *iLine*, p and 1_p, respectively. Thus, architecture for obtaining *iSamp*, q and 1_q is given in Figure 9.11.

  As shown in Figure 9.11, three adders, seven multiplexers (MUX21) and nine flipflops, three of which are shared when *iLine*, p and 1_p are used, are required. In addition, three MUXs are public. After the processing of the whole subtract_mod, *iSamp*, q, 1_q, *iLine*, p and 1_p is acquired with the done signal in the same clock cycle. When obtaining these variables, *iSamp* and *iLine* are sent to the next submodule to retrieve the address of gray

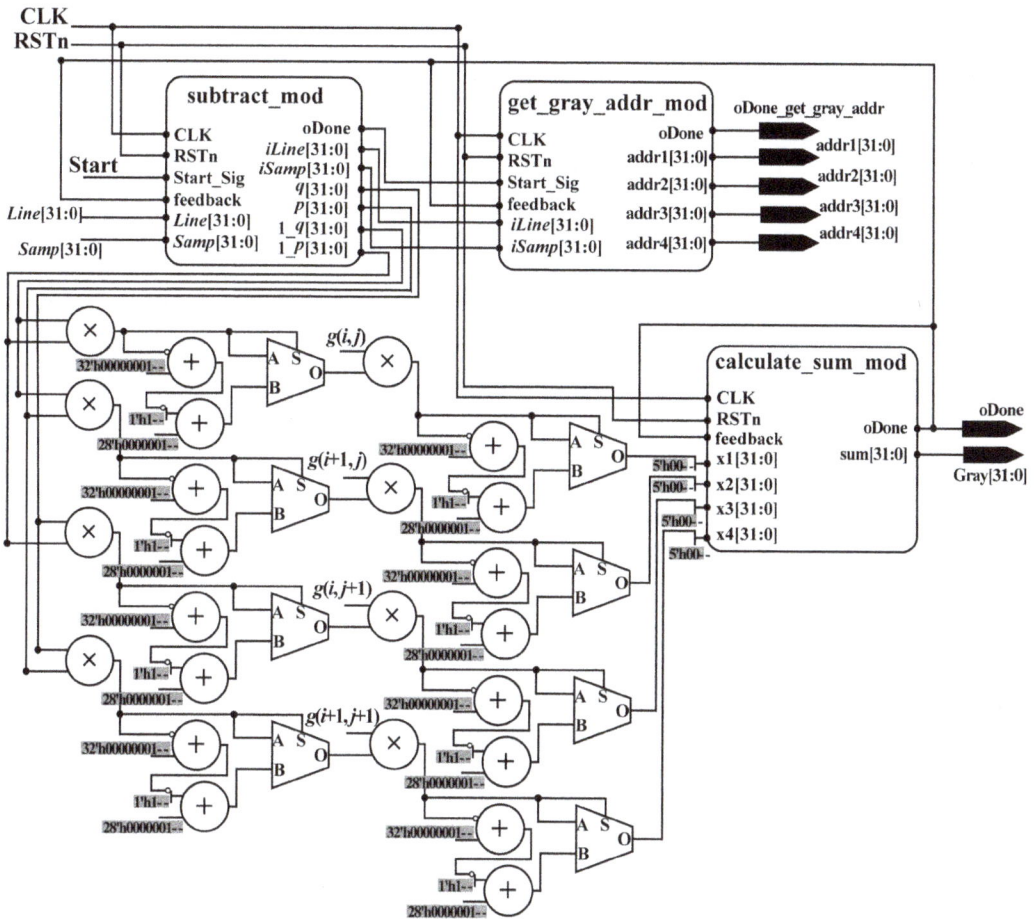

FIGURE 9.10 Architecture of bilinear gray interpolation module.

for four neighbors in RAM. Meanwhile, q, 1_q, $iLine$, p and 1_p is sent to another part to perform multiplication.

- **Get_gray_addr_mod**

  In order to obtain the gray values of four neighbors around the obtained column and row indexes in parallel, a parallel-computing hardware architecture is proposed in Figure 9.12, called "*get_gray_addr_mod*", where 3 LESS-THAN comparators, 9 adders, 10 MUX21, 12 flipflops and 70 MUX are applied. After being processed for the get_gray_addr_mod, four addresses are obtained with the done signal in the same clock cycle. According to the obtained addresses, the gray values can be acquired from RAM. Then, they are sent to the multiplication part to perform the multiplication.

- **Calculate_sum_mod**

  As shown in Figure 9.10, after the multiplication process, four variables x_1, x_2, x_3 and x_4 are obtained in the same clock cycle. To implement the addition for four variables, two levels of additions are needed. Each addition corresponds to an extended addition that has an architecture that is similar to Figure 9.7. The details can be found in Section 9.2.2.2 and Figure 9.7. After being processed for the calculate_sum_mod, the result of gray interpolation in ($Samp$, $Line$) can be obtained finally.

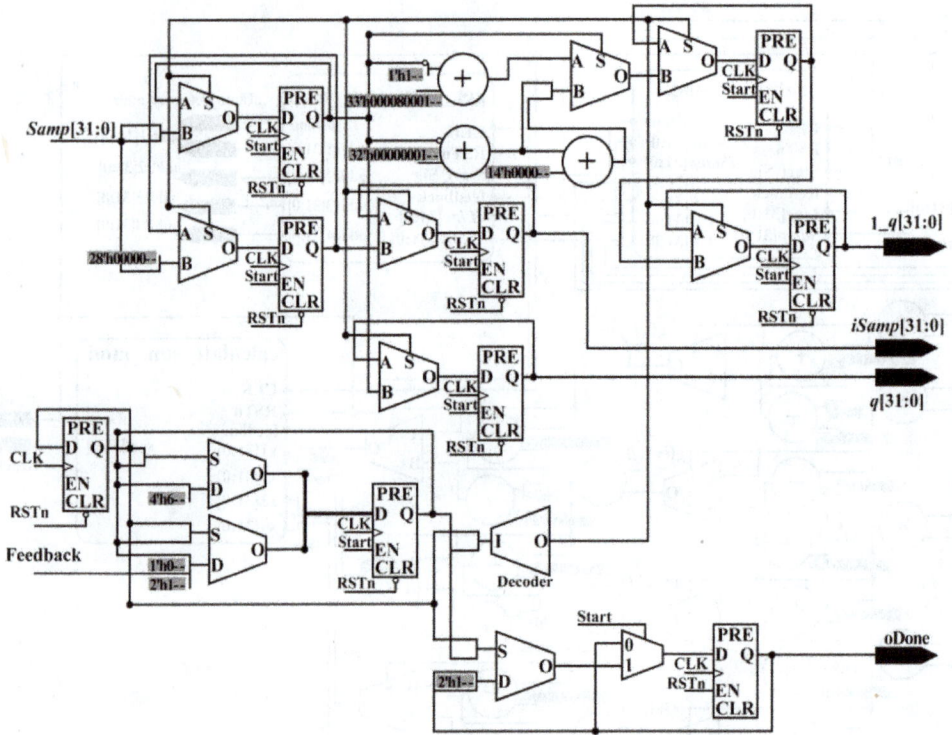

FIGURE 9.11   Architecture of the subtract_mod module.

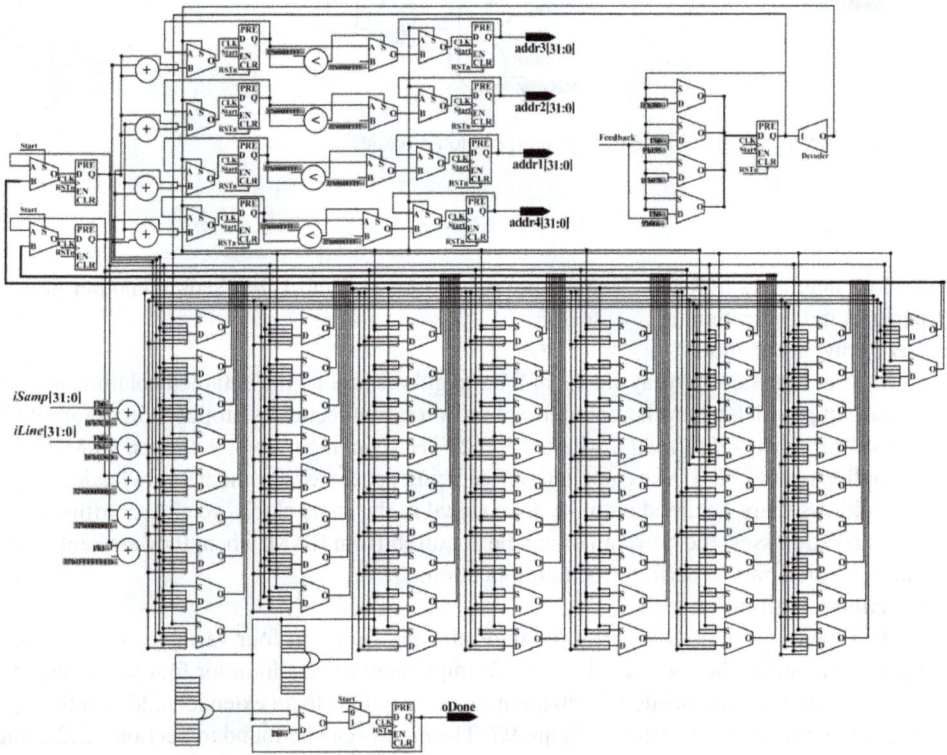

FIGURE 9.12   Architecture of the get_gray_addr_mod module.

FIGURE 9.13   The FPGA hardware.

9.3   VALIDATIONS

9.3.1   SOFTWARE AND HARDWARE ENVIRONMENT

An FPGA with Kintex-7 XC7K325TFFG900-1 is used (see Figure 9.13). The design tool is Vivado-2016.4, and the simulation tool is ModelSim SE10.1d. The PC with Windows 7 (64-bit) operating system, Intel(R) Core (TM) i7-4790 CPU @ 3.6 GHz processor with 8 GB RAM is used. To validate the proposed method, the ortho-rectification algorithm is also implemented using MATLAB 2012a (MathWorks, 1 Apple Hill Drive, Natick, MA 01760-2098, USA).

9.3.2   DATASET

Two test data sets are used in this study (Figure 9.14). The first study area is located in San Diego, California, US. The IKONOS-2 PAN image with the resolution of 1.0 meter and the wavelength of 450–900 nm collected on February 7, 2000, is used. The second study area is located in Genhe,

(a)   (b)

FIGURE 9.14   (a) The original IKONOS image. (b) The original SPOT6 image.

TABLE 9.2

The Normalized Parameters

#	1st Area	2nd Area
$Line_{off}$ (pixels)	1135	24,874.5
$Samp_{off}$ (pixels)	2548	17,962.5
Lat_{off} (degrees)	32.718700	50.737358
Lon_{off} (degrees)	-117.13340	121.44648
H_{off} (meters)	36.000	500
$Line_{scale}$ (pixels)	1829	24,874.5
$Samp_{scale}$ (pixels)	6570	17,962.5
Lat_{scale} (degrees)	0.01710000	0.41058849
Lon_{scale} (degrees)	0.07090000	0.45794952
H_{scale} (meters)	223	500

Inner Mongolia, China. The SPOT-6 PAN image with the resolution of 1.5 meter acquired on September 29, 2013, and the wavelength of 450–745 nm is used.

According to the proposed method, input parameters should be transformed into fixed-point data. As shown in Tables 9.2–9.4, the values of parameters of two study areas are in different range. To ensure computation accuracy, all parameters are transformed into fixed-point of 32 bits using different scale factor, τ. The details are given as follows. In addition, the clock frequency is 100 MHz.

Figures 9.15(a) and 9.16(a) are the orthorectified DOM using the proposed FPGA-based method. To validate the accuracy and the speed from the proposed method, the comparison analysis using the PC-based MATLAB at 32-bits floating-point format and 32-bits fixed point format was conducted. The results are shown in Figures 9.15(b),(c) and 9.16(b),(c). In addition, for the 32-bits fixed-point

TABLE 9.3

The Rational Function Polynomial Coefficients for the First Study Area

#	Values	#	Values	#	Values	#	Values
a_1	-7.52883250E-4	b_1	1	c_1	-9.23491680E-4	d_1	1
a_2	4.60115225E-3	b_2	-1.68736561E-3	c_2	1.01134804	d_2	-1.68736561E-3
a_3	-1.03642070	b_3	1.88384395E-3	c_3	3.57115249E-4	d_3	1.88384395E-3
a_4	-3.93943040E-2	b_4	-6.55340329E-4	c_4	-1.17541741E-2	d_4	-6.55340329E-4
a_5	1.75874570E-3	b_5	2.11928788E-7	c_5	1.71162003E-3	d_5	2.11928788E-7
a_6	2.25762210E-4	b_6	-2.15886792E-7	c_6	-2.77384658E-4	d_6	-2.15886792E-7
a_7	6.47497342E-4	b_7	6.60194370E-8	c_7	-4.67286564E-5	d_7	6.60194370E-8
a_8	-1.22344418E-3	b_8	1.29969058E-6	c_8	-1.70712175E-3	d_8	1.29969058E-6
a_9	-1.95386510E-3	b_9	-6.96485750E-7	c_9	7.61699396E-7	d_9	-6.96485750E-7
a_{10}	2.70799645E-5	b_{10}	3.41030606E-7	c_{10}	2.98181047E-6	d_{10}	3.41030606E-7
a_{11}	5.10672113E-7	b_{11}	5.17265975E-10	c_{11}	8.68077245E-7	d_{11}	5.17265975E-10
a_{12}	2.05965613E-6	b_{12}	2.71171743E-10	c_{12}	1.42413537E-6	d_{12}	2.71171743E-10
a_{13}	-2.18726634E-7	b_{13}	-1.47633205E-10	c_{13}	-1.11312088E-6	d_{13}	-1.47633205E-10
a_{14}	5.40855097E-8	b_{14}	3.59570414E-10	c_{14}	2.35665930E-7	d_{14}	3.59570414E-10
a_{15}	-3.96996732E-6	b_{15}	2.28588675E-10	c_{15}	5.40107978E-7	d_{15}	2.28588675E-10
a_{16}	7.19308892E-7	b_{16}	-1.11864088E-10	c_{16}	-1.10872161E-10	d_{16}	-1.11864088E-10
a_{17}	-3.89372910E-7	b_{17}	-1.37823694E-10	c_{17}	-4.86264967E-10	d_{17}	-1.37823694E-10
a_{18}	-4.18443985E-6	b_{18}	-3.32951199E-9	c_{18}	-8.43531861E-7	d_{18}	-3.32951199E-9
a_{19}	4.50802285E-8	b_{19}	6.33689691E-10	c_{19}	-3.64346611E-8	d_{19}	6.33689691E-10
a_{20}	-1.57227534E-8	b_{20}	-5.49482473E-11	c_{20}	-2.38643375E-9	d_{20}	-5.49482473E-11

TABLE 9.4

The Rational Function Polynomial Coefficients for the Second Study Area

#	Values	#	Values	#	Values	#	Values
a_1	0.00207581	b_1	1	c_1	-0.01727220	d_1	1
a_2	0.05939323	b_2	5.06562471E-9	c_2	1.01955596	d_2	-2.43637767E-6
a_3	-1.06139835	b_3	-2.23329117E-9	c_3	0.00149223	d_3	1.76928700E-6
a_4	0.00300505	b_4	-2.21235982E-11	c_4	-0.00498582	d_4	-1.29701506E-7
a_5	9.99447200E-5	b_5	-3.85166222E-8	c_5	-0.01544990	d_5	-4.50015117E-5
a_6	4.48210655E-6	b_6	7.22875706E-11	c_6	0.00070933	d_6	1.03746933E-6
a_7	-0.00011601	b_7	1.96276656E-9	c_7	-0.00021711	d_7	-2.16896140E-6
a_8	-0.00285285	b_8	1.03082157E-8	c_8	0.01454522	d_8	2.27253718E-5
a_9	-0.00025849	b_9	4.58273834E-8	c_9	0.00146642	d_9	2.71848287E-5
a_{10}	8.66439267E-8	b_{10}	4.12786890E-12	c_{10}	-2.61417544E-6	d_{10}	-1.20465997E-7
a_{11}	1.94846265E-7	b_{11}	-8.77450090E-11	c_{11}	-2.38142758E-5	d_{11}	-8.45078827E-9
a_{12}	7.22766464E-7	b_{12}	-1.25925035E-9	c_{12}	4.37422992E-5	d_{12}	-2.25373082E-9
a_{13}	-2.45655192E-5	b_{13}	-1.41399102E-9	c_{13}	-0.00012373	d_{13}	1.66156835E-7
a_{14}	-2.15776820E-10	b_{14}	1.09626013E-12	c_{14}	3.1682470454E-7	d_{14}	-2.87815787E-10
a_{15}	3.53191253E-5	b_{15}	1.82192638E-9	c_{15}	0.00022374	d_{15}	-1.35493121E-7
a_{16}	3.58935305E-5	b_{16}	-8.74976255E-10	c_{16}	1.88815906E-5	d_{16}	-2.31068042E-8
a_{17}	-3.00220333E-9	b_{17}	2.05074275E-13	c_{17}	-1.37072926E-7	d_{17}	5.66675066E-10
a_{18}	-3.32028434E-7	b_{18}	8.02116204E-11	c_{18}	1.95362054E-5	d_{18}	-2.37161908E-9
a_{19}	-1.44250161E-7	b_{19}	-1.15845372E-10	c_{19}	2.68897003E-6	d_{19}	6.41890372E-9
a_{20}	1.88636696E-12	b_{20}	5.66352056E-16	c_{20}	-1.15446432E-9	d_{20}	-7.23372539E-12

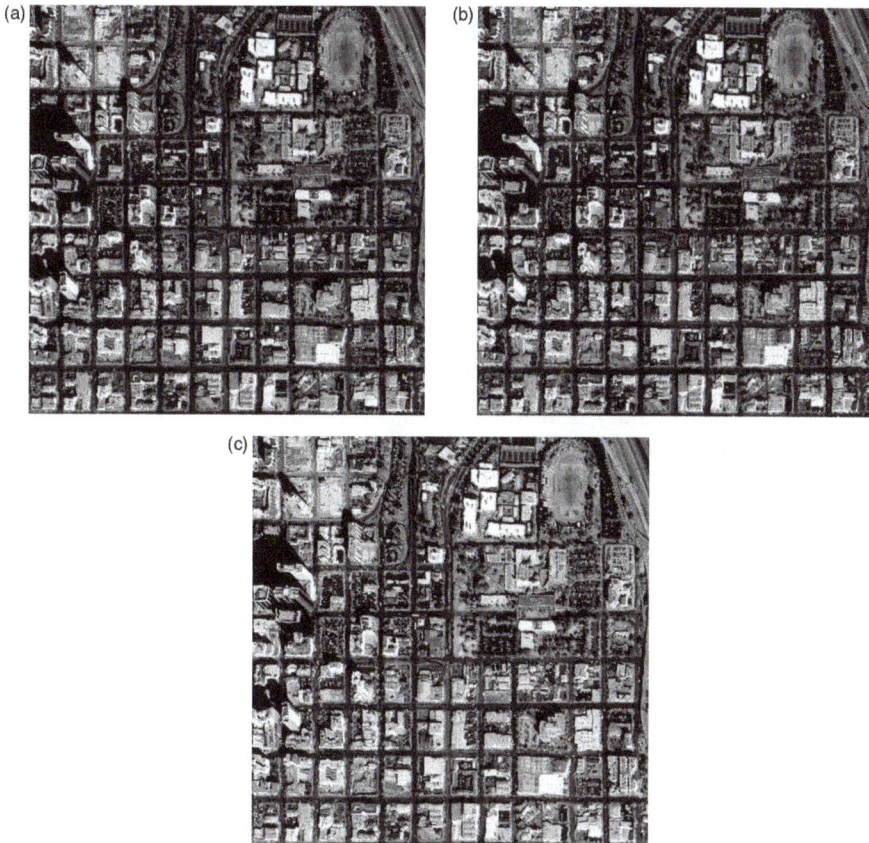

FIGURE 9.15 The DOM for the first study area generated by (a) the proposed FPGA-based method; (b) the PC MATLAB-based with the float point format; (c) the PC MATLAB-based with the fixed-point format method.

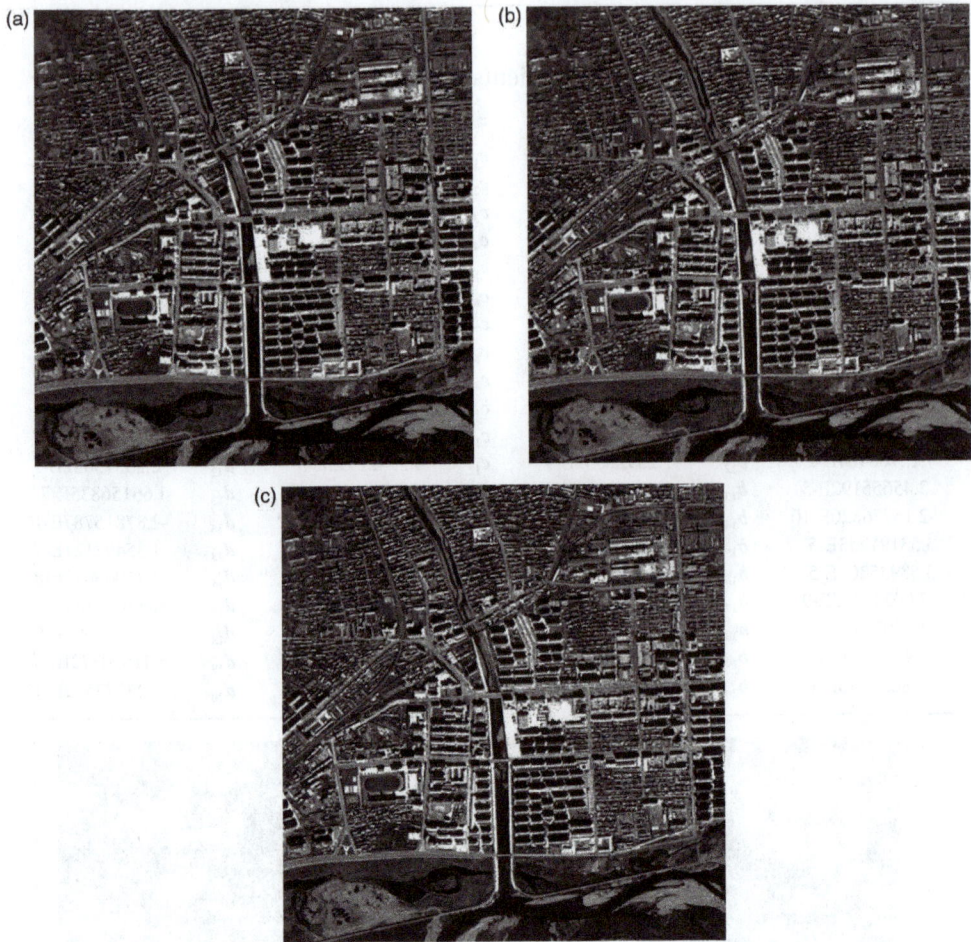

FIGURE 9.16 The DOM for the second study area generated by (a) the proposed FPGA-based method; (b) the PC MATLAB-based with the float point format; (c) the PC MATLAB-based with the fixed-point format method.

PC-based method, the scale factors are the same as used in FPGA-based method, and the results are shown in Tables 9.5 and 9.6.

In order to verify the accuracy of the different methods, the differences of the resulted DOM image gray for three method, in which the DOM generated by FPGA-based method is taken as a base, are shown in Figures 9.17 and 9.18. As observed from Figures 9.17(b) and 9.18(b), no significant difference between the proposed FPGA-based method and the fixed-point PC-based method, which

TABLE 9.5

Scale Factors for Parameters for the First Study Area

	τ	Range	Accuracy
Lat', Lon', Hei'	23	(-256, 255.999999881)	0.000000119
Lat'_{off}, Lat'_{scale}, Lon'_{off}, Lon'_{scale}	23	(-256, 255.999999881)	0.000000119
H'_{off}, H'_{scale}	23	(-256, 255.999999881)	0.000000119
$Line'_{off}$, $Line'_{scale}$, $Samp'_{off}$, $Samp'_{scale}$	18	(-8192, 8191.999996185)	0.000003815
a'_i, b'_i, c'_i and d'_i (i=1 to 20)	30	(-2, 1.999999999)	0.000000001

TABLE 9.6

Scale Factors for Parameters for the Second Area

	τ	Range	Accuracy
Lat′, Lon′, Hei′	23	(-256, 255.999999881)	0.000000119
Lat′$_{off}$, Lat′$_{scale}$, Lon′$_{off}$, Lon′$_{scale}$	23	(-256, 255.999999881)	0.000000119
H′$_{off}$, H′$_{scale}$	21	(-1024, 1023.999999523)	0.000000477
Line′$_{off}$, Line′$_{scale}$, Samp′$_{off}$, Samp′$_{scale}$	16	(-32768, 32767.999984741)	0.000015259
a'_i, b'_i, c'_i and d'_i (i=1 to 20)	30	(-2, 1.999999999)	0.000000001

demonstrates that the two methods can reach the same accuracy. However, the difference of the DOM image gray in Figures 9.17(a) and 9.18(a) indicates that many pixels have a significant difference, which should be mainly caused by the used bit wide and scale factor (French and Balster 2014).

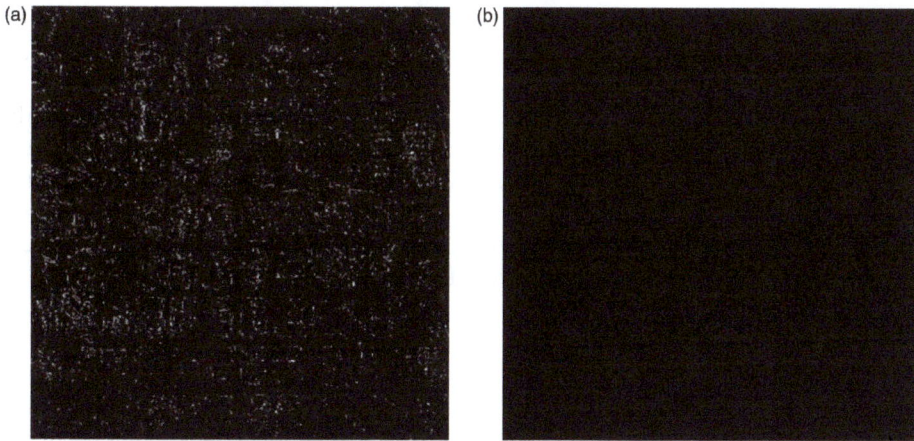

FIGURE 9.17 The difference of DOM image grays for the first study area (a) between the FPGA-based method; the PC MATLAB-based with the float point format method; (b) between the FPGA-based method and the PC MATLAB-based fixed-point RPC method.

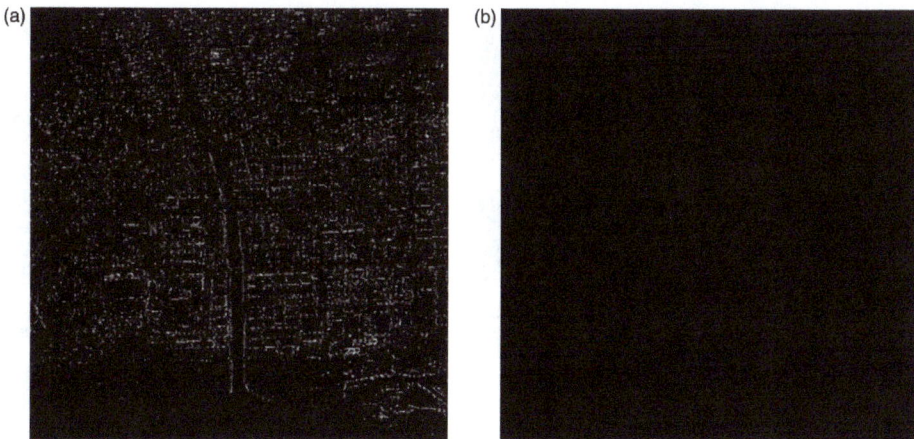

FIGURE 9.18 The difference of DOM image grays for the second study area (a) between the FPGA-based method; the PC MATLAB-based with the float point format method; (b) between the FPGA-based method and the PC MATLAB-based fixed-point method.

9.4 DISCUSSIONS

9.4.1 ERROR ANALYSIS

To quantitatively evaluate the accuracy of the proposed FPGA-based ortho-rectification method, the root-mean-square error (RMSE) (Shi and Shaker 2003; Reinartz et al. 2006) was utilized. Mathematically, the RMSEs of the image coordinates along the vertical axis (ΔI) and horizontal axis (ΔJ) and distance (ΔS) can be calculated by, respectively,

$$\Delta I = \sqrt{\frac{\sum_{h=1}^{n}(I'_h - I_h)^2}{n-1}} \quad \Delta J = \sqrt{\frac{\sum_{h=1}^{n}(J'_h - J_h)^2}{n-1}} \tag{9.11}$$

$$\Delta S = \sqrt{\frac{\sum_{h=1}^{n}((I'_h - I_h)^2 + (J'_h - J_h)^2)}{n-1}} \tag{9.12}$$

where I'_h and J'_h are the image coordinates rectified by the proposed ortho-rectification method; I_h and J_h are the reference image coordinates; and n is the number of check points.

To compute the RMSEs, 40 check points for each study area were selected randomly (see Figure 9.19). The RMSEs (ΔI, ΔJ and ΔS) are 0.35, 0.30 and 0.46 pixels, respectively, for the first study area; 0.27, 0.36 and 0.44 pixels, respectively, for the second study area. The other statistics analysis is calculated in Table 9.7.

With the Table 9.5, the DOM generated using the proposed FPGA-based method is accepted because the RMSEs are less than one pixel (Richards and Jia 1999; Schowengerdt 2007b; Schowengerdt 2007a).

9.4.2 COMPARISON ANALYSIS FOR PROCESSING SPEED

The processing speed is one of the most importance indicators for evaluating on-board computation. To this end, the throughput, which is a normalized metric, is used. The processing time have averagely been calculated from 10-time running of the DOM generation for each image. The

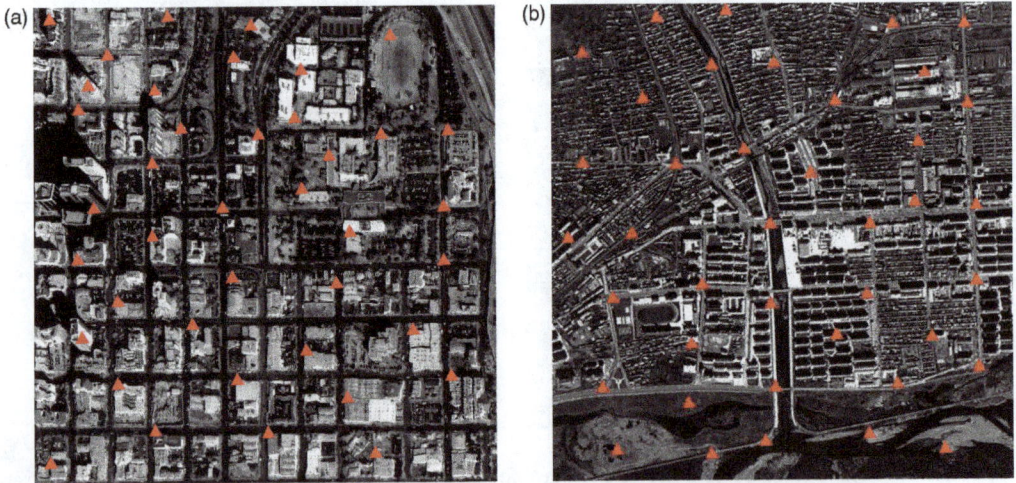

FIGURE 9.19 Check-point distribution in (a) the first study area; (b) the second study area.

TABLE 9.7

Statistical Analysis for the Different Image Coordinates Obtained by MATLAB and FPGA (Unit: Pixel)

Study Area No.		Maximum	Minimum	Mean	STD
1st	$|I'-I|$	0.65	0.02	0.29	0.19
	$|J'-J|$	0.68	0.01	0.25	0.18
2nd	$|I'-I|$	0.72	0.01	0.20	0.16
	$|J'-J|$	0.73	0.01	0.26	0.24

average processing time for different sizes of image are listed in Table 9.8. The average throughput is approximately 675.67 Mpixels/s for the proposed FPGA-based method, approximately 61677.49 pixels/s for the PC MATLAB-based method. This means that the proposed FPGA-based method has higher processing capacity than the MATLAB-based method does.

The speed-up is defined as the consuming time by PC MATLAB-based divided by the consuming time by FPGA-based method (Senthilnath et al. 2014). The results are listed in Table 9.8 also. As observed from Table 9.8, the maximum speed-up is for a size of 1024 pixels×1024 pixels, which reaches 11095.8709. It can therefore be demonstrated that the speed-up increases with the size of image.

9.4.3 Resource Consumption

Besides the speed of processing, the utilization ratio of each type of resource is also a key indicator when assessing the quality of a proposed method. Usually, if the utilization ratio of a type of resource reaches 60%–80%, the selected device satisfies the requirement of the design scheme. Thus, a few important resources, such as look-up tables (LUTs), registers and total pins are compared and analyzed. The results are listed in Table 9.9. As observed from Table 9.9, the utilization ratios of LUTs and registers are 44.42% and 5.59%, respectively. The utilization of input and output (IO) is 368, which is 73.60% of the total IOs.

Therefore, it can be concluded that the resources of the proposed FPGA-based DOM generation method can meet the requirement of FPGA resources.

TABLE 9.8

Average Processing Time for the FPGA-Based and PC MATLAB-Based DOM Generation

No.	Image Size (Pixels)	MATLAB Time (s)	FPGA Time (ms)	Speed-up
1	256×256	1.0515	0.09686	10,855.8745
2	512×512	4.2461	0.3875	11,008.8151
3	1024×1024	17.1986	1.5500	11,095.8709

TABLE 9.9

Utilization Ratio of Resources for the Proposed FPGA-Based Ortho-Rectification

#		Utilization	Available	Utilization Ratio (%)
Slice logic	Slice LUTs	90,634	203,800	44.42
	Slice registers	22,798	407,600	5.59
IO		368	500	73.60

9.5   CONCLUSIONS

This chapter presents a named, "field-programmable gate array (FPGA)-based fixed-point (FP) rational polynomial coefficient (RPC) method" (briefly, "FPGA-based FP-RPC method") for on-board generation of DOM. The proposed FPGA-based FP-RPC method contains three main sub-modules: Read_parameter_mod, Coordinate_transform_mod and Interpolation_mod.

To validate the accuracy from the proposed FPGA-based FP-RPC method, the PC-based (MATLAB 2012a) method was used as a reference, and two data sets, IKONOS and SPOT-6 images, were used as data resources. The RMSE, which is associated with the maximum, minimum, standard deviation (STD), and mean of row and column coordinates' differences, was used as evaluation index. The experimental results show that the STD of the row and column coordinates' differences are 0.19 and 0.18 pixels, respectively, for the first study area, and 0.16 and 0.24 pixels, respectively, for the second study area. The RMSE of the row coordinate (ΔI) and column coordinate (ΔJ) and the distance ΔS are 0.35, 0.30 and 0.46 pixels, respectively, for the first study area; are 0.27, 0.36 and 0.44 pixels, respectively, for the second study area. It can therefore be concluded that the proposed method can meet the demand of on-board DOM generation.

In addition, a comparison analysis for the processing speed was also performed. The throughput of the FPGA-based FP-RPC method and the PC-based RPC method are 675.67 megapixels per second and 61,070.24 pixels/s, respectively, which demonstrates that the processing speed of the FPGA-based FP-RPC method is faster (by approximately 11,000 times) than the processing speed of the MATLAB-based RPC method. In addition, it can be found that the utilization ratios of ALUTs, registers and IO are 44.42%, 5.59% and 73.60%, respectively.

REFERENCES

Aguilar, M.A., Saldaña, M.D.M., Aguilar, F.J., Assessing geometric accuracy of the ortho-rectification process from GeoEye-1 and WorldView-2 panchromatic images. *International Journal of Applied Earth Observation and Geoinformation*, 2013, 21: 427–435.

Balster, E.J., Fortener, B.T., Turri, W.F., Integer computation of lossy JPEG2000 compression. *IEEE Transactions Image Processing*, 2011, 20: 2386–2391.

Dai, C., Yang, J., Research on ortho-rectification of remote sensing images using GPU-CPU cooperative processing. *In International Symposium on Image and Data Fusion*, Tengchong, Yunnan, China, 9–11 Aug. 2011, IEEE: US, 1–4.

Eadie, D., Shevlin, F., Nisbet, A., Correction of geometric image distortion using FPGAs. *In Proceedings of SPIE – The International Society for Optical Engineering*, Society of Photo-Optical Instrumentation Engineers (SPIE), Galway, Ireland, 19 March 2003.

Escamilla-Hernández, E., Kravchenko, V., Ponomaryov, V., Robles-Camarillo, D., Ramos, L.E., Real time signal compression in radar using FPGA. *Científica*, 2008, 12: 131–138.

Ferrer, M.A., Alonso, J.B., Travieso, C.M. Offline geometric parameters for automatic signature verification using fixed-point arithmetic. *IEEE Transactions on Pattern Analysis & Machine Intelligence*, 2005, 27: 993–997.

Fraser, C. S., Hanley, H. B., Yamakawa, T., Three-dimensional geopositioning accuracy of IKONOS imagery. *Photogrammetric Record*, 2002, 17: 465–479.

French, J.C., Balster, E.J., A fast and accurate ortho-rectification algorithm of aerial imagery using integer arithmetic. *IEEE Journal of Selected Topics Applied Earth Observations and Remote Sensing*, 2014, 7: 1826–1834.

Grodecki, J., Dial, G., Block adjustment of high-resolution satellite images described by rational polynomials. *Photogrammetric Engineering & Remote Sensing*, 2003, 69: 59–68.

Habib, A., Xiong, W., Yang, F., HeH, L., Crawford, M., Improving ortho-rectification of UAV-based push-broom scanner imagery using derived orthophotos from frame cameras. *IEEE Journal of Selected Topics Applied Earth Observations and Remote Sensing*, 2017, 10: 262–276.

Halle, W., Venus, H., Skrbek, W., Thematic data processing on board the satellite BIRD. *In Proceedings of SPIE 4132, Imaging Spectrometry VI*, 15 November 2000, SPIE, pp. 412–419.

Hoja, D., Schneider, M., Müller, R., Lehner, M., Reinartz, P., Comparison of ortho-rectification methods suitable for rapid mapping using direct georeferencing and RPC for optical satellite data. *Presented at the ISPRS Conference 2008*, Peking (China), 3–11 July 2008, pp. 1617–1624.

Huang, J., Zhou, G. On-board detection and matching of feature points. *Remote Sensing*, 2017, 9: 601–618.

Huang, J., Zhou, G., Zhou, X., Zhang, R., A new FPGA architecture of fast and BRIEF algorithm for on-board corner detection and matching. *Sensors*, 2018, 18: 1014–1031.

Kate, D. Hardware implementation of the Huffman encoder for data compression using altera DE2 board. *International Journal of Advances in Engineering*, 2012, 2: 11–15.

Kumar, P. R., Sridharan, K., VLSI-efficient scheme and FPGA realization for robotic mapping in a dynamic environment. *IEEE Transactions on Very Large-Scale Integration (VLSI) Systems*, 2007, 15: 118–123.

Liu, H., Yang, J., Liu, H., Zhang, J., A new parallel ortho-rectification algorithm in a cluster environment. *In Third International Congress on Image and Signal Processing*, Yantai, China, 16–18 Oct. 2010, IEEE: US, pp. 2080–2084.

Li, H. H., Liu, S., Piao, Y., Snow removal of video image based on FPGA. *In Proceedings of the 5th International Conference on Electrical Engineering and Automatic Control*, Huang, B., Yao, Y., Eds., Berlin, Heidelberg: Springer Berlin Heidelberg, 2016a, pp. 207–215. ISBN: 978-3-662-48768-6.

Li, H., Xiang, F., Sun, L., Based on the FPGA video image enhancement system implementation. *In 2016 International Conference on Electronic Information Technology and Intellectualization*, Guangzhou, China, 18–19 June 2016, 2016b, pp. 427–434.

Marsetič, A., Oštir, K., Fras, M.K., Automatic ortho-rectification of high-resolution optical satellite images using vector roads. *IEEE Transactions Geoscience Remote Sensing*, 2015, 53: 6035–6047.

Ontiveros-Robles, E., Gonzalez-Vazquez, J.L., Castro, J. R., Castillo, O., A hardware architecture for real-time edge detection based on interval type-2 fuzzy logic, *In 2016 IEEE International Conference on Fuzzy Systems (FUZZ-IEEE)*, Vancouver, BC, Canada, 24–29 July 2016, IEEE: US, 804–810.

Ontiveros-Robles, E., Vázquez, J.G., Castro, J.R., Castillo, O., A FPGA-based hardware architecture approach for real-time fuzzy edge detection. *In Nature-Inspired Design of Hybrid Intelligent Systems*, Melin, P., Castillo, O., Kacprzyk, J., Eds., Cham: Springer International Publishing, 2017, pp. 519–540. ISBN: 978-3-319-47054-2.

Pal, C., Kotal, A., Samanta, A., Chakrabarti, A., Ghosh, R., An efficient FPGA implementation of optimized anisotropic diffusion filtering of images. *International Journal of Reconfigurable Computing*, 2016, 2016: 1–17.

Quan, J., Wang, P., Wang, H. Ortho-rectification of optical aerial images by GPU acceleration. *Optics and Precision Engineering*, 2016, 24: 2863–2871.

Reguera-Salgado, J., Calvino-Cancela, M., Martin-Herrero, J., GPU geocorrection for airborne pushbroom imagers. *IEEE Transactions Geoscience Remote Sensing*, 2012, 50: 4409–4419.

Reinartz, P., Müller, R., Lehner, M., Schroeder, M., Accuracy analysis for DSM and orthoimages derived from SPOT HRS stereo data using direct georeferencing. *ISPRS Journal of Photogrammetric Remote Sensing*, 2006, 60: 160–169.

Richards, J. A., Jia, X., Remote sensing digital image Analysis: An introduction. *In Remote Sensing Digital Image Analysis: An Introduction*, Richards, J. A., Jia, X. Eds., Berlin, Heidelberg: Springer Berlin Heidelberg, 1999, pp. 39–74. ISBN: 978-3-642-30062-2.

Schowengerdt, R.A., CHAPTER 7 – Correction and calibration. *In Remote Sensing, 3rd* ed., Burlington: Academic Press, 2007a, pp. 285–XXIII. ISBN: 978-0-12-369407-2.

Schowengerdt, R.A., CHAPTER 8 – Image registration and fusion. *In Remote Sensing (Third* edition), Burlington: Academic Press, 2007b, pp. 355–XXVI. ISBN: 978-0-12-369407-2.

Senthilnath, J., Sindhu, S., Omkar, S.N., GPU-based normalized cuts for road extraction using satellite imagery. *Journal of Earth System Science*, 2014, 123: 1759–1769.

Shi, W., Shaker, A., Analysis of terrain elevation effects on IKONOS imagery rectification accuracy by using non-rigorous models. *Photogrammetric English Remote Sensing*, 2003, 69: 1359–1366.

Tomasi, M., Vanegas, M., Barranco, F., Diaz, J., Ros, E., Real-time architecture for a robust multi-scale stereo engine on FPGA. *IEEE Transactions, VLSI SYST*, 2012, 20: 2208–2219.

Wang, H., Ellis, E.C., Spatial accuracy of orthorectified IKONOS imagery and historical aerial photographs across five sites in China. *International Journal of Remote Sensing*, 2005, 26: 1893–1911.

Wang, E., Yang, F., Tong, G., Qu, P., Pang, T., Particle filtering approach for GNSS receiver autonomous integrity monitoring and FPGA implementation. *TELKOMNIKA (Telecommunication Computing Electronics and Control)*, 2016, 14: 1321–1328.

Warpenburg, M. R., Siegel, L. J., SIMD image resampling. *IEEE Transactions on Computing*, 1982, 31: 934–942.

Wittenbrink, C.M., Somani, A.K., 2D and 3D optimal parallel image warping. *In Proceedings Seventh International Parallel Processing Symposium*, Newport, CA, USA, 13–16 April 1993, pp. 331–337.

Yang, G.D., Zhu, X., Ortho-rectification of SPOT 6 satellite images based on RPC models. *Applied Mechanics & Materials*, 2013, 392: 808–814.

Yang, G., Xin, X., Wu, Q., A study on ortho-rectification of SPOT6 image. *Presented at the 2017 International Conference on Mechanical and Mechatronics Engineering (ICMME 2017)*, Bangkok Thailand, 26–17 March 2017, pp. 26–27.

Zhang, C., Liang, T., Mok, P.K.T., Yu, W., FPGA implementation of the coupled filtering method. *In 2016 IEEE International Conference on Bioinformatics and Biomedicine (BIBM)*, Shenzhen, China, 15–18 Dec 2016, IEEE: US, pp. 435–442.

Zhang, G., Qiang, Q., Luo, Y., Zhu, Y., Gu, H., Zhu, X., Application of RPC model in ortho-rectification of spaceborne SAR imagery. *Photogrammetric Record*, 2012, 27: 94–110.

Zhang, R., Zhou, G., Zhang, G., et al., RPC-based orthorectification for satellite images using FPGA[J]. *Sensors*, 2018, 18(8): 2511–2534.

Zhang, R., GA-RLS-RFM ortho-rectification Algorithm of Satellite Images and Its Hardware Implementation using FPGA. *Dissertation,* Tianjin University, August 2019.

Zhou, G., Near real-time ortho-rectification and mosaic of small UAV video flow for time-critical event response. *IEEE Transactions Geoscience Remote Sensing*, 2009, 47: 739–747.

Zhou, G., Jezek, K., Wright, W., Rand, J., Granger, J., Ortho-rectification of 1960s satellite photographs covering Greenland. *IEEE Transactions Geoscience Remote Sensing*, 2002, 40: 1247–1259.

Zhou, G., Baysal, O., Kaye, J., Habib, S., Wang, C., Concept design of future intelligent earth observing satellites. *International Journal of Remote Sensing*, 2004, 25: 2667–2685.

Zhou, G., Chen, W., Kelmelis, J., A comprehensive study on urban true ortho-rectification, *IEEE Transactions on Geoscience and Remote Sensing*, 2005, 43(9): 2138–2147. doi: 10.1109/TGRS.2005.848417.

Zhou, G., Jiang, L., Huang, J., Zhang, R., Liu, D., Zhou, X., Baysal, O., FPGA-based on-board geometric calibration for linear CCD array sensors. *Sensors*, 2018, 18: 1794–1812.

Zhou, G., Zhang, Rongting, Huang, Jingjin, et al., Real-time ortho-rectification for remote sensing images. *International Journal of Remote Sensing*, 2019, 40(5–6): 2451–2465.

Zhou, G., Zhang, Rongting, Liu, N., On-board ortho-rectification for images based on an FPGA[j]. *Remote Sensing*, 2017, 9(9): 874–895.

10 On-Board Flood Change Detection Using SAR Images

10.1 INTRODUCTION

A change detection refers to the use of images from the same region acquired by multiple phases to determine and analyze changes. Since synthesis aperture sensors (SAR) are able to acquire data in all weather conditions, SAR image-based change detection has become a magnet in remote sensing research and is widely used in the field of natural disaster monitoring and evaluation. The existing change detection methods can be divided into three categories, pixel-based change detection (PBCD), object-based change detection (OBCD) and the hybrid change detection (HCD). The PBCD method is simple, fast and effective, but it is susceptible to noise (Lu et al. 2004). OBCD first performs image classification and changes detection. The detection accuracy is affected by the classification accuracy, and there is error accumulation phenomenon (Chen et al. 2012). HCD combines two methods of PBCD and OBCD, but there are still inherent errors in the two methods (Xiong et al. 2012). The algorithms of OBCD and HCD are high in complexity and time-consuming, and are still in the development stage. Currently, PBCD is the most commonly used method for automated change detection (Bovolo and Bruzzone 2005).

Since on-board flood monitoring requires high timeliness and accuracy performance. A PC-based change detection implementation was not appropriated owing to its limited speed. field programmable gate array (FPGA)-based method was faster and lower power-consumption comparing with PC-based ones (Ghofrani et al. 2014). In addition to the on-board implementations, including image denoising, image correction, image compression, etc. using based on FPGA chip (Shu 2019; Shu et al. *2019*), Zhou et al. (2004) and Huang and Zhou (2016) proposed an FPGA-based implementation for on-board features detection. many similar achievements have been reached. This chapter focus on the description on FPGA-based on-board implementation for flood change detection from SAR image with good precision.

10.2 THE PROPOSED CHANGE DETECTION METHOD

To implement the on-board change detection hardware for SAR flood image, this chapter adopts the method of log-ratio based image composition combining with multi-scale analysis and CFAR (Constant false alarm detection) thresholding method. The log-ratio based difference map construction method turns the inherent multiplicative speckle noise in SAR image into additive noise. The multi-scale fusion method based on stationary wavelets have good performance in noise inhibition and detail preserving (Carvalho et al. 2001; Yu et al. 2008; Celik and Ma 2010). In addition, the CFAR automatic threshold segmentation method is used to obtain the change detection result. The FPGA implementation was resource-conserving and less time consuming by means of hardware migration optimization, whose details can be referenced to Zhou (2020), Shu (2019) and Shu et al. (2019).

Two SAR flood images acquired at different times in the same area are taken as input images, which are respectively labeled as X1 and X2. A change detection, which can be boiled down to a two-category problem, is to extract the change and non-change regions from the two images. The flowchart is shown in Figure 10.1:

The basic steps of the proposed algorithm are summarized as follows (Shu 2019):

a. Generation of a log-ratio difference map from two SAR images;
b. Stationary wavelet-based decomposition and fusion of different scale;
c. The change detection result obtained by CFAR thresholding method.

DOI: 10.1201/9781003319634-10

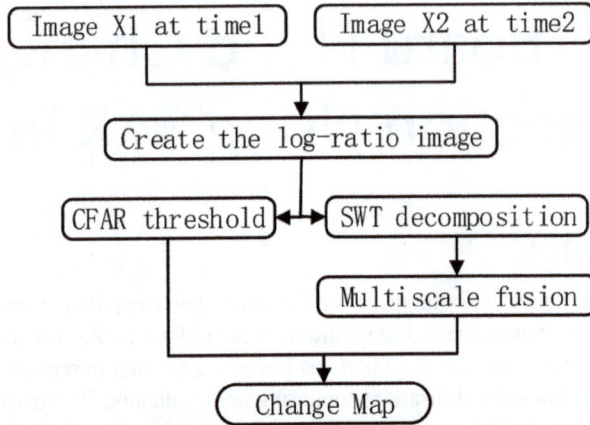

FIGURE 10.1 General scheme of the proposed approach.

10.2.1 LOG-RATIO

After the X_1 and X_2 of SAR images are calibrated and registered, the ratio operation is carried out pixel by pixel, and the absolute value of the natural logarithm of the ratio result is taken. The log-ratio method turns the multiplicative noise into additive noise, which is convenient for subsequent noise reduction processing in the next step, the difference image is defined as:

$$DI = \left|\log^{\frac{X_1}{X_2}}\right| = \left|\log^{X_1} - \log^{X_2}\right| \tag{10.1}$$

10.2.2 WAVELET DECOMPOSITION AND FUSION

Wavelet-based multi-scale decomposition and fusion are performed on the difference image (DI). In the process of multi-scale decomposition, two-dimensional stationary wavelet transform (2D-SWT) (Nason and Silverman 1995) is adopted.

SWT avoids the image downsampling and upsampling with the convolutional operator, which makes the size of the sub-image of each resolution level the same as the original image and avoids the inverse wavelet transform. Since the spatial detail of the next-resolution level is included in the current level, and the scale zero contains both low-frequency and high-frequency information, only the low-frequency stationary wavelet decomposition is performed here. The set X_{DS} of images with multiscale resolution is defined as:

$$X_{DS} = \{X_{RL}^0, \ldots, X_{RL}^n, \ldots, X_{RL}^{N-1}\} \tag{10.2}$$

where $n = 0, 1, \ldots, N-1$ indicates the scale and the DI $= X_{RL}^0$. Weight-based image fusion is taken considering the different trade-offs between spatial detail preservation and speckle reduction (Bovolo and Bruzzone 2005).

$$X_{FS} = \sum_{r=0}^{N-1} X_{RL}^r * \frac{1}{r} \tag{10.3}$$

where N represents optional, whereas when N increases, the image Xfs contains a smaller amount of both special details and speckle noises.

10.2.3 CFAR THRESHOLDING

Since the histogram distribution of the processed SAR image difference map is close to Rayleigh distribution, The Rayleigh distribution based Constant false alarm detection (CFAR) thresholding method is used to detect the changed and non-changed regions. The distribution density function of the Rayleigh distribution is defined as:

$$f(x) = \begin{cases} \dfrac{x}{\sigma^2}\exp\left(-\dfrac{x^2}{2\sigma^2}\right), x > 0 \\ 0, x \leq 0 \end{cases} \tag{10.4}$$

where σ is the Rayleigh parameter, and the threshold calculation formula for CFAR detection based on Rayleigh distribution is:

$$T = \frac{\sqrt{-2logP_{fa}} - \sqrt{\pi/2}}{\sqrt{2 - \pi/2}} \tag{10.5}$$

The change detection result M after the threshold segmentation is defined as:

$$M(i,j) = \begin{cases} 1, X_{FS}(i,j) > T \\ 0, X_{FS}(i,j) \leq T \end{cases} \tag{10.6}$$

10.3 FPGA-BASED HARDWARE ARCHITECTURE

The FPGA-based hardware architecture of on-board detection and matching is depicted in Figure 10.2. The operation principle is described as follows (Shu et al. 2019):

First, the two SAR images are input and stored in the RAM module, which is built by the LUT with a depth of 262,144 (512 × 512) and a width of 8 bits. Then the two images are read to the log-ratio module at the same time.

The obtained difference image from the log-ratio module is transmitted to the CFAR threshold determination module to obtain a threshold for automatic segmentation. At the same time, the difference image is input into the SWT module, and the resolution sub-image set X_{DS} is output to the image fusion module to obtain a new difference image. Finally, the obtained difference image and the threshold T obtained by CFAR transmitted to the threshold segmentation module, and the final change map is obtained by threshold segmentation.

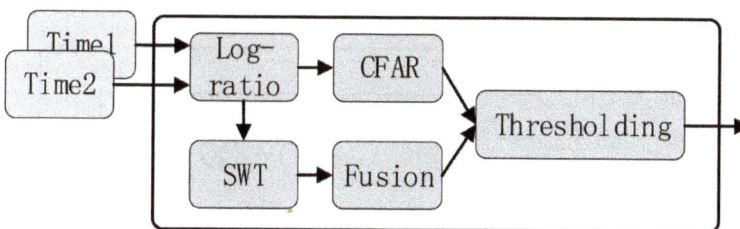

FIGURE 10.2 FPGA-based hardware architecture for on-board detection.

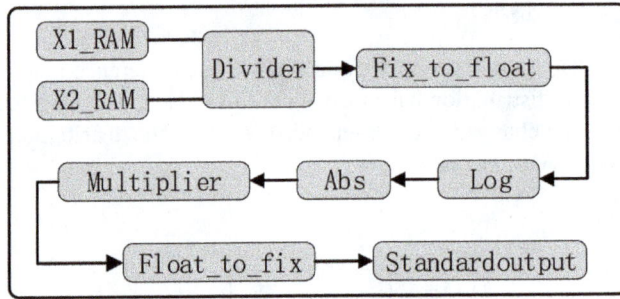

FIGURE 10.3 FPGA-based log-ratio implementation.

10.3.1 IMPLEMENTATION OF IMAGE COMPOSITION (LOG-RATIO)

First, use the divider module to calculate the value of X divided by X2, and then input the result to the logarithmic module after passing the fixed point to the floating point. The module adopts the 32-bit floating point type, and the obtained result is obtained by the absolute value obtaining module and the standardization module, the log result range 0-ln (255) is extended to the gray level range of 0-255 (just multiply by 46, can be completed by shift addition). Finally, it is converted to 8-bit fixed-point number output through floating-point to fixed-point module (Figure 10.3).

10.3.2 IMPLEMENTATION OF WAVELET DECOMPOSITION

In this section, an 8-layer wavelet resolution is used to obtain sub-image sets X. Considering the characteristics of FPGA: comparing with the dividing operation, the speed of shift and summation operation is faster and the resource consumption is smaller than traditional operation does. Smooth db wavelet is obtained using the proposed method. Since only low-pass filter convolution operation is needed, the low-pass convolution calculation of smooth db wavelet is performed. The lpd is (0.5, 0.5) and its upsampling such as {0.5,0,0.5}. etc. Multiplying by 0.5 is equal to shifting 1 bit to the right, and the convolution calculation is simplified to a shift summation operation of two numbers. Therefore, the wavelet decomposition operation can be simplified as follows: the row-based convolution operation, and the result is transposed and stored, and then the row-based convolution calculation is performed (this is equivalent to the convolution operation of the column), and thus the completion of one-layer convolution operation. The storage is again transposed, and the subsequent convolution operation is performed, and the result of the 8-layer wavelet resolution sub-image sets is obtained in this cycle (see Figure 10.4).

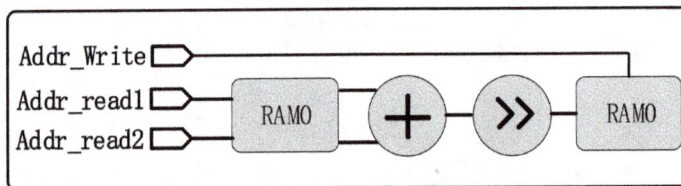

FIGURE 10.4 Implementation for FPGA-based convolution operation.

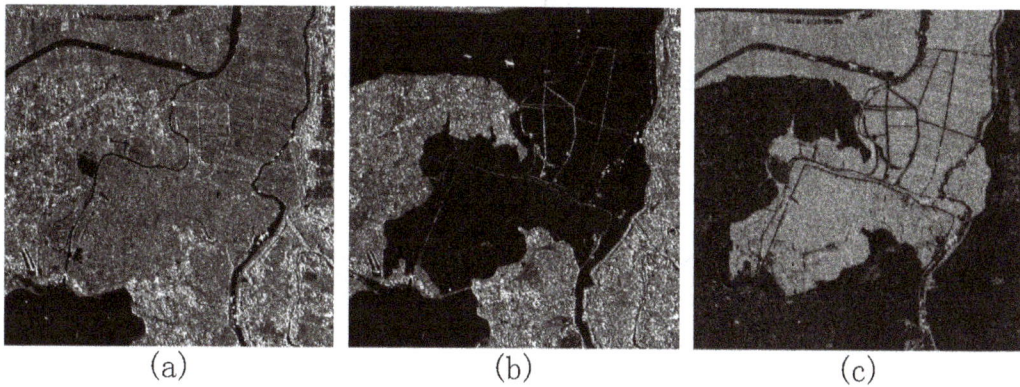

FIGURE 10.5   Sentinel-1 SAR images. (a) Reference image. (b) Flood image. (c) Ground truth

10.4   EXPERIMENT AND DISCUSSION

10.4.1   HARDWARE ENVIRONMENT

The proposed FPGA-based hardware architecture is a custom-designed board that contains a Xilinx XC7VH580T FPGA. The detailed resource utilization of the selected FPGA can be referenced to Chapter 2. In addition, the system environment is Windows, the design tool is Vivado2014.2, the simulation tool is Vivado Simulator, and the hardware design language is Verilog HDL.

10.4.2   EXPERIMENTAL RESULTS

The two images obtained in this chapter are part of the Huaihe River flood disaster in Anhui Province on June 11, 2016 and July 5, 2016, and the image size is 512×512. The data is Sentine-1A multi-temporal, VV polarized SAR image (see Figure 10.5).

The detection results of these methods are shown in Figure 10.6. There are a lot of noises in the result when using the logarithmic mean-based thresholding change detection method (Sumaiya and Kumari 2016), and the result of Xiong et al. (2012) method encounters the over-smoothing problem.

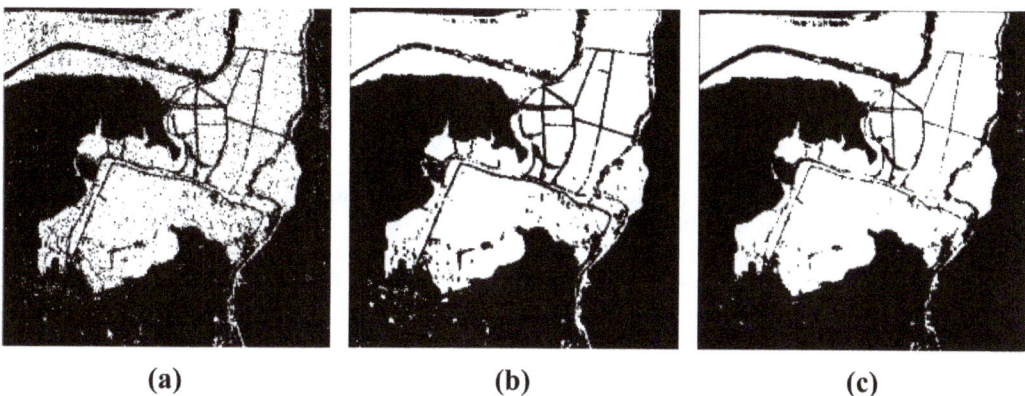

FIGURE 10.6   The experimental results:(a) the change detection (CD) map from the logarithmic mean-Based Thresholding change detection (Sumaiya and Kumari 2016). (b) the change detection map from Xiong et al. (2012). (c) the change detection map from the proposed method in this chapter.

TABLE 10.1

Quantitative Evaluation of Different Results

Method	False Alarm	Missed Alarm	Overall Error
LMCD (2)	1359	13,030	14,389
Xiong (3)	3734	9649	13,383
FPGA-based	2505	4442	6947

TABLE 10.2

The Resource Utilization (XC7VH580T FPGA)

Resources	Estimation	Available	Utilization (%)
FlipFlop	3524	725,600	0.48
Logic LUT	17,553	362,800	4.84
Memory LUT	98,304	141,600	69.42
BRAM	2	1880	0.11

The method proposed in this paper obtained a better performance both in geometric details and speckle denoising.

False alarm, missed alarm, and overall error are used as the measurements. The results are listed in Table 10.1.

In addition, the detection speed of SAR image with size of 512×512 is about 50 ms under the clock period is 100 MHz (see Table 10.2), which is ten times faster than PC-based performance.

10.5 CONCLUSION

In order to validate the timeliness on-board flood monitoring, this chapter proposes a SAR flood image change detection using FPGA implementation. The log-ratio difference map construction method and multi-scale wavelet decomposition and fusion are adopted. The experimental results show that the algorithm not only can achieve a good level of accuracy, but the processing speed is greatly improved.

REFERENCES

Bovolo, F., Bruzzone, L., A detail-preserving scale-driven approach to unsupervised change detection in mult-itemporal SAR images, *IEEE Transactions Geoscience Remote Sensing*, 43(12): 2963–2972, Dec. 2005.

Carvalho, L.M.T., et al., Digital change detection with the aid of multiresolution wavelet analysis. *International Journal of Remote Sensing*, 2001, 22(18): 3871–3876.

Celik, T., Ma, K.K., Unsupervised change detection for satellite images using dual-tree complex wavelet trans-form. *IEEE Transactions on Geoscience & Remote Sensing*, 2010, 48(3): 1199–1210.

Chen, G., et al., Object-based change detection. *International Journal of Remote Sensing*, 2012, 33(14): 4434–4457.

Ghofrani, Z., et al., Evaluating coverage changes in national parks using a hybrid change detection algorithm and remote sensing. *Journal of Applied Remote Sensing*, 2014, 8(1): 083646.

Huang, J., Zhou, G., A FPGA-based implementation for onboard features detection in remote sensing image, *2016 Instrument Science & Biomedical Engineering PhD Forum of China*, Beijing, China, 27–29 May 2016.

Lu, D., et al., Change detection techniques. *International Journal of Remote Sensing*, 2004, 25(12): 2365–2401.

Nason, G.P., Silverman, B.W., The stationary wavelet transform and some statistical applications. *Wavelets and Statistics*, Springer, New York, NY, 1995, pp. 281–299

Shu, Lei. FPGA Based SAR Flood Image Change Detection Algorithm Design and Hardware Implementation, Thesis, Tianjin University, October 2019.

Shu, Lei, Zhou, G., Liu, D., Huang, J., On-Board Wavelet Based Change Detection Implementation of SAR Flood Image. *IGARSS 2019 – 2019 IEEE International Geoscience and Remote Sensing Symposium*. IEEE, 2019.

Sumaiya, M.N., Kumari, R.S.S., Logarithmic mean-based thresholding for SAR image change detection. *IEEE Geoscience & Remote Sensing Letters*, 2016, 13(11): 1726–1728.

Xiong, B.l., Chen, J.M., Kuang, G., A change detection measure based on a likelihood ratio and statistical properties of SAR intensity images. *Remote Sensing Letters*, 2012, 3(3): 267–275.

Yu, G., Vladimirova, T., Sweeting, M.N., An efficient onboard lossless compression design for remote sensing image data, *IEEE International Geoscience and Remote Sensing Symposium, IGARSS 2008*, IEEE, Boston, MA, USA, 2008, pp. II-970–II-973.

Zhou, G., *Urban High-Resolution Remote Sensing: Algorithms and Modelling*, Taylor & Francis/CRC Press, 2020. ISBN: 978-03-67-857509, 465 pp.

Zhou, G., et al., Concept design of future intelligent earth observing satellites. *International Journal of Remote Sensing*, 2004, 25(14): 2667–2685.

Index

Note: Page numbers with *italics* refer to the figure and **bold** refer to the table.

For Product Safety Concerns and Information please contact our EU
representative GPSR@taylorandfrancis.com
Taylor & Francis Verlag GmbH, Kaufingerstraße 24, 80331 München, Germany

www.ingramcontent.com/pod-product-compliance
Lightning Source LLC
Chambersburg PA
CBHW061408210326
41598CB00035B/6139